D0261463

NUMERICAL ANALYSIS

NUMERICAL ANALYSIS

AR Mitchell 75th Birthday Volume

Editors

DF Griffiths & GA Watson

Department of Mathematics and Computer Science
University of Dundee
Dundee, Scotland

Published by

World Scientific Publishing Co. Pte. Ltd.

P O Box 128, Farrer Road, Singapore 912805

USA office: Suite 1B, 1060 Main Street, River Edge, NJ 07661

UK office: 57 Shelton Street, Covent Garden, London WC2H 9HE

British Library Cataloguing-in-Publication Data
A catalogue record for this book is available from the British Library.

NUMERICAL ANALYSIS: A R MITCHELL 75TH BIRTHDAY VOLUME

ISBN 981-02-2719-1

This book is printed on acid-free paper.

Printed in Singapore by Uto-Print

PREFACE

This Volume is intended as a small tribute to Ron Mitchell on the occasion of his 75th birthday, which falls on 22 June 1996. It consists of a collection of articles, written by people with links with Ron, as students, as colleagues, as co-workers, or just as regular visitors to Dundee. This set is of course very large, and we were obliged to restrict our requests for contributions to a subset, which was chosen in a somewhat arbitrary manner. Some of those who were not asked for contributions might feel disappointed or even offended: we very much regret if that is the case. Of course, some of those we did approach were, for various perfectly legitimate reasons, unable to provide us with a contribution within the given timescale. Since it will not be clear into which of these groups any particular person falls, we hope therefore that this venture will not lose us too many friends.

Some details of Ron's work and career are given elsewhere in this book but an explanation for its existence might be expected in a preface such as this. We could do this by drawing attention to the fact that Ron has made many substantial contributions to Numerical Analysis, both at a national and an international level, through his papers, his books, his conference talks, his visits, his supervision of students and of research fellows, his encouragement of young people, friends and colleagues. But that formal list does not do full justice to Ron. Because what is not included is the range of human qualities, the immense personal magnetism, the marvellous sense of humour. No attempt to characterise Ron would be complete without mention of these, because they are essential parts of a full picture. They are a major part of the reason why Ron is regarded in such high esteem, and with so much affection, by his friends and colleagues.

We are grateful to the World Scientific Publishing Company for giving us the opportunity to produce this book, and we would like to thank them, and in particular Dr Anju Goel for help in the pre-publication process. We have been privileged to have been members of the Numerical Analysis Group in Dundee, led by Ron, for over 25 years. We join with the other contributors to this volume in saying: "Happy birthday, Ron, and many more of them".

David F. Griffiths
G. Alistair Watson

Dundee
January, 1996

CONTENTS

A. R. MITCHELL: SOME BIOGRAPHICAL AND MATHEMATICAL NOTES

D. F. GRIFFITHS, J. D. LAMBERT, G. A. WATSON
Department of Mathematics and Computer Science,
University of Dundee, Dundee DD1 4HN, Scotland
E-mail: dfg@mcs.dundee.ac.uk, gawatson@mcs.dundee.ac.uk

and

G. FAIRWEATHER
Department of Mathematical and Computer Sciences,
Colorado School of Mines, Golden, Colorado 80401–1887, USA
E-mail: gfairwea@mines.edu

1. The Period up to 1967

Andrew Ronald Mitchell was born on 22 June 1921. He went to school at Morgan Academy, Dundee, and in 1938 went on to read mathematics at the old University College, Dundee (at the time a college of the University of St Andrews) where E. T. Copson held the Chair of Mathematics; Ron graduated with First Class Honours in 1942. Partly due to the war, student numbers were low, and Ron was the only Honours student in mathematics. On graduating, he was called up and sent to the wartime Ministry of Aircraft Production in London, where he remained until after the end of the war. His duties included the interrogation of captured Luftwaffe pilots, in an attempt to get information about their aircraft: some years later he met one of them at a conference. Ron had shown some prowess as a soccer player, and during this period he turned out a few times for Chelsea.

In October 1946, Ron decided to take some time out to do a PhD, and returned to Dundee to see if this might be possible. There was no available supervisor in University College, but he made contact with D. E. Rutherford, who was then a Lecturer in Mathematics and Applied Mathematics at St Andrews University. Lecturing staff were in short supply at that time (they were badly paid even in those days), and Dan Rutherford agreed to act as supervisor in return for Ron taking an Assistant Lectureship for the duration of his PhD.

Although Dan Rutherford was responsible for the Applied Mathematics part of the Mathematics Department in St Andrews, his main research interest was in Lattice Theory, so the supervision must have been fairly nominal, particularly as Ron's thesis was concerned with relaxation methods in compressible flow. Deciding that University life was not so bad after all, Ron stayed in St Andrews after being awarded his PhD in 1950, and was appointed Lecturer, later Senior Lecturer and eventually Reader. Some of Dan Rutherford's later work concerning the eigenvalues of certain banded matrices (having constant diagonals, apart from contributions from boundaries) was almost certainly influenced by Ron's interest in relaxation methods. They did some joint work in this area and, as well as writing two joint papers, actually discovered an early form of SOR before the famous paper of David Young, although this was never published,.

During much of this time, Ron continued his active interest in football. No doubt looking to supplement his salary, he signed as a part-time professional footballer with a number of Scottish clubs. During the period 1946-1955, he played with St Johnstone, East Fife, Brechin City and Berwick Rangers in that order. With typical modesty, Ron insists that his playing days at the last-named club were extended beyond their sell-by date because of his vital role as interpreter; five of the Berwick side came from Glasgow and five from Newcastle.

Ron had developed an interest in Numerical Analysis, initially as a means of tackling fluid dynamics problems using Southwell's relaxation methods. Jack Lambert was a member of the Senior Honours class in 1953–54 when Ron taught an Honours special topic in Numerical Analysis. This was the first time Numerical Analysis had been taught in St Andrews.

Ron's first PhD student was Jim Murray, who started in 1953 on a topic in boundary layer fluid dynamics. In these days the Air Ministry published a list of their top ten problems in fluids. Number 6 at that time was flow into a pitot tube: was the speed of flow which was registered the correct speed of the aircraft? This was Jim Murray's PhD problem. As is well known, he went on to an illustrious career, which included an FRS and the Chair of Mathematical Biology at Oxford. Ron's second PhD student was J. Y. Thompson who started in 1954 working on numerical aspects of fluid dynamics. Following the award of his PhD, he went on to a Lectureship in Applied Mathematics in Liverpool, where he married a widowed medical doctor with a large number of children, retrained into the medical profession, and was lost to Mathematics.

In 1959, Ron married Ann, and took up a one year post of Senior Research Fellow in the Mathematics Department at California Institute of Technology. Jack Lambert was appointed as a Lecturer at St Andrews in the same year, and became Ron's third PhD student, working for the degree part–time. He worked on an idea of Ron's of incorporating higher derivatives into methods for ODE's—apparently one of the few times Ron strayed away from PDE's to ODE's. In 1963, Ron and Jack Lambert jointly took on the PhD supervision of Graeme Fairweather and Brian Shaw, but this arrangement resolved itself into 2 pairings with Ron and Graeme Fairweather working on PDE's and the other two on ODE's. In September 1964, the group attended a Workshop in Perugia in Italy on Alcune Questioni di Analisi Numerica. Travel was by car, boat and train, and was a very complicated process. There Ron met Vlastimil Ptak, who visited St Andrews the following year; also present were Peter Wynn, Walter Gautschi and F. L. Bauer. A local newspaper published a picture of the St Andrews contingent listening intently to one of the lectures.

A joint Mitchell/Fairweather paper, published in Numerische Mathematik in 1964, was the first in a series on high order alternating direction finite difference methods for elliptic PDE's. It caused surprise in some quarters, where it had been believed that such higher order methods did not exist.

St Andrews got its first computer in 1964, an IBM 1620 with 64K (or was it 32K?) memory. The machine was capable of solving Laplace's equation in a cube using an optimal alternating direction finite difference method with a $5 \times 5 \times 5$ mesh in 15 minutes—on a good day. In a square, a 20×20 mesh could be tackled. The computer was housed in the Observatory, over a mile from the Department of Mathematics, and hands-on access was provided for an hour each morning and afternoon, with no exceptions, even when the printer ribbon wrapped itself around the type bar, a frequent occurrence. Batch jobs could be run at other times. The method published in the Numerische Mathematik paper was not completely reliable, although it did well on problems with homogeneous Dirichlet boundary conditions on at least two sides of the square. Contrary to Ron's belief that there was an error in the program, it turned out that there was a problem with the handling of the boundary conditions in high–order methods. An elegant way round the problem was obtained by Ron and Graeme Fairweather in 1966, and was published in SINUM the following year. This paper also described how to deal with problems in L–shaped regions. Earlier joint work by the same authors involved the use of a difference scheme based on the Schwarz alternating procedure, published in the Computer Journal in 1966: this paper may have been the first to give numerical results obtained using a domain decomposition method.

Graeme Fairweather had completed his PhD in 1965, and the following year went to Rice University in Texas, a visit set up as a consequence of correspondence between Ron and Jim Douglas Jr. Sandy Gourlay started a PhD with Ron in 1964, as did Pat Keast, followed one year later by John Morris. A good discussion of the work done by Ron and his students around this time is given by Lapidus and Pinder in their book *Numerical Solution of Partial Differential Equations in Science and Enginering* published by Wiley in 1982.

Throughout his time at the University of St Andrews, Ron continued to live in Dundee. On a typical day, he would catch the 8.02 train from Dundee to St Andrews which got him into his office around 8.40. At 9 he met his students, and he then had a lecture at 10. At 11 he would have coffee in the Staff Club in the Younger Hall, and after possibly seeing his students again he would gather his things together to catch the train back to Dundee at 12.40. An overlong discussion could result in a rush for the train, and a jogging party to the station which would leave the students hanging on to the railings to recover. Ron kept himself extremely fit, and at that time was one of the best squash players in the University. Only a very special event, such as an important visitor, would keep Ron in St Andrews for an afternoon. Seminars were comparatively rare in those days and the relatively large group of research students at St Andrews supervised by Ron and Jack Lambert was likely to turn up anywhere: they became known (to others) as the "all purpose colloquium audience". Trips were made as far as Newcastle for a one–hour seminar.

Mike Osborne came to Edinburgh in July 1963 as Assistant Director (to Sidney Michaelson) of the University Computer Unit and he was joined by Donald Kershaw, together with some students including, in October 1964, Alistair Watson. "Computer Unit" was, in fact, a bit of a misnomer, as the Unit had no computer, and Atlas Autocode programmes were sent to the Manchester Atlas, but that is another story. John Todd and Olga Tausky were visiting Arthur Erdelyi in Edinburgh, and E. T. Copson (who had moved from Dundee to St Andrews) invited John Todd over to give a talk. Mike Osborne made a fourth in Arthur Erdelyi's car and there was a lively conversation in which the driver was an active participant. One result was that a particular signpost "St Andrews 10 miles" was passed several times, and not always in the correct direction.

After the talk, Ron and Mike Osborne swapped concerns about the need for more interaction, and this led to the idea of a "do it ourselves" seminar. Ron claimed a friendship with the warden of a particular Hall of Residence, and undertook to see if he could get a good rate (ie student rate) after the June 1965 examinations but before end of term. His persuasive powers were equal to the task, and so the idea was pushed forward, and Ron and Jack Lambert agreed to run the meeting. The Department of Mathematics was very happy to be involved on the basis that if the meeting made a profit, it belonged to the Department, and if it made a loss it came out of the pockets of the organisers. The main organizational arrangements were made by Jack Lambert, and no conference has ever had its financial estimates done more meticulously.

Particular encouragement was given to participation by students (remembered rather differently by the students as a form of coercion), and a surprisingly good response was obtained. In fact the Edinburgh and St Andrews contingents made up only around half those attending, and John Mason from Oxford, Ken Wright from Newcastle, Will McLewin from Manchester and Garry Tee from Lancaster appear to have travelled furthest. If there were any records of the meeting, they seem not to have survived; the best estimates are that there were about 30 attendees and a program extending over two days. Ron's group talked about their work on ADI and high accuracy discretizations.

This model has been used in many other places, but its origins are recorded here in some detail because it is now recognised as the first "Dundee" meeting. Emboldened by the success of this meeting, a second one was held in St Andrews from 26–30 June 1967, attracted 85 participants, and established the biennial pattern.

Around 1965–66, Ron went to evening classes in Dundee to learn Russian. Having long since lost his School Leaving Certificate, he experienced some difficulty in persuading the organisers that he had an appropriate level of general education to allow him entry to the course; apparently a PhD was not an acceptable alternative. During Graeme Fairweather's thesis work, it had been realised that some Russians, in particular Samarskii, Andreyev and D'Yakonov were also working on high order difference methods for PDEs. Indeed a method, essentially that of the 1964 Numerische Mathematik paper, had been published in Russian at about the same time, and D'Yakonov had also discovered the loss of accuracy referred to earlier. A knowledge of Russian not only allowed Ron to keep up with the Russian literature as soon as it appeared, but was invaluable when he attended the ICM Meeting in Moscow in 1966. There he met D'Yakonov and, as a result, the latter visited Dundee in the late sixties. In Moscow, Ron was able to indulge his love of soccer: he played for The Rest of the World against the USSR in a soccer match which was organised in the stadium of Moscow Dynamo. The home team, who had been in training for several weeks, won 5–2.

The work of Mitchell/Fairweather lay somewhere between the classical ADI approach of Douglas, Peaceman, Rachford and Gunn, and that of D'Yakonov. The former would not handle the loss of accuracy at the boundary, while the latter would, but was cumbersome. A byproduct was that people in the West became much more aware of the activity in the USSR concerning split operator techniques.

2. The Period from 1967 onwards

In 1965 D. S. Jones was appointed to the Ivory Chair of Mathematics in Queen's College (formerly University College), Dundee. With a level of priority which was atypical of a classical Applied Mathematician in those days, he decided to build up Numerical Analysis, which he was far–sighted enough to see as a growth area. For example he started an MSc course in Numerical Analysis in 1965. There was a numerical analyst already in Dundee, R. P. Pearce, a Senior Lecturer, who had collaborated with Ron and Jack Lambert while they were at St Andrews. However, he left to fill a Chair at the University of Reading at the end of the 1966/67 academic year. Meantime, Douglas Jones obtained funds to establish a Chair of Numerical Analysis in Dundee, and Ron was appointed in 1967, the year in which Queen's College Dundee formally severed its links with St Andrews and became the University of Dundee. The same year, Jack Lambert, who had moved to Aberdeen in 1965, joined Ron in Dundee as a Senior Lecturer, and Sandy Gourlay came from St Andrews as a Lecturer.

Ron continued to attract research students and, with funding from NCR and the Ministry of Defence obtained largely through the efforts of Douglas Jones, other numerical analysts were appointed to post–doctoral positions. John Morris came from St Andrews to a Research Fellowship, and Sean McKee took up a similar position after completing a PhD with Ron in 1970. Other Research Fellows who came to Dundee from elsewhere at that time were Nancy Nichols and Alistair Watson. Sandy Gourlay left in 1970 to join IBM, and Alistair Watson was appointed to the vacant post. David Griffiths joined the Mathematics Department as a Lecturer in the same year, and he and Ron have worked closely ever since.

The 3rd Biennial Dundee Conference had by now been held in 1969, this time actually in Dundee. It attracted 148 participants. In the following year, Ron obtained substantial Science Research Council funding for a Numerical Analysis Year lasting from September 1970 to September 1971. This was an important and high profile period which went a long way to putting Dundee on the Numerical Analysis map. The year began with a Symposium on the Theory of Numerical Analysis from 15-23 September, 1970 with speakers Gene Golub, Vidar Thomée, Gene Wachspress and Olof Widlund. There was a Conference on the Applications of Numerical Analysis from 23–26 March 1971 with 177 participants, a Conference on Numerical Methods for Nonlinear Optimisation from 28 June–1 July, 1971 with 198 participants, a Seminar on Ritz–Galerkin and the Finite Element Method from 8–9 July, 1971 and finally a Conference on the Numerical Solution of ODEs from 5–6 August, 1971.

In addition to those already mentioned, about 34 other numerical analysts of international repute visited Dundee during the Numerical Analysis Year as Senior Visiting Fellows, some for short periods and others for longer periods up to the full year. Available records list those as C. Bardos, F. L. Bauer, R. Bellman, G. D. Birkhoff, J. Bramble, H. Brunner, J. C. Butcher, L. Collatz, G. Dahlquist, P. J. Davis, J. Douglas Jr., C. W. Gear, W. Gragg, J. L. Greenstadt, P. Henrici, A. S. Householder, T. E. Hull, E. Isaacson, R. E. Kalaba, H. B. Keller, H. O. Kreiss, P. Lascaux, J. L. Lions, M. R. Osborne, M. J. D. Powell, V. Ptak, J. R. Rice, R. D. Richtmyer, J. B. Rosen, I. J. Schoenberg, H. J. Stetter, G. Strang, R. Temam, R. S. Varga.

As those who have spent time with Ron know, unusual and remarkable things are always likely to happen. During his visit, Vidar Thomée was persuaded by Ron to investigate his ancestry by inquiring at a local Tartan shop which claimed to be able to find a tartan associated with any given surname: in this case it turned out to be MacDonald. Ron also managed to persuade Mike Osborne and Gene Wachspress to attend a Burns' supper in Dundee resplendent in kilts.

Although Ron and Jack Lambert were back as colleagues, they had an understanding that they would pursue their own interests in an attempt to keep the Numerical Analysis base wide. A major contribution to the widening of that base was the appointment in 1973 of Roger Fletcher, already a leading figure in optimization, to a Senior SRC Fellowship; he moved to a permanent appointment in 1976. Also in 1973, a Conference on The Numerical Solution of Differential Equations was held from 3–6 July 1973, and attracted 234 participants. The 1975 Conference was on general Numerical Analysis, and this pattern continues to this day, reflecting the wider Numerical Analysis base referred to above. Further information about all the Dundee meetings can be found on the World Wide Web, starting from the home page of the Department of Mathematics and Computer Science: http://www.mcs.dundee.ac.uk:8080/. Many eminent numerical analysts have had long associations with Dundee, and have strongly supported the Dundee conferences. Two of those, Lothar Collatz and Gene Golub, have been awarded Honorary Degrees by the University of Dundee. Gene

Golub, in addition, presented the inaugural A. R. Mitchell lecture at the 1991 meeting which celebrated Ron's 70th birthday. Two other staunch supporters of the Dundee conferences have been Mike Powell and Bill Morton, who gave the A. R. Mitchell lectures in 1993 and 1995, respectively.

The MSc course in Numerical Analysis has been successful in attracting many good students, and once again Ron must take much of the credit for this. Many students went on to do PhD's with Ron, for example Dick Wait who was the first to work with Ron in the newly emerging field of finite elements, and who continued on to a Research Fellowship and further collaboration, including a book. Among others who took the MSc course and have gone on to carve out academic careers for themselves are Mehi Al-Baali, Peter Alfeld, Ken Brodlie, Dugald Duncan, Julian Hall, Per Skafte Hansen, Chus Sanz–Serna, Sven Sigurdsson, Ian Stewart and Rob Womersley.

Ron's interests changed in the late 1960's to finite elements, a move allegedly instigated by Dick Wait who turned up at Ron's doorstep and announced that he would like to do a PhD in the area. This was virgin territory for numerical analysts and Ron did much pioneering work during the next five years with George Phillips, Gene Wachspress, Bob Barnhill and his students Dick Wait, Robin McLeod and Jim Marshall, work which focussed mainly on the treatment of boundaries—the approximation of curved boundaries and the exact matching of boundary data using blending interpolants (see the Addendum).

In the booklet commemorating Lothar Collatz and published by the University of Hamburg, a lecture given by Ron in 1973 is fondly recalled in which he was dealing with a finite element with one curved side. A crude reproduction of Ron's diagram is shown below where he was explaining how to approximate the curved side (solid line) by a "parabolic arc" (dashed curve). There were some gasps from the audience which caused him to explain—"Of course a parabola does not behave in this way, except it is a Scottish parabola!"

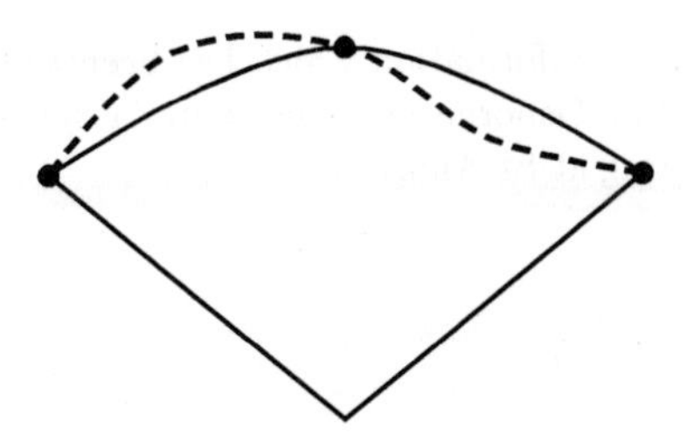

The next change of direction occurred as a consequence of a lecture given by Olec Zienkiewicz at the second MAFELAP conference organized by John Whiteman at Brunel University in 1975. In this talk Olec Zienkiewicz described instabilities they had experienced in converting their successful finite element codes for structural problems into codes for solving the Navier–Stokes and related equations in fluid dynamics. Finite difference practitioners had known for many years that the instability could be overcome by the use of "upwind differencing" and Ron was immediately intrigued to know how this type of stabilization could be applied to the finite element situation.

On his return to Dundee, he and David Griffiths attacked this problem with some gusto over the next few weeks and the end result was upwind-biased test functions, and what is now known as the Petrov–Galerkin finite element method (this term was coined, we believe, in a joint paper Ron wrote with Bob Anderssen). Ian Christie, who was on the MSc course at that time, then developed the ideas further in both his MSc and PhD dissertations.

There followed several fruitful years working on convection–diffusion problems until, through his interest in diffusion and dispersion effects and his collaboration with Brian Sleeman, he became interested in nonlinear effects in the early 1980's. Some of the problems arose from Mathematical Biology, on which "Mano" Manoranjan did much of his PhD work, but Ron was also interested in solitons, particularly those arising from the Korteweg–de Vries and Schrödinger equations. He was instrumental in bringing the subject of spurious solutions to the fore. Nonlinearity continues to be his abiding passion as he currently wrestles with the Korteweg–de Vries–Burgers equation.

As in most of the areas in which Ron has worked, he has had the uncanny knack of alighting on fundamental issues which, through his many papers and conference talks, have drawn others to the subject. He has a long and illustrious list of publications, and a complete bibliography is given as Appendix 1. Equally if not more impressive is the list of his PhD students, given in Appendix 2. One of his great strengths is the way he has been able to motivate and encourage his students, and this is borne out by the many who have gone on to great things; he has a truly outstanding talent for getting the best out of research students and for instilling self-confidence in them.

Acknowledgement

This article is based on information which has been collected from a number of people. In particular Mike Osborne provided material for Section 1 concerning the origins of the 1965 meeting at St Andrews.

ADDENDUM: THE FINITE ELEMENT METHOD AND COMPUTER AIDED GEOMETRIC DESIGN

ROBERT E. BARNHILL
Office of the Vice President for Research,
Box 872703, Arizona State University, Tempe, AZ 85287-2703
E-mail: barnhill@asu.edu

She bends most gracefully who resists the bending.

This quotation, attributed to Rayleigh, in connection with the calculus of variations, elegantly expresses the ever–present search for æsthetic optimization. The Finite Element Method (FEM), or Rayleigh–Ritz–Galerkin method, is a particular instance of optimization.

Computer Aided Geometric Design (CAGD) is the representation and approximation of free form curves, surfaces, and volumes in an interactive computer graphics environment. CAGD has roots both in classical approximation theory and in practical problems. Approximation theory precedents for CAGD can be seen in Philip Davis-Interpolation and Approximation(1963) where he refers to the potential usefulness of Bernstein approximations for shape preservation: this led to what is called Bezier approximations. Davis also discussed two–point Hermite interpolation, which, applied to surfaces, is the foundation for Coons Patches. The early practical situations which called for CAGD methods included the automobile and airplane industries: Bezier and Coons Patches were developed for and first used in the former. Bicubic patches formed as tensor products of univariate cubic Hermite interpolants were used by Ferguson in the aircraft industry. Both of these sets of developments first occurred during the 1960s. CAGD involves either Design or Representation. The automobile and aircraft applications are those of Design. Representation examples include modelling given data, such as from physical or biological measurements. In this case interpolation is usually desired and, for data in general locations, leads to "scattered data interpolation".

The Finite Element Method provides a solution to variational principles corresponding to differential equations. The variational principle is solved approximately, in that a finite family of functions replaces the infinite set of functions that satisfy the boundary conditions and are sufficiently smooth for the appropriate integration by parts. The creative step is to determine this finite family suitably. For concreteness, consider the domain to be a nice region in the plane, a polygon, for example. Then the region is divided into pieces and, over the network of pieces, suitable approximating functions are defined. Typically, piecewise defined interpolants are used for the approximating functions. In the FEM, the data for the interpolants are not known: determining these values is in fact the point of the FEM. In the FEM these values are determined by the solution of a least squares approximation in the energy seminorm appropriate for the variational principle. For example, Poisson's Equation corresponds to the Sobolev seminorm involving first derivatives.

The basis functions from FEM are frequently used in CAGD, because of their optimal properties. A very basic example is the cubic interpolatory spline for curves; cubic interpolatory splines satisfy the æsthetic variational principle of minimizing the linearized curvature. (Although some refer by "splines" to any piecewise defined function, I prefer "splines" to mean that the function satisfies a variational principle.) CAGD uses these FEM basis functions in the following two fundamental situations:

Design: a curve, surface or volume is constructed that conforms to a designer's geometric ideas. Examples range from automobiles and airplanes to computer animation.

Representation: data from an applications area are interpolated exactly. Examples of such modelling include brain mapping and oil exploration.

Personal notes: Early CAGD–FEM connection: I helped give CAGD its name in its defining conference at The University of Utah in the spring of 1974. The idea was to delineate the more mathematical aspects of Computer Aided Design. (CAD). Then I spent the 1974–75 school year at the University of Dundee in order to work with Ron Mitchell there. Ron and I, and his students Jim Marshall and Jim Brown, combined CAGD and FEM by studying the effect of exactly matching boundary conditions on the quality of the approximations. Various aspects of this research continued thereafter.

Ron gave me two other great gifts during that year: a sharpened sense of humor and a love for soccer. However, he disclaims responsibility for my implementation of the latter! Happy birthday, Ron.

APPENDIX 1: LIST OF PUBLICATIONS OF A. R. MITCHELL

BOOKS

1. *Computational Methods in Partial Differential Equations.* J. Wiley & Sons, London, 1969.
2. *Finite Element Methods in Partial Differential Equations.* J. Wiley & Sons, London, 1976. (with R. Wait).
3. *The Finite Difference Method in Partial Differential Equations.* J. Wiley & Sons, London, 1980. (with D. F. Griffiths).
4. *Finite Element Analysis and Applications.* J. Wiley & Sons, London, 1985. (with R. Wait).

PAPERS

1. Application of relaxation methods to compressible flow past a double wedge. *Proc. Roy. Soc. Edinburgh*, 63:139–154, 1951. (with D. E. Rutherford).
2. Application of relaxation to the rotational flow behind a bow shock wave. *Quart. J. Mech. and Appl. Math.*, 4:371–383, 1951.
3. The rotational field behind a bow shock wave in axially symmetric flow using relaxation methods. *Proc. Roy. Soc. Edinburgh*, 63:371–380, 1952.
4. Stability of difference relations in the solution of ordinary differential equations. *Math. Tables and other Aids to Comp.*, 7:127–129, 1953. (with J. W. Craggs).
5. Round–off errors in the solution of the heat conduction equation by relaxation methods. *Appl. Sci. Res.*, 3:109–119, 1953.
6. Round–off errors in relaxation solutions of Poisson's equation. *Appl. Sci. Res.*, 4:456–454, 1954.
7. On the theory of relaxation. *Proc. Glasgow Math. Assoc.*, 1:101–110, 1955. (with D. Rutherford).
8. Two dimensional flow with constant shear past cylinders with various cross sections. *ZAMP*, 6:223–235, 1957. (with J. D. Murray).
9. Round-off errors in implicit finite difference methods. *Quart. J. Mech. and Appl. Math.*, 9:111–121, 1956.
10. Flow with variable shear past circular cylinders. *Quart. J. Mech. and Appl. Math.*, 10:13–23, 1957. (with J. D. Murray).
11. Boundary value techniques for initial value problems in ordinary differential equations. *Quart. J. Mech. and Appl. Math.*, 10:232–243, 1957. (with L. Fox).
12. The application of finite difference methods to the solution of problems in boundary layer flow. *Proc. Int. Appl. Math. Congress, Brussels*, 1957. (with J. Y. Thompson).

13. Finite difference methods of solution of the Von Mises boundary layer equation with special reference to conditions near a singularity. *ZAMP*, 9:26–37, 1958. (with J. Y. Thompson).
14. The influence of critical boundary conditions on finite difference solutions of two point boundary problems. *Math. Tables and other Aids to Comp.*, 13:252–260, 1959.
15. The effect of boundary condition and mesh size on the accuracy of finite difference solutions of two point boundary value problems. *ZAMP*, 10:221–232, 1959. (with D. Borwein).
16. Solution of the Von Mises boundary layer equation using a high speed computer. *Math. Comp.*, 15:238–242, 1961.
17. On the solution of $y' = f(x,y)$ by a class of high accuracy difference formulae. *ZAMP*, 13:223–232, 1962. (with J. D. Lambert).
18. On finite difference methods of solution of the transport equation. *Math. Comp.*, 16:155–169, 1962. (with R. P. Pearce).
19. High accuracy difference formulae for the numerical solution of the heat conduction equation. *The Computer Journal*, 5:142–146, 1963. (with R. P. Pearce).
20. The use of higher derivatives in quadrature formulae. *The Computer Journal*, 5:322–327, 1963. (with J. D. Lambert).
21. Repeated quadrature using derivatives of the integrand. *ZAMP*, 15:84–90, 1964. (with J. D. Lambert).
22. Explicit difference methods for solving the cylindrical heat conduction equation. *Math. Comp.*, 17:426–432, 1963. (with R. P. Pearce).
23. A generalised ADI method of Douglas–Rachford type for solving the biharmonic equation. *The Computer Journal*, 7:242–245, 1964. (with G. Fairweather).
24. Improved forms of the ADI methods of Douglas, Peaceman and Rachford for solving parabolic and elliptic equations. *Numer. Math.*, 6:285–292, 1964. (with G. Fairweather).
25. A new ADI method for parabolic equations in three space variables. *SIAM J.*, 13:957–965, 1965. (with G. Fairweather).
26. A high accuracy ADI method for the wave equation. *J. Inst. Maths Applics.*, 1:309–316, 1965. (with G. Fairweather).
27. Some computational results of an improved ADI method for the Dirichlet problem. *The Computer Journal*, 9:9–15, 1966. (with G. Fairweather).
28. Two level difference schemes for hyperbolic systems. *SIAM J. Numer. Anal.*, 3:474–485, 1966. (with A. R. Gourlay).
29. A stable implicit difference method for hyperbolic systems in two space variables. *Numer. Math.*, 8:367–375, 1966. (with A. R. Gourlay).
30. Alternating direction methods for hyperbolic systems. *Numer. Math.*, 8:137–

149, 1966. (with A. R. Gourlay).
31. On the instability of the Crank–Nicholson formula under derivative boundary conditions. *The Computer Journal*, 9:110–114, 1966. (with P. Keast).
32. Split operator methods for hyperbolic systems in p space variables. *Math. Comp.*, 21:351–354, 1967. (with A. R. Gourlay).
33. Intermediate boundary corrections for split operator methods in three dimensions. *BIT*, 7:31–38, 1967. (with A. R. Gourlay).
34. Finite difference solution of the third boundary problem in elliptic and parabolic equations. *Numer. Math.*, 10:67–75, 1967. (with P. Keast).
35. A new computational procedure for ADI methods. *SIAM J. Numer. Anal.*, 4:163–170, 1967. (with G. Fairweather).
36. Some high accuracy difference schemes with a splitting operator for equations of parabolic and elliptic type. *Numer. Math.*, 10:56–66, 1967. (with G. Fairweather and A. R. Gourlay).
37. High accuracy ADI methods for parabolic equations with variable coefficients. *Numer. Math.*, 12:180–185, 1968. (with A. R. Gourlay).
38. The equivalence of certain alternating direction and locally one–dimensional difference methods. *SIAM J. Numer. Anal.*, 6:37–46, 1969. (with A. R. Gourlay).
39. A classification of split difference methods for hyperbolic equations in several space dimensions. *SIAM J. Numer. Anal.*, 6:62–71, 1969. (with A. R. Gourlay).
40. Alternating direction methods for parabolic equations in two space dimensions with a mixed derivative. *The Computer Journal*, 13:81–86, 1970. (with S. McKee).
41. Splitting methods in partial differential equations. *Abhandlungen aus dem Mathematischen Seminar der Universität Hamburg*, 36:45–56, 1971.
42. The solution of time dependent problems by Galerkin methods. *J. Inst. Maths Applics.*, 7:241–250, 1971. (with R. Wait).
43. The finite element method in partial differential equations. *Methoden und Verfahren der Mathematischen Physik*, 5:101–115, 1971.
44. Corner singularities in elliptic problems by finite element methods. *J. Comp. Phys.*, 8:45–52, 1971. (with R. Wait).
45. Alternating direction methods for parabolic equations in three space dimensions with mixed derivatives. *The Computer Journal*, 14:295–300, 1971. (with S. McKee).
46. Forbidden shapes in the finite element method. *J. Inst. Maths Applics.*, 8:260–269, 1971. (with G. M. Phillips and E. L. Wachspress).
47. Variational principles and the finite element method in partial differential equations. *Proc. Roy. Soc. London*, A323:211–217, 1971.
48. On the structure of ADI and LOD difference methods. *J. Inst. Maths Applics.*,

9:80–90, 1972. (with A. R. Gourlay).
49. Construction of basis functions in the finite element method. *BIT*, 12:81–89, 1972. (with G. M. Phillips).
50. Basis functions for curved elements in the finite element method. In *Proc. Second Manitoba Conf. on Num. Math.*, pages 35–50, 1972.
51. An introduction to the mathematics of the finite element method. In J. R. Whiteman, editor, *The Mathematics of Finite Elements and Applications I*, pages 37–58. Academic Press, 1972.
52. Variational principles and the finite element method. *J. Inst. Maths Applics.*, 9:378–389, 1972.
53. The construction of basis functions for curved elements in the finite element method. *J. Inst. Maths Applics.*, 10:382–393, 1972. (with R. J. Y. McLeod).
54. An exact boundary technique for improved accuracy in the finite element method. *J. Inst. Maths Applics.*, 12:355–362, 1973. (with J. A. Marshall).
55. Variational principles—a survey. In J. G. Gram, editor, *Numerical Solution of Partial Diffential Equations*, NATO Advanced Study Series, pages 17–23. D. Reidel Press, 1973.
56. Element types and base functions. In J. G. Gram, editor, *Numerical Solution of Partial Diffential Equations*, NATO Advanced Study Series, pages 107–150. D. Reidel Press, 1973.
57. Curved elements in the finite element method. In G. A. Watson, editor, *Numerical Solution of Differential Equations*, pages 89–104. Springer Verlag, 1973. (with R. J. Y. McLeod).
58. Numerical analysis. In *Encyclopaedia Brittanica*, pages 381–392. 1974.
59. Matching of essential boundary conditions in the finite element method. In J. J. H. Miller, editor, *Topics in Numerical Analysis II*, pages 109–120. Academic Press, 1974. (with J. A. Marshall).
60. Curved boundaries in the finite element method. In J. Albrecht and L. Collatz, editors, *Finite Elemente und Differenzverfahren*, volume 28 of *ISNM*, pages 71–90. Birkhauser Verlag, 1975.
61. The use of parabolic arcs in matching curved boundaries in the finite element method. *J. Inst. Maths Applics.*, 16:139–246, 1975. (with R. J. Y. McLeod).
62. Basis functions for curved elements in the mathematical theory of finite elements. In J. R. Whiteman, editor, *The Mathematics of Finite Elements and Applications II*, pages 43–58. Academic Press, 1975.
63. Finite element methods for second order differential equations with significant first derivatives. *Int. J. Numer. Meth. Engng.*, 10:1389–1396, 1976. (with I. Christie, D. F. Griffiths and O. C. Zienkiewicz).
64. Finite element methods in conduction–convection problems. In J. Albrecht and L. Collatz, editors, *Numerische Behandlung von Differentialgleichungen*, volume 31 of *ISNM*, pages 171–179. Birkhauser Verlag, 1976.

65. Exact control of a parabolic differential equation using an implicit Runge–Kutta method. *J. Inst. Maths Applics.*, 18:9–14, 1976. (with P. J. Harley).
66. Finite element methods in time dependent problems. In D. A. H. Jacobs, editor, *The State of the Art in Numerical Analysis*, pages 671–692. Academic Press, 1976.
67. Generalized Galerkin methods for second order equations with significant first derivatives. In G. A. Watson, editor, *Numerical Analysis, Dundee*, pages 90–104. Springer, 1977. (with D. F. Griffiths).
68. An upwind finite element scheme for two–dimensional convective transport equation. *Int. J. Numer. Meth. Engng.*, 11:131–143, 1977. (with J. C. Heinrich, P. S. Huyakorn and O. C. Zienkiewicz).
69. A finite element collocation method for the exact control of a parabolic problem. *Int. J. Numer. Meth. Engng.*, 11:345–353, 1977. (with P. J. Harley).
70. Computable finite element error bounds for Poisson's equation. *Int. J. Numer. Meth. Engng.*, 11:593–603, 1977. (with R. E. Barnhill, J. H. Brown and N. McQueen).
71. Upwinding of high order Galerkin methods in conduction–convection problems. *Int. J. Numer. Meth. Engng.*, 12:1764–1771, 1978. (with I. Christie).
72. Blending interpolants in the finite element method. *Int. J. Numer. Meth. Engng.*, 12:77–83, 1978. (with J. A. Marshall).
73. The parabolic–hyperbolic interface in conduction–convection problems. In I. Marek, editor, *Proc. Fourth Symposium on Basic Problems in Numerical Mathematics*, pages 141–151. Czech Academy of Sciences, 1978. (with F. Lawlor).
74. Semi–discrete generalised Galerkin methods for time–dependent conduction–convection problems. In J. R. Whiteman, editor, *The Mathematics of Finite Elements and Applications III*, pages 19–35. Academic Press, 1978.
75. Finite difference/finite element methods at the parabolic–hyperbolic interface. In P. W. Hemker and J. J. H. Miller, editors, *Numerical Analysis of Singular Perturbation Problems*, pages 339–361. Academic Press, 1978. (with I. Christie).
76. Blending interpolants in the finite element method. *Int. J. Numer. Meth. Engng.*, 12:77–83, 1978. (with J. A. Marshall).
77. An analysis of a cubic isoparametric transformation. *Int. J. Numer. Meth. Engng.*, 12:1587–1595, 1978. (with G. Woodford and R. J. Y. McLeod).
78. Upwinding of high order Galerkin methods in conduction–convection problems. *Int. J. Numer. Meth. Engng.*, 12:1764–1771, 1978. (with I. Christie).
79. Semi–discrete generalized Galerkin methods for time–dependent conduction–convection problems. In J. R. Whiteman, editor, *The Mathematics of Finite Elements and Applications III*, pages 19–34. Academic Press, 1979. (with D. F. Griffiths).

80. On generating upwind finite elements. In T. J. R. Hughes, editor, *Finite Elements for Convection Dominated Flows, AMD vol 34*, pages 91–105, New York, 1979. ASME. (with D. F. Griffiths).
81. A piecewise parabolic C^1 approximation technique for curved boundaries. *Computers and Math. with Applics.*, 5:277–284, 1979. (with R. J. Y. McLeod).
82. Advantages of cubics for approximating element boundaries. *Computers and Math. with Applics.*, 5:321–327, 1979.
83. Analysis of generalised Galerkin methods in the numerical solution of elliptic equations. *Math. Meth. in the Appl. Sci.*, 1:3–15, 1979. (with R. S. Anderssen).
84. Analysis of error growth for explicit difference schemes in conduction convection problems. *Int. J. Numer. Meth. Engng.*, 15:1075–1081, 1980. (with I. Christie and D. F. Griffiths).
85. Upwinding by Petrov–Galerkin methods in convection–diffusion problems. *J. Comp. and Appl. Math.*, 6:219–228, 1980. (with D. F. Griffiths).
86. Finite element Galerkin methods for convection–diffusion and reaction–diffusion. In L. S. Frank O. Axelsson and A. van der Sluis, editors, *Analytical and Numerical Approaches to Asymptotic Problems in Analysis.* North Holland, 1981. (with D. F. Griffiths and A. F. Meiring).
87. A comparison of finite element error bounds for Poisson's equation. *IMA J. Num. Anal.*, 1:95–103, 1981. (with R. E. Barnhill and J. H. Brown).
88. Product approximation for non–linear problems in the finite element method. *IMA J. Num. Anal.*, 1:253–266, 1981. (with I. Christie, D. F. Griffiths and J. M. Sanz–Serna).
89. The stability of a Petrov–Galerkin method for the periodic convection–diffusion problem. *Math. Num. Sinica*, 4:138–146, 1982. (with D. F. Griffiths and Guo Pen–Yu).
90. Finite element studies in reaction–diffusion. In J. R. Whiteman, editor, *The Mathematics of Finite Elements and Applications IV*, pages 17–36. Academic Press, 1982. (with V. S. Manoranjan).
91. A moving Petrov–Galerkin method for transport equations. *Int. J. Numer. Meth. Engng.*, 18:1321–1336, 1982. (with B. M. Herbst and S. W. Schoombie).
92. A numerical study of the Belousov–Zabotinskii reaction using Galerkin finite element methods. *J. Math. Biol.*, 16:251–260, 1983. (with V. S. Manoranjan).
93. The stability of the Petrov–Galerkin method for the initial boundary value problem for the convection-diffusion equation. *Acta Math Sinica*, 26:54–64, 1983. (with Guo Ben Yu).
94. Equidistributing principles in moving finite element methods. *J. Comp. and Appl. Math.*, 9:377–389, 1983. (with B. M. Herbst and S. W. Schoombie).
95. A self-adaptive difference scheme for the nonlinear Schrödinger equation. *Arab Gulf J. of Scientific Research*, 1:461–473, 1983. (with J. Ll. Morris).
96. A numerical study of the Belousov–Zhabotinskii reaction using Galerkin finite

element methods. *J. Math Biology*, 16:251–260, 1983. (with V. S. Manoranjan).

97. Recent developments in the application of finite element methods to current problems in applied mathematics. In R. Alani, editor, *First Int. Conf. in Math.*, pages 1–26. Arab Bureau of Education for the Gulf States, 1983.
98. A numerical study of the nonlinear Schrödinger equation. *Comp. Meth. Appl. Mech. Engng.*, 45:177–216, 1984. (with D. F. Griffiths and J. Ll. Morris).
99. Finite element studies of solitons. In R. W. Lewis, P. Bettess, and E. Hinton, editors, *Numerical Methods in Coupled Systems*, pages 465–489. John Wiley & Sons, 1984. (with S. W. Schoombie).
100. Generalized Petrov–Galerkin methods for the numerical solution of Burger's equation. *J. Comp. Phys.*, 20:1273–1289, 1984. (with B. M. Herbst, D. F. Griffiths and S. W. Schoombie).
101. Numerical solutions of the good Boussinesq equation. *SIAM J. Sci. Stat. Comp.*, 5:946–957, 1984. (with V. S. Manoranjan and J. Ll. Morris).
102. Curved elements. In D. F. Griffiths, editor, *The Mathematical Basis of the Finite Element Method*, IMA Conference Series, pages 157–168. Oxford University Press, 1984.
103. Nonconforming elements. In D. F. Griffiths, editor, *The Mathematical Basis of the Finite Element Method*, pages 41–70. Oxford University Press, 1984. (with D. F. Griffiths).
104. Recent developments in the finite element method. In *Computational Techniques and Applications*, pages 2–14. North Holland Press, 1984.
105. Bifurcation studies in reaction-diffusion. *J. Computational and Applied Mathematics*, 11:27–37, 1984. (with V. S. Manoranjan, B. D. Sleeman and Guo Ben Yu).
106. Numerical studies of bifurcation and pulse evolution in mathematical biology. In J. R. Whiteman, editor, *The Mathematics of Finite Elements and Applications V*, pages 175–191. Academic Press, 1985. (with V. S. Manoranjan).
107. On the stability of the nonlinear Schrödinger equation. *J. Comp. Phys.*, 60:263–281, 1985. (with B. M. Herbst and J. A. C. Weideman).
108. Numerical experience with the nonlinear Schrödinger equation. *J. Comp. Phys.*, 60:282–305, 1985. (with B. M. Herbst and J. Ll. Morris).
109. A numerical study of chaos in a reaction–diffusion equation. *Num. Methods PDEs.*, 1:13–23, 1985. (with J. C. Bruch Jr).
110. Adaptive grids in Petrov–Galerkin computations. In *Accuracy Estimates and Adaptive Refinements in Finite Element Computations*, pages 315–324. John Wiley & Sons, Chichester, 1986. (with B. M. Herbst).
111. Analysis of a nonlinear difference scheme in reaction-diffusion. *Numer. Math.*, 49:511–527, 1986. (with Guo Ben–Yu).
112. Beyond the linearised stability limit in nonlinear problems. In D. F. Griffiths

and G. A. Watson, editors, *Numerical Analysis Proceedings, Dundee*, pages 140–156. Pitman Research Notes in Mathematics Series, Vol 140, 1986. (with D. F. Griffiths).

113. On nonlinear instabilities in leap-frog finite difference schemes. *J. Comp. Phys.*, 67:372–395, 1986. (with D. M. Sloan).
114. Spatial effects in a two–dimensional model of the budworm–balsam fir ecosystem. *Comp. and Math. with Applics.*, 12B:1117–1132, 1986. (with Guo Ben–Yu and B. D. Sleeman).
115. On a one–dimensional difference scheme in reaction diffusion. *J. Comp. Math.*, 5:191–202, 1987. (with Guo Ben–Yu and Chen Sui–Yang).
116. Period doubling bifurcations in non–linear difference equations. *SIAM J. Sci. Stat. Comp.*, 9:543–557, 1988. (with B. D. Sleeman, D. F. Griffiths and P. D. Smith).
117. Periodic structure beyond a Hopf bifurcation. *Commun. Appl. Numer. Meth.*, 4:263–272, 1988. (with G. Stein and M. Maritz).
118. Nonlinear instability in long time calculations of a partial difference equation. In *Comp. Meth in Water Resources VII*, pages 153–161. Elsevier, 1988.
119. Nonlinear diffusion and stable period 2 solutions of a discrete reaction-diffusion model. *J. Comp. and Appl. Math.*, 25:363–372, 1989. (with S. W. Schoombie).
120. Nonlinear diffusion and stable period 2 solutions of a discrete reaction-diffusion model. *J. Comp. and Appl. Math.*, 25:363–372, 1989. (with S. W. Schoombie).
121. Spurious behaviour and nonlinear instability in discretised partial differential equations. In D. S. Broomhead and A. Iserles, editor, *Proceedings of Dynamics of Numerics and Numerics of Dynamics*, pages 215–242. Cambridge University Press, 1990. (with D. F. Griffiths).
122. Spatial patterning of the Spruce budworm in a circular region. *Science in China (Series A)*, 34:676–688, 1991. (with Guo Ben–Yu and B. D. Sleeman).
123. Numerical solution of Hamiltonian systems in reaction–diffusion by symplectic difference schemes. *J. Comp. Phys.*, 95:339–358, 1991. (with B. A. Murray and B. D. Sleeman).
124. Attractors in long time solutions of discretised nonlinear systems. In T. F. Russell and R. E. Ewing, editors, *Numerical Methods in Water Resources*, pages 107–120. Elsevier Applied Science, 1992.
125. Bifurcation studies in long time solutions of discrete mode in reaction diffusion. *University of Dundee Report* NA/136, 1992. (with A. R. Gardiner).
126. Periodic bifurcations of an explicit discretisation of the Ginzburg–Landau equation. *University of Dundee Report* NA/137, 1992. (with A. R. Gardiner).
127. Unstable attractors of a discretised inviscid Burgers equation. *University of Dundee Report* NA/139, 1992. (with A. R. Gardiner).
128. The influence of dispersion on the inviscid Burgers equation. *University of Dundee Report* NA/142, 1992. (with A. R. Gardiner).

APPENDIX 2: PhD STUDENTS OF A. R. MITCHELL

Student	Date of Degree Award	Known/Most Recent Location
J. D. Murray	1955	University of Washington
J. Y. Thompson	1957	
J. D. Lambert	1963	University of Dundee
G. Fairweather	1965	Colorado School of Mines
A. R. Gourlay	1966	IBM
P. Keast	1967	Dalhousie University
J. Ll. Morris	1967	University of Dundee
J. S. McKee	1970	University of Strathclyde
R. Wait	1971	University of Liverpool
R. J. Y. MacLeod	1972	Saltire Software Inc.
K. M. Vine	1973	Royal Navy, Shrivenham
P. J. Harley	1974	University of Sheffield
G. Woodford	1975	
J. A. Marshall	1975	
J. H. Brown	1976	
D. P. Laurie	1977	University of Potchefstroom
I. Christie	1977	West Virginia University
I. M. Snyman*	1979	University of South Africa
F. M. M. Lawlor	1979	
A. F. Meiring*	1980	University of Pretoria
V. S. Manoranjan	1982	Washington State University
R. Schofield	1982	
S. Herring	1987	
P. Cumber*	1988	
Y. Tourigny*	1988	University of Bristol
P. M. John–Charles	1988	
A. R. Gardiner	1991	University of Sussex

* Jointly supervised.

FIXED POINTS AND SPURIOUS MODES OF A NONLINEAR INFINITE-STEP MAP

MARK A. AVES*
E-mail: maves@mcs.dundee.ac.uk

PENNY J. DAVIES
E-mail: pdavies@mcs.dundee.ac.uk

and

DESMOND J. HIGHAM*
E-mail: dhigham@mcs.dundee.ac.uk
Department of Mathematics and Computer Science,
University of Dundee, Dundee, DD1 4HN, Scotland

ABSTRACT

We study discretisations of an integro-differential equation with convolution kernel that arises in population dynamics. This leads to infinite-step variations of the discrete logistic map (which is perhaps the canonical example of a simple, nonlinear map that gives rise to complicated behaviour). We focus attention on the existence and stability of correct, approximate and spurious steady states, and examine how these properties are affected by the choice of quadrature rule.

1. Background

Time-stepping plays an essential role in the numerical solution of initial-value differential systems. Traditionally, analysis of time-stepping methods has been restricted to either

- convergence (as some mesh parameter tends to zero) over a compact time interval; or
- stability for linear problems.

More recently, however, attention has turned to a scenario that lies outside these two extremes—the long term behaviour of numerical methods on nonlinear problems. Results in this area have direct relevance to applications where steady state behaviour of a differential system is to be investigated by numerical simulation.

Consider a nonlinear evolution equation

$$u'(t) = f(u), \quad u(0) = u_0, \quad f : \mathbb{R}^m \to \mathbb{R}^m, \tag{1}$$

*These authors acknowledge funding by the Engineering and Physical Sciences Research Council of the U.K. under research grant number GR/H94634.

(which may represent a spatially discretised partial differential equation). With a fixed time-step Δt any standard k-step numerical method will produce an iteration of the form

$$U_{n+1} = F_{\Delta t}(U_n, U_{n-1}, \ldots, U_{n-k}), \qquad F_{\Delta t} : \mathbb{R}^m \times \mathbb{R}^m \times \cdots \times \mathbb{R}^m \to \mathbb{R}^m, \tag{2}$$

where U_n is the numerical approximation of $u(n\Delta t)$. We are concerned with long term behaviour, and hence the appropriate limits are $t \to \infty$ in (1) and $n \to \infty$, with Δt fixed, in (2). In studying the long term behaviour of the map (2) and the underlying flow (1), an important first step is to examine the fixed points. In particular, it is desirable that stable fixed points of the flow are preserved by the numerical method, and that no extra (spurious) stable fixed points are introduced by the discretisation.

Results about spurious fixed points have been obtained on a range of nonlinear equations. Often, problems are chosen for which there is known to be a stable steady state $u(t) \equiv u^\star$. After linearising (1) about $u^\star$ and asking for the numerical method to behave appropriately on the linearised problem, we obtain a *linearised stability interval* $(0, \Delta t^\star)$ for the time-step. It is then possible to investigate

1. Spurious behaviour that occurs inside the linear stability interval, $\Delta t \in (0, \Delta t^\star)$.

2. Spurious solutions that bifurcate from $u^\star$ at $\Delta t = \Delta t^\star$.

3. Spurious solutions that exist beyond the linear stability interval, $\Delta t > \Delta t^\star$.

Case 1 is clearly of relevance, since it applies when the time-step is small, in the linearised sense. However, cases 2 and 3 are also worthy of study since, on complicated problems, the size of $\Delta t^\star$ may be difficult to determine in practice.

It is appropriate to mention that many results in this area have been produced by Ron Mitchell and his co-workers at Dundee, and we conclude this section by citing a sample of relevant references.

In Mitchell et al.[16] a connection was made between simple discretisatons of nonlinear problems and discrete maps that have been widely studied in the context of dynamical systems. Finite difference discretisations of nonlinear partial differential equations (PDEs) have also been studied[5,15], where the analysis focuses on bifurcations and spurious solutions for time-steps that exceed the linear stability limit. Griffiths and Mitchell[6] considered a number of issues relating to long term dynamics on nonlinear PDEs, including spurious solutions, blow up and travelling waves. Gardiner and Mitchell[3,4] studied spuriosity arising from reaction-diffusion problems with homogeneous Dirichlet and periodic boundary conditions. Here, the aims were to exhibit solutions that are periodic in time and/or space and to illustrate the effect of varying the amount of dissipation. It was shown that the long term behaviour of the solution can be extremely sensitive to the initial conditions, and that unstable periodic solutions may have an impact on the basins of attractions of the stable solutions.

Mitchell and Gardiner[13,14] performed extensive numerical tests on the inviscid Burgers' equation and investigated known theoretical results concerning the development of shock wave and soliton solutions.

2. The Continuous Problem

Perhaps the simplest example of a nonlinear ordinary differential equation (ODE) is the logistic equation

$$u'(t) = u(t)\,(1 - u(t)) \;, \quad \text{for } t > 0 \;, \tag{3}$$

with $u(0)$ given. This equation (in its non-normalised form) has been proposed as a basic model for a population that is susceptible to overcrowding. There are two steady states: the unstable $u(t) \equiv 0$ and the stable $u(t) \equiv 1$, which attracts all positive initial data.

Equation (3) has proved useful for illustrating the potential for spurious behaviour in numerical methods. For example, Iserles[9], Mitchell et al.[16] and Yee et al.[17] consider discretisations of (3), and Griffiths and Mitchell[5] and Higham and Owren[8] use the right-hand side as a reaction term in a time-dependent PDE.

A variation on the logistic ODE is the delay differential equation

$$u'(t) = u(t)\,(1 - u(t - T)) \;, \quad \text{for } t > 0 \;, \tag{4}$$

with $u(t)$ given for $-T \leq t \leq 0$, where $T > 0$ is the constant *delay.* In this case the growth rate depends upon the population size at a time T units earlier. It can be shown that the fixed point $u(t) \equiv 0$ is always unstable, while the fixed point $u(t) \equiv 1$ is linearly stable when $T < \pi/2$ (see, for example, May[12]). A discretisation of (4) is studied by Higham[7], and the dynamics are compared with those arising from the non-delay problem (3).

For a population model, it has been argued that it is more realistic to replace the single delayed value $u(t-T)$ in (4) by an average over past values[2,12]. This reasoning leads to the integro-differential equation (IDE)

$$u'(t) = u(t)\left(1 - \int_{-\infty}^{t} \mathrm{k}(t-\tau)\,u(\tau)\,d\tau\right) \;, \quad \text{for } t > 0 \;, \tag{5}$$

with $u(t)$ given for $t \leq 0$. Here, the kernel $\mathrm{k}(t)$ determines the weighting that is attached to the delayed values. We consider the case

$$\mathrm{k}(t) = t\; \mathrm{e}^{-t/T} / T^2 \;, \quad \text{where } T > 0 \text{ is constant} \;. \tag{6}$$

(An alternative kernel, $\mathrm{k}(t) = \mathrm{e}^{-t/T}/T$, is studied by Aves et al.[1].) Note that with the choice (6), the weighting $\mathrm{k}(t - \tau)$ in (5) is zero at $\tau = t$, maximum at $\tau = t - T$ and decays exponentially as $\tau \to -\infty$. Hence we may continue to regard T as the delay parameter.

The IDE (5) has fixed points $u(t) \equiv \{0,1\}$, whose linear stability can been studied via the Laplace transform[2,12]. For the non-zero fixed point we let $u(t) = 1 + \epsilon(t)$, linearise and truncate the integral to obtain

$$\epsilon'(t) = \int_0^t \mathrm{k}(t-\tau)\,\epsilon(\tau)\,d\tau \ .$$

For this linear convolution problem stability is equivalent to the condition

$$z - \widehat{\mathrm{k}(z)} \neq 0 \ , \quad \text{for all } \Re\{z\} \geq 0 \ , \tag{7}$$

where $\widehat{\mathrm{k}(z)}$ is the Laplace transform

$$\widehat{\mathrm{k}(z)} = \int_0^\infty \mathrm{k}(s)\,\mathrm{e}^{-zs}\,ds \ .$$

For the kernel (6) we have $\widehat{\mathrm{k}(z)} = -(zT+1)^{-2}$ and, by applying the Routh-Hurwitz criterion[10], the stability condition (7) reduces to $T < 2$.

3. The Discretisation

We discretise the IDE (5) by using Euler's Method for the time derivative and a quadrature rule for the integral. Applying Euler's Method produces the intermediate recurrence

$$U_{n+1} = U_n + \Delta t\, U_n \left(1 - \sum_{j=0}^{\infty} \int_{(n-j-1)\Delta t}^{(n-j)\Delta t} \mathrm{k}(n\Delta t - s)\, U(s)\, ds\right) \ , \tag{8}$$

where $U(s) \approx u(s)$ and we use the integral identity

$$\int_{-\infty}^{n\Delta t} \mathrm{k}(nh-\tau)\, U(\tau)\, d\tau = \sum_{j=0}^{\infty} \int_{(n-j-1)\Delta t}^{(n-j)\Delta t} \mathrm{k}(n\Delta t - s)\, U(s)\, ds \ .$$

When quadrature based on the discrete values $\{U_j\}$ is applied to the convolution integral in equation (8), we have a recurrence of the general form

$$U_{n+1} = U_n + \Delta t\, U_n \left(1 - \sum_{j=0}^{\infty} \omega_j\, U_{n-j}\right) \ , \tag{9}$$

where the weights ω_j are determined by the quadrature formula. For example, assuming a piecewise linear (PL) variation of the solution over a time-step produces the infinite discrete map

$$U_{n+1} = U_n + \Delta t\, U_n \left(1 - \sum_{j=0}^{\infty} [(j+1)W_j - V_j]\, U_{n-j} + [V_j - jW_j]\, U_{n-j-1}\right) \ , \tag{10}$$

where the quadrature weights are equal to or approximate the kernel integrals:

$$W_j \approx \int_{j\Delta t}^{(j+1)\Delta t} \mathrm{k}(s)\, ds \ , \quad V_j \approx \frac{1}{\Delta t}\int_{j\Delta t}^{(j+1)\Delta t} s\,\mathrm{k}(s)\, ds \ . \tag{11}$$

Typically, for complicated kernels, the weights (11) are also calculated by a quadrature formula. We consider two cases, obtained by approximating the kernel by piecewise constants and piecewise linears. Since the kernel (6) can be integrated explicitly, we also consider exact integration. The other recurrences that we analyse are generated by taking $U(s)$ to be piecewise constant (PC). This simplifies (8) further, with the weights in the map (9) reducing to $\omega_j = W_j$.

4. Fixed Points

The fixed points of the map (9) are found by setting the solution U_j to be a constant for all j. We then find that (for $\Delta t \neq 0$) the map has two fixed points, $U_j \in \{0, U^\star\}$, where

$$U^\star = \left(\sum_{j=0}^{\infty} \omega_j\right)^{-1} \ . \tag{12}$$

For both the PC and PL solution approximations the non-zero fixed point (12) is given by $U^\star = \left(\sum_{j=0}^{\infty} W_j\right)^{-1}$. Note also that if the integrals in (11) are evaluated exactly then $U^\star$ is the true fixed point of the continuous equation (5).

For the kernel (6), a piecewise constant approximation for the weights W_j in (11) gives

$$W_j = \Delta t\, \mathrm{k}(j\Delta t) = r^2 j\, q^j \ , \tag{13}$$

where $r = \Delta t/T$ and $q = \mathrm{e}^{-r}$. If the kernel is instead approximated by piecewise linear sections then

$$W_j = r^2 q^j \left(j + (j+1)\, q\right)/2 \ . \tag{14}$$

Substituting these into equation (12) then gives the non-zero fixed point of the infinite discrete map (9) as

$$U^\star = \begin{cases} (1-q)^2\, r^{-2} q^{-1}, & \mathrm{k}(t) \text{ piecewise constant,} \\ (1-q)^2\, r^{-2} q^{-1}, & \mathrm{k}(t) \text{ piecewise linear,} \\ 1, & \mathrm{k}(t) \text{ exact.} \end{cases} \tag{15}$$

The fixed points obtained by the two approximate evaluations of the integrals in (11) are identical (this is not true in general, see Aves et al.[1]) and converge to the true fixed point of the continuous equation as $\Delta t \to 0$. Also note that if the integrals are approximated then the fixed point $U^\star$ depends upon both the time-step and the value taken for the delay.

5. Stability

It is important to note that (9) is an infinite-step recurrence: U_{n+1} depends on all previous iterates. In this respect, the map is fundamentally different from those that arise from standard discretisations of the ODE (3), or the fixed-delay equation (4). With regard to linear stability, Lubich[11] has derived an analogue of the theory for fixed-step maps that can be applied in our case. The following theorem gives the key result.

Theorem (Lubich[11])
Consider the scalar recurrence

$$y_n = f_n + \sum_{j=0}^{n} a_{n-j} y_j \ , \quad n \geq 0 \ , \tag{16}$$

where $\{a_n\}_{n=0}^{\infty}$ *belongs to* l^1. *Then*

$$y_n \to 0 \quad \textit{whenever} \quad f_n \to 0 \ , \quad \textit{as } n \to \infty \ , \tag{17}$$

if and only if

$$\sum_{n=0}^{\infty} a_n \xi^n \neq 1 \ , \quad \textit{for } |\xi| \leq 1 \ . \tag{18}$$

We denote the non-zero fixed point of the recurrence (9) by $U^\star$, so that $U^\star = (\sum_{j=0}^{\infty} \omega_j)^{-1}$. Then letting $U_n = U^\star + \epsilon_n$ in (9) we obtain

$$\epsilon_{n+1} = \epsilon_n - \Delta t \, (U^\star + \epsilon_n) \sum_{j=0}^{\infty} \omega_j \epsilon_{n-j} \ .$$

Linearising this gives

$$\epsilon_{n+1} = \epsilon_n - \Delta t \; U^\star \sum_{j=0}^{\infty} \omega_j \epsilon_{n-j} \ , \tag{19}$$

which we write in the form

$$\epsilon_{n+1} = \gamma_n + \epsilon_n - \Delta t \; U^\star \sum_{j=0}^{n} \omega_{n-j} \epsilon_j \ , \quad \text{where } \gamma_n := -\Delta t \; U^\star \sum_{-\infty}^{1} \omega_{n-j} \epsilon_j \ . \tag{20}$$

For each of the types of quadrature defined in section 3, it can be shown that the weights ω_j are in l^1. Further, if we assume that the perturbations of the initial data are bounded in the sense that

$$\sup_{-\infty \leq j \leq 0} |\epsilon_j| = \epsilon_{\max} < \infty \ ,$$

then the tail of the summation $\gamma_n \to 0$ as $n \to \infty$. Hence, applying the above theorem we obtain the condition for linear stability as

$$(1 - \Delta t\, \omega_0\, U^\star)\xi - \Delta t\; U^\star \sum_{n=2}^{\infty} \omega_{n-1}\xi^n \neq 1\;, \quad \text{for } |\xi| \leq 1\;.$$

Setting $z = 1/\xi$, we write this as

$$1 - \Delta t\; U^\star\; \widehat{\omega}(z) \neq z\;, \quad \text{for } |z| \geq 1\;, \tag{21}$$

where $\widehat{\omega}(z)$ is the z-transform of the sequence $\{\omega_n\}$, i.e.

$$\widehat{\omega}(z) = \sum_{n=0}^{\infty} \omega_n z^{-n}\;.$$

It follows that stability of the non-zero fixed point for the PC solution approximation is governed by the equation

$$z = 1 - \Delta t\, U^\star \sum_{j=0}^{\infty} W_j z^{-j}\;, \tag{22}$$

whilst in the PL case the stability equation is

$$z = 1 - \Delta t\, U^\star \sum_{j=0}^{\infty} \left[(j+1)W_j - V_j\right] z^{-j} + \left[V_j - jW_j\right] z^{-j-1}\;. \tag{23}$$

(Conditions (22) and (23) imply that the zero fixed point is unstable for all $\Delta t > 0$.) In general, stability of the fixed points (15) depends upon both the approximation taken for the solution and how the quadrature weights are evaluated.

We first consider the stability of the non-zero fixed point of the PC approximation. When a piecewise constant approximation is used to calculate the quadrature weights in (11), the stability equation (22) reduces to the cubic

$$p(z) \equiv z^3 - (1 + 2\,q)\, z^2 + \left((1-q)^2\, \Delta t + q\,(q+2)\right) z - q^2 = 0\;, \tag{24}$$

where $q = \mathrm{e}^{-\Delta t/T}$. For a given value of the delay T (< 2), equation (24) may then be shown to be a Schur polynomial for values of the time-step satisfying the set of transcendental inequalities (see, for example, Lambert[10])

$$0 < (1-q)^2\, \Delta t + 2\,(1+q)^2\;, \tag{25}$$
$$0 < (1-q)\,(4\,(1+q) - (1-q)\,\Delta t)\;, \tag{26}$$
$$0 < (1-q)^2\,(2 - \Delta t)\;, \tag{27}$$
$$0 < (1-q)^2\,\Delta t\;, \tag{28}$$
$$0 < (1-q)^2\left(1 - \Delta t - q^2\right)\;. \tag{29}$$

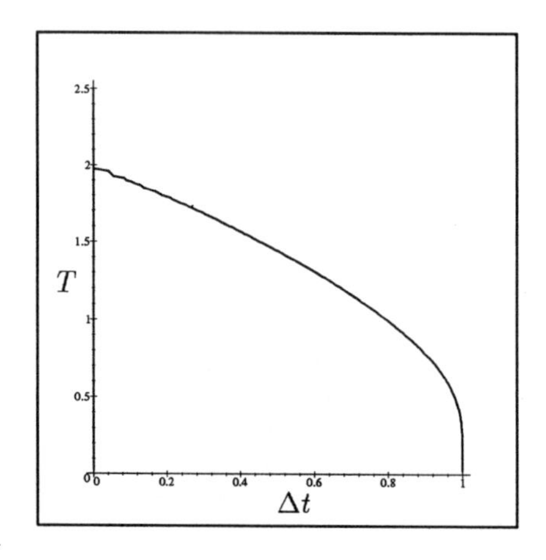

Fig. 1. Stability boundary for the PC map with a piecewise constant kernel approximation.

(Note that these conditions depend implicitly upon the delay, via the time-step ratio $\Delta t/T$.) Condition (28) requires $0 < \Delta t$, whilst (29) is the most restrictive of the other conditions and determines the stability limit, $\Delta t^\star$, such that the fixed point is stable for all time-steps $\Delta t \in (0, \Delta t^\star)$. Figure 1 graphs the stability region, obtained by solving (29) at equality numerically (using the computer algebra package `Maple`). The stability limit $\Delta t^\star$ tends to 1 as $T \to 0$, and to zero as $T \to 2$. We discuss the way in which the fixed point for the PC solution and piecewise constant kernel approximation loses stability after examining the stability equations and regions for the other cases of interest.

When the weights are generated from piecewise linear kernel sections, stability of the fixed point is governed by the equation

$$2\,z^3 + \Big((1-q)^2\,\Delta t - 2\,(1+2\,q)\Big)\,z^2 + \Big((1-q)^2\,\Delta t + 2\,q\,(2+q)\Big)\,z - 2\,q^2 = 0\;. \tag{30}$$

The stability limit is now determined by the equation

$$2\,(1-q^2) - (1+q^2)\,\Delta t = 0\;, \tag{31}$$

and the stability region is enlarged to that shown in Figure 2. The stability limit tends to 2 as $T \to 0$, and to zero as $T \to 2$.

The stability region is further extended when the weights are computed by exact integration of the kernel. The stability equation in this case is

$$\begin{aligned} T\,z^3 &+ \Big(-q\,\Delta t^2 + T\,(1-q)\,\Delta t - T\,(1+2\,q)\Big)\,z^2 \\ &+ \Big(q\,\Delta t^2 - T\,q\,(1-q)\,\Delta t + T\,q\,(2+q)\Big)\,z - T\,q^2 = 0\;, \end{aligned} \tag{32}$$

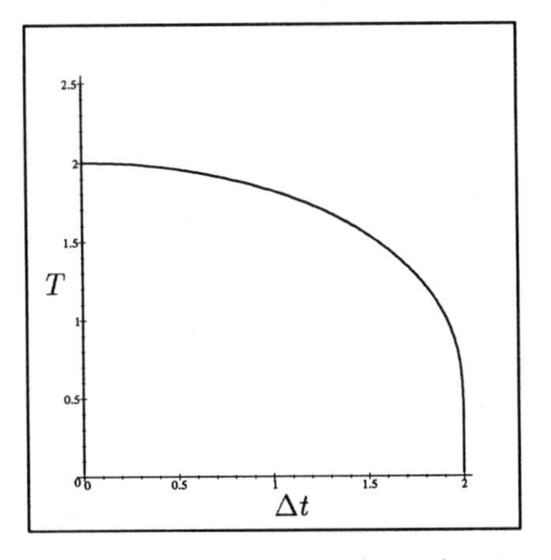

Fig. 2. Stability boundary for the PC map with a piecewise linear kernel approximation.

and (for fixed $T < 2$) the stability limit is given by the solution of the equation

$$2\,q\,\Delta t^2 - \left(1 - q^2\right) T\,\Delta t + 2\,(1+q)^2\,T = 0\ . \tag{33}$$

Now the stability limit $\Delta t^\star$ is greater than 2 for all $T < 2$, and the map has the stability region graphed in Figure 3. Note, however, that the map has become marginally "over-stable" in the sense that it permits a stable non-zero fixed point for some $T > 2$.

When the PL solution approximation is used in conjunction with approximate evaluation of the quadrature weights the stability regions are smaller than those obtained from the corresponding PC approximations. For the piecewise constant and piecewise linear evaluation of the weights the stability limits are bounded by $\Delta t^\star < 2\,(\sqrt{2} - 1)$ and $\Delta t^\star < 1{\cdot}2$, respectively, and the stability limit shrinks to zero as $T \to 2$. The stability region for the case when exact integration is used for the weights is shown in Figure 4.

We now return to the PC case with piecewise constant kernel approximation, and examine the way in which the non-zero fixed point loses stability at $\Delta t^\star$. Recall that $\Delta t^\star$ is the non-zero solution of (29) at equality, i.e.

$$\Delta t^\star = 1 - \exp(2\Delta t^\star/T)\ . \tag{34}$$

We know that $|z| = 1$ when $\Delta t = \Delta t^\star$ for at least one of the roots of $p(z) = 0$ (equation (24)). First note that ± 1 are not roots of $p(z)$ for any $\Delta t > 0$, since

$$p(1) = (1-q)^2 \Delta t > 0, \text{ and}$$
$$p(-1) = -2(1+q)^2 - (1-q)^2 \Delta t < 0.$$

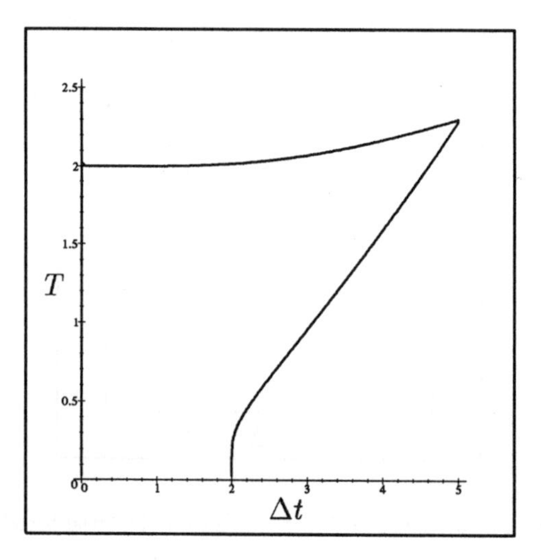

Fig. 3. Stability boundary for the PC map with exact kernel integration.

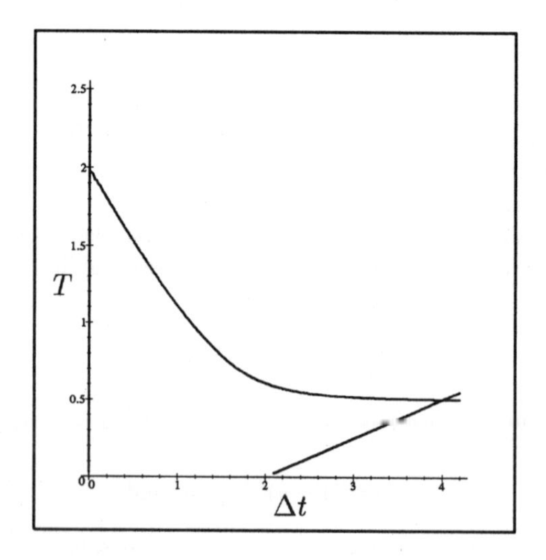

Fig. 4. Stability boundary for the PL map with exact kernel integration.

Hence there must be a root $z = e^{i\theta}$ for some θ with $\sin\theta \neq 0$. The coefficients of $p(z)$ are real and so $z = e^{-i\theta}$ must also be a root. Thus

$$p(z) = (z^2 - 2z\cos\theta + 1)(z - a)$$

for some θ and a. Equating coefficients and using (34) gives $a = q^2$ and

$$\begin{aligned} \cos\theta &= (1 - q^2 + 2q)/2 \\ &= \frac{\Delta t^\star + 2\sqrt{1 - \Delta t^\star}}{2} . \end{aligned} \tag{35}$$

If $\theta = 2\pi/m$ for some integer m then it is likely that the non-zero fixed point will lose stability at $\Delta t^\star$ to a period-m solution of (9) (which will only be seen if it is stable). When $\theta = 2\pi/m$, equation (35) can be inverted to give

$$\Delta t^\star = 4\sin(\pi/m)(1 - \sin(\pi/m)) ,$$

and we denote the value of the delay T which gives this $\Delta t^\star$ by T_m. It follows from (34) that

$$T_m = \frac{-2\Delta t^\star}{\ln(1 - \Delta t^\star)},$$

and it can be shown numerically that there is a value T_m of T for which $\theta = 2\pi/m$ for each integer $m > 6$. Corresponding values of T_m and $\Delta t^\star$ are tabulated below for $m = 7 : 15$.

Table 1. Values of T_m and $\Delta t^\star$ (see text for details).

m	T_m	$\Delta t^\star$
7	0.4856	0.9825
8	0.6518	0.9449
9	0.7813	0.9002
10	0.8874	0.8541
11	0.9765	0.8094
12	1.0525	0.7673
13	1.1180	0.7282
14	1.1752	0.6920
15	1.2254	0.6587

6. Spurious Periodic Solutions

It has been shown that discretisations of the logistic equation and delay differential equation models admit spurious periodic and chaotic solutions over a range of parameter and time-step values[7,12]. We now investigate the various discretisations of the integro-differential equation model and look for spurious periodic behaviour.

We first consider period-2 motion and look for solutions of the form

$$U_n = \begin{cases} u & n \text{ even.} \\ v & n \text{ odd.} \end{cases} \tag{36}$$

Substituting this into the map (9) yields the recurrence

$$\left. \begin{aligned} v &= u + u\Delta t\,(1 - u\,\Sigma_0 - v\,\Sigma_1) \\ u &= v + v\Delta t\,(1 - v\,\Sigma_0 - u\,\Sigma_1) \end{aligned} \right\} , \tag{37}$$

where the even and odd sums are defined by $\Sigma_k = \sum_{l=0}^{\infty} \omega_{2l+k}$. Solving this recurrence relation for u gives the quartic polynomial

$$u\,(u\,(\Sigma_0 + \Sigma_1) - 1)\left(u^2 \Delta t^2\, \Sigma_0\,(\Sigma_0 - \Sigma_1) - u\Delta t\,(\Delta t + 2)\,(\Sigma_0 - \Sigma_1) + \Delta t + 2\right) = 0\ .$$

Two of the solutions are given by the (period-1) fixed points of the map (9). If real period-2 solutions are to be admitted by the map then the discriminant of the quadratic factor must be positive, which gives

$$(\Sigma_0 - \Sigma_1)\,((\Sigma_0 - \Sigma_1)\,\Delta t - 2\,(\Sigma_0 + \Sigma_1)) > 0\ . \tag{38}$$

Further, if equation (5) represents a population model (with necessarily positive solutions) then we must also have $\Sigma_0 > \Sigma_1$. Note that period-2 solutions are prevented altogether if $\Sigma_0 \equiv \Sigma_1$.

When the weights are approximated from piecewise constants or linears then real positive period-2 solutions are suppressed for all $\Delta t, T > 0$, as seen from the following table.

kernel approximation	U_n approximation	
	PC	PL
piecewise constant	$\Sigma_0 < \Sigma_1$	$\Sigma_0 \equiv \Sigma_1$
piecewise linear	$\Sigma_0 \equiv \Sigma_1$	$\Sigma_0 < \Sigma_1$

Real positive period-2 solutions are however possible when the weights are evaluated exactly. In this case there are parts of the stability region for which one of the conditions (38) and $\Sigma_0 > \Sigma_1$ is satisfied (for both the PC and PL solution approximations). However, these conditions are only satisfied *simultaneously* for the PL approximation outwith the stability region, as indicated by (the shaded part of) Figure 5. The figure shows that there are real positive period-2 solutions in a region touching the stability boundary for small delays and hence, in this range, it is likely that the solution loses stability at $\Delta t = \Delta t^*$ by period doubling.

The situation is more complicated when the PC approximation is used with exact kernel integration. Then period-2 solutions are permitted within the stability region graphed in Figure 3. However, a stability analysis of the recurrence (37) reveals that all period-2 solutions are unstable within the stability region, and the fixed point solutions bifurcate into period-2 behaviour at the stability limit for all delays $T < 2$.

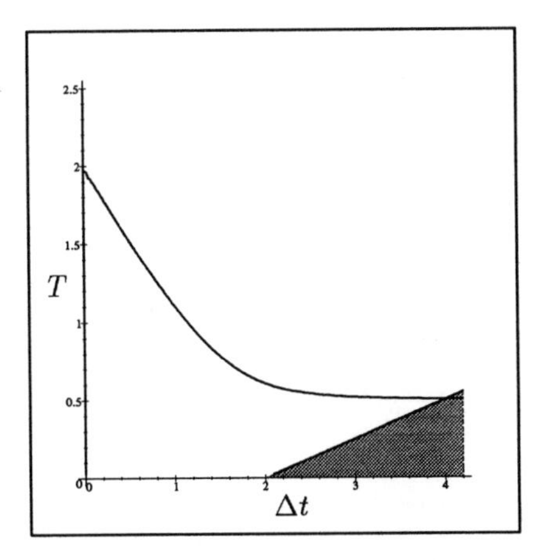

Fig. 5. Stability region (bounded by the axes and solid lines) and region for which period-2 solutions are possible (shaded) for the PL map with exact kernel integration.

We now turn our attention to the next mode of oscillation. Period-3 solutions have the form

$$U_n = \begin{cases} u, & n = 3l \\ v, & n = 3l+1 \\ w, & n = 3l+2 \end{cases} , \tag{39}$$

and substituting these solutions into the map (9) gives

$$\left.\begin{aligned} v &= u + u\Delta t\,(1 - u\,\Sigma_0 - w\,\Sigma_1 - v\,\Sigma_2) \\ w &= v + v\Delta t\,(1 - v\,\Sigma_0 - u\,\Sigma_1 - w\,\Sigma_2) \\ u &= w + w\Delta t\,(1 - w\,\Sigma_0 - v\,\Sigma_1 - u\,\Sigma_2) \end{aligned}\right\} , \tag{40}$$

with the sums now defined as $\Sigma_k = \sum_{l=0}^{\infty} \omega_{3l+k}$. Solving equations (40) for u gives the condition

$$u\,[u\,(\Sigma_0 + \Sigma_1 + \Sigma_2) - 1]\,F(u, \Delta t, \Sigma_0, \Sigma_1, \Sigma_2) = 0\ , \tag{41}$$

where the factor $F(u, \Delta t, \Sigma_0, \Sigma_1, \Sigma_2)$ is a sixth-order polynomial in u with real coefficients. The two linear factors correspond to the fixed points 0 and U^* of the map (9).

The roots of the sixth-order factor in equation (41) will, in general, be complex and cannot be determined explicitly. However, using the technique of Aves et al.[1] it is possible to search for repeated roots. Period-3 solutions of multiplicity two exist when the polynomial

$$s(x) = x^6 + g_5x^5 + g_4x^4 + g_3x^3 + g_2x^2 + g_1x + g_0 \tag{42}$$

can be written as a cubic squared: $p(x) = (x^3 + ax^2 + bx + c)^2$. This imposes the constraints

$$g_0 = c^2\ , \qquad g_1 = 2bc\ , \qquad g_2 = 2ac + b^2\ ,$$
$$g_3 = 2ab + 2c\ , \qquad g_4 = a^2 + 2b\ , \qquad g_5 = 2a\ .$$

The coefficients in the cubic are then given by

$$a = g_5/2, \quad b = (4g_4 - {g_5}^2)/8, \quad c = (8g_3 - 4g_4g_5 + {g_5}^3)/16\ ,$$

and the additional conditions on the coefficients can be shown to hold when

$$16\, {g_0}^3\, {g_5}^2 - {g_1}^4 + 8\, {g_1}^2\, g_0\, g_2 - 16\, {g_0}^2\, {g_2}^2 = 0\ . \tag{43}$$

We now consider this condition for the PC approximation with exact kernel integration. The period-3 sums then have the form

$$\Sigma_0 = \frac{T(q^2+q+1) - 2\,q^2\Delta t - q\Delta t}{T\,(q^2+q+1)^2}\ ,$$
$$\Sigma_1 = \frac{(T(q^2+q+1) + \Delta t - q^2\Delta t)\,q}{T\,(q^2+q+1)^2}\ ,$$
$$\Sigma_2 = \frac{(T(q^2+q+1) + q\Delta t + 2\,\Delta t)\,q^2}{T\,(q^2+q+1)^2}\ ,$$

where $q = \mathrm{e}^{-\Delta t/T}$. Equation (43) can be shown to contain two 14-th order polynomials in q; one factor corresponds to two period-3 solutions consisting of one real root and a complex conjugate pair, and the other factor to real period-3 solutions. The numerically computed roots of this real factor are shown in Figure 6 along with the stability boundary.

We finish this section with some numerical results that illustrate the long-term behaviour of the PC map with weights given by exact integration of the kernel. We cannot perform numerical experiments using an infinite map and therefore approximate (9) by the finite map

$$U_{n+1} = U_n + \Delta t\, U_n \left(1 - \sum_{j=0}^{N-1} U_{n-j}\, W_j\right)\ , \tag{44}$$

where the quadrature weights are $W_j = q^j\,(1 + jr + q\,(1 + (j+1)r))$, with $q = \mathrm{e}^{-r}$ and $r = -\Delta t/T$. Here N is fixed for each choice of the time-step. The non-zero fixed point of (44) is then

$$U_N^\star = \left(1 - q^N\,(1 + Nr)\right)^{-1}\ , \tag{45}$$

and $U_N^\star \to U^\star$ (the fixed point of the infinite map) as $N \to \infty$.

For each fixed time-step we chose the number of points in the finite sum so that $\mathrm{k}(j\Delta t) \leq 5{\cdot}0{\times}10^{-4}$ for all $j > N$. We iterated for 500 steps and plotted the last

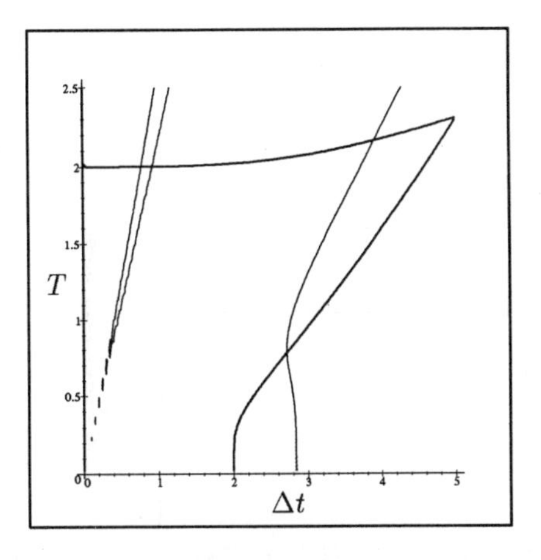

Fig. 6. Stability boundary and period-3 solutions of the PC map with exact kernel integration.

10 values, $\{U_j\}_{j=491}^{500}$. Figure 7 shows typical long-term behaviour of the map (44) using constant time-steps $\Delta t \in [0{\cdot}1, 4{\cdot}0]$ and constant initial data $\{U_j\}_{j=1-N}^{0} \in [0, 2]$. The fixed points obtained by the search procedure are given by equation (45) until the stability limit is reached at $\Delta t \approx 3{\cdot}05$. The fixed point then becomes unstable and the solution bifurcates to period-2. Note that period-3 solutions are also present and occur at the approximate location predicted by the infinite analysis $\Delta t \approx 2{\cdot}8$ (see Figure 6). However, the above analysis only predicts the location of period-3 solutions of multiplicity two. Two branches of period-3 should originate from the multiplicity-two solution (only the stable branch is visible). The main point to note is that, in contrast to period-2 solutions, the spurious period-3 solutions exist below the linear stability limit. The same phenomena occur for the PL approximation, as shown in Figure 8: there is bifurcation to period-2 at the stability limit and period-3 solutions are visible below the stability limit.

7. Summary

Our aim in this study has been to investigate the fixed points and spurious low period solutions of a discrete version of an integro-differential equation with the exponential convolution-type integral kernel $\mathrm{k}(t) = t\ \mathrm{e}^{-t/T}/T^2$.

We have examined the behaviour of a single explicit time discretisation of the IDE, using several different quadrature formulae for the convolution integral. Specifically, we have used a piecewise constant (PC) and a piecewise linear (PL) solution approximation with the quadrature weights evaluated both exactly and approximately. In all

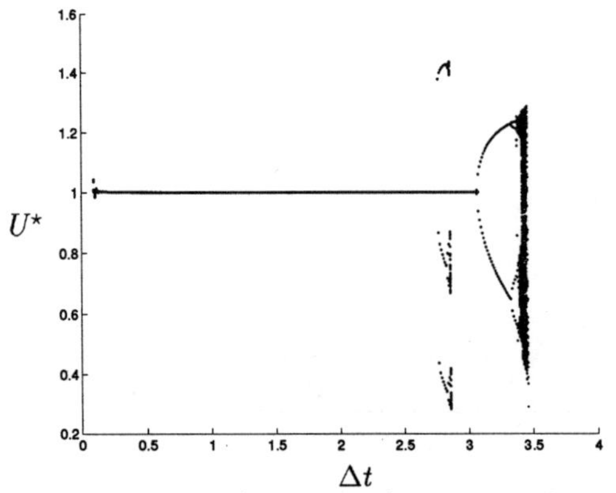

Fig. 7. Numerical solution of PC map (finite version) for $T = 1$ with exact kernel integration.

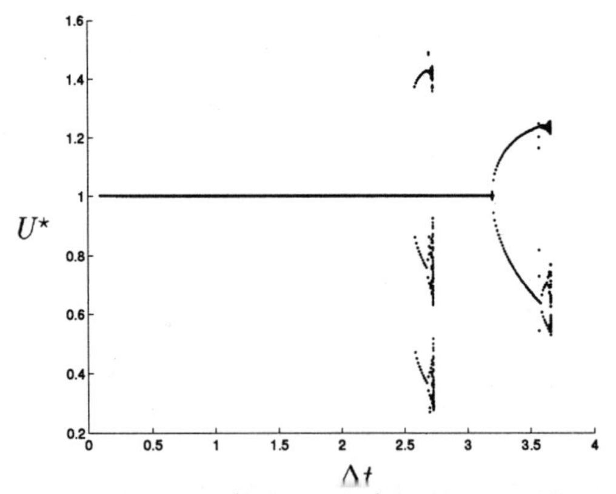

Fig. 8. Numerical solution of PL map (finite version) for $T = 0.3$ with exact kernel integration.

cases the discrete map possesses a single non-zero fixed point that either equals or converges to (as $\Delta t \to 0$) the non-zero fixed point of the continuous equation (5). If the quadrature weights are evaluated exactly then the fixed point is exact, whilst approximating the weights gives a fixed point which depends upon both the time-step Δt and the delay T. As for the IDE with simpler exponential kernel $\mathrm{k}(t) = \mathrm{e}^{-t/T}/T$ investigated in Aves et al.[1], the additional weights V_j associated with the PL approximation do not affect the non-zero fixed point $U^\star$ (the PC and PL solution approximations have the same fixed points).

Our stability analysis of the fixed point of the infinite-step map (9) makes use of a result that Lubich[11] derived for Volterra integral equations, which can be couched in terms of the z-transform. When approximate kernel integration is used the stability region of the PC map is always larger than that of the corresponding PL map. We have shown that the stability intervals typically shrink to zero as the delay tends towards the limit of valid fixed point solutions at $T = 2$. However, when the PC map is used in conjunction with exact kernel integration, the stability interval is non-zero for $T = 2$ and stable fixed point solutions exist for delays above this value.

None of the maps we have investigated can have positive period-2 solutions inside the linear stability interval, for various reasons:

- approximate kernel integration—$\Sigma_0 \geq \Sigma_1$;
- piecewise linear approximation and exact kernel integration—no positive period-2 solutions for $\Delta t < \Delta t^\star$; and
- piecewise constant approximation and exact kernel integration—positive period-2 solutions unstable for $\Delta t < \Delta t^\star$.

In contrast, stable period-3 solutions are supported in all cases and exist within the linear stability interval for certain values of the delay.

References

1. M.A. Aves, P.J. Davies, and D.J. Higham. The effect of quadrature on the dynamics of a discretised nonlinear integro-differential equation. Technical Report NA/166, University of Dundee, 1995.
2. J.M. Cushing. *Integrodifferential Equations and Delay Models in Population Dynamics.* Springer-Verlag, Lecture Notes in Biomathematics, 1977.
3. A.R. Gardiner and A.R. Mitchell. Bifurcation studies in long time solutions of discrete models in reaction diffusion. Technical Report NA/136, University of Dundee, 1992.
4. A.R. Gardiner and A.R. Mitchell. Periodic bifurcations of an explicit discretisation of the Ginzburg-Landau equation. Technical Report NA/137, University of Dundee, 1992.

5. D.F. Griffiths and A.R. Mitchell. Stable periodic bifurcations of an explicit discretization of a nonlinear partial differential equation in reaction diffusion. *IMA J. Numer. Anal.*, 8:435–454, 1988.
6. D.F. Griffiths and A.R. Mitchell. Spurious behaviour and nonlinear instability in discretised partial differential equations. In A. Iserles and D. Broomhead, editors, *Proceedings of the 1990 IMA Conference on Dynamics of Numerics and Numerics of Dynamics*, pages 215–242. Oxford University Press, 1992.
7. D.J. Higham. The dynamics of a discretised nonlinear delay differential equation. In D.F. Griffiths and G.A. Watson, editors, *Proceedings of the 1993 Dundee Conference on Numerical Analysis*, pages 167–179. Pitman Research Notes in Mathematics, 1994.
8. D. J. Higham and B. Owren. Non-normality effects in a discretised, nonlinear, reaction-convection-diffusion equation. *J. Comp. Phys., to appear.*
9. A. Iserles. Stability and dynamics of numerical methods for nonlinear ordinary differential equations. *IMA J. Numer. Anal.*, 10:1–30, 1990.
10. J.D. Lambert. *Numerical Methods for Ordinary Differential Systems.* Wiley, 1991.
11. Ch. Lubich. On the stability of linear multistep methods for Volterra convolution equations. *IMA J. Num. Anal.*, 3:439–465, 1983.
12. R.M. May. Time-delay verses stability in population models with two and three trophic levels. *Ecology*, 54:315–325, 1973.
13. A.R. Mitchell and A.R. Gardiner. The influence of dispersion on the inviscid Burgers' equation. Technical Report NA/142, University of Dundee, 1992.
14. A.R. Mitchell and A.R. Gardiner. Unstable attractors of a discretised inviscid Burgers' equation. Technical Report NA/139, University of Dundee, 1992.
15. A.R. Mitchell and D.F. Griffiths. Beyond the linearised stability limit in nonlinear problems. In D.F. Griffiths and G.A. Watson, editors, *Proceedings of the 1985 Dundee Conference on Numerical Analysis*, pages 140–156. Pitman Research Notes in Mathematics, 1986.
16. A.R. Mitchell, P. John-Charles, and B.D. Sleeman. Long time calculations and non-linear maps. In R.W. Lewis, E. Hinton, P. Bettess, and A. Schrefler, editors, *Numerical Methods for Transient and Coupled problems*, pages 199–211. John Wiley and Sons, 1987.
17. H.C. Yee, P.K. Sweby, and D.F. Griffiths. Dynamical systems approach study of spurious steady state numerical solutions of nonlinear differential equations. 1. The ODE connection and its implications for algorithm developments in computational fluid dynamics. *J. Comp. Phys.*, 97:249–310, 1991.

RUNGE-KUTTA METHODS AS MATHEMATICAL OBJECTS

J. C. BUTCHER
Mathematics Department, The University of Auckland
Private Bag 92019, Auckland, New Zealand
E-mail: butcher@math.auckland.ac.nz

ABSTRACT

Sets of Runge-Kutta methods, closely related in certain specific ways, are regarded as comprising equivalence classes. Using these classes as mathematical objects, an algebraic system is built up that can be used to represent the essential numerical properties of Runge-Kutta methods and of sequences of methods applied sequentially over several steps. By generalizing the formulation slightly, the algebraic system is able to represent other computationally significant quantities. The principal application of this theory is in the analysis of multistage-multivalue-multiderivative methods. However, even for the narrow class of Runge-Kutta methods, it suggests a generalization of the order concept to include "effective order". A new Runge-Kutta method of effective order 5 is given, as is a new method which achieves order 5 behaviour in a different way: by moving additional information from step to step.

1. Introduction

Runge-Kutta methods have become very popular both as computational techniques and as thesis subjects. To survey all the wealth of applications and theoretical knowledge that has come into existence, even within the last 30 years, would be a formidable task and it will be avoided in this paper.

Rather than present a Michelin guide to Runge-Kutta methods, we will here look at some aspects of these methods that make them worthy of study as mathematical objects. What could possibly justify doing this? The answer is that the sort of objects that will be constructed have a wider role as a means of studying more general numerical methods. Thus, while we will shun abstraction for its own sake, we will certainly not shun abstraction for the sake of insight and practical application.

I would like to think that this approach, of using interesting and demanding, but above all appropriate, mathematics as a means of analysing and understanding the solution of practical problems and practical numerical methods, is in the spirit of Scottish applied mathematics. I think it is certainly within the spirit of the work of Ron Mitchell who exemplifies all that is best in the applied mathematical traditions of his country.

The objects that will be discussed in this paper are nothing more than families of Runge-Kutta methods, where members of each family have, as their claims to kinship, an identification of some of their algebraic and numerical properties. Nothing more? Perhaps a little more, because we will allow ourselves the indulgence of generalizing slightly the type of method that we regard as a Runge-Kutta method. The description of these methods and families of methods will constitute Section 2, with generalizations introduced in Section 5.

When representative members of two families are applied in turn to numerical data, the composition of these two members always lies in a single family, which can be thought of as the product of the original families. We will discuss this in Section 3 and lead on to Section 4, where a group, whose elements are represented by mappings from rooted trees to real numbers, is introduced as a homomorphic image of this product.

Also in Section 4, we will discuss some interesting subgroups and factor groups along with their numerical significance. This is in preparation for Section 6, where an interpretation of order of Runge-Kutta methods, will be discussed. An interesting generalization, first introduced at an early Dundee conference [1], will be described.

The generalized Runge-Kutta methods of Section 5 also have algebraic ramifications; the group introduced in Section 4 is now extended so that it now has vector space properties with a subset (consisting of members of the original group) acting as a set of linear operators.

Just as the algebraic quantities themselves have an interpretation as representing numerical methods, so do they also act like members of a field over which vector spaces can be constructed. In this role, they can be used to represent the results of carrying out various numerical operations and they can be manipulated and operated on. This leads, in Section 7, to a convenient method of defining the truncation error order of a wide family of numerical methods, of which Runge-Kutta methods are a very small sub-class. An example of the use of the calculus built up from these ideas will be discussed. This leads to the construction of a new numerical method which, in a sense, breaks the order 5 barrier for Runge-Kutta methods. There may be prospects here for solving non-stiff problems more efficiently than is possible with genuine fifth order Runge-Kutta methods: something that should please a canny numerical analyst.

Although the algebraic ideas which are at the centre of this paper date from the author's 1972 paper [2], the algebraic approach is less well known than the more analytic emphasis inherent in the paper of Hairer and Wanner [4]. Nevertheless, the author hopes that this present gentle introduction to his early paper, and to the relevant parts of his monograph [3], will make the algebraic approach accepted as a complementary view to that of the B-series work of Hairer and Wanner.

2. Families of Runge-Kutta methods

At the background to discussions of Runge-Kutta methods is a generic differential equation system which might be of the form

$$y'(x) = f(x, y(x)), \tag{1}$$

or of the simpler, but essentially just as general, autonomous form

$$y'(x) = f(y(x)). \tag{2}$$

We will use both representations of a general differential equation. In discussions of how numerical methods might be applied to solve a problem, (1) seems to be more appropriate. However, for the purposes of analysis, (2) leads to simpler looking formulae.

There are senses in which two or more Runge-Kutta methods might possess such similarities that they can be looked upon as being really the same method in a different form. We will look at this in terms of three example methods, as represented by the tableaus

$$\text{Method A: } \begin{array}{c|cc} 0 & 0 & 0 \\ \frac{1}{2} & \frac{1}{2} & 0 \\ \hline & 0 & 1 \end{array}, \quad \text{Method B: } \begin{array}{c|ccc} 0 & 0 & 0 & 0 \\ \frac{1}{3} & -\frac{1}{4} & \frac{5}{6} & -\frac{1}{4} \\ \frac{1}{2} & \frac{1}{2} & 0 & 0 \\ \hline & 0 & 0 & 1 \end{array}, \quad \text{Method C: } \begin{array}{c|ccc} 0 & 0 & 0 & 0 \\ \frac{1}{2} & \frac{1}{2} & 1 & -1 \\ \frac{1}{2} & \frac{1}{2} & -2 & 2 \\ \hline & 0 & \frac{3}{7} & \frac{4}{7} \end{array}.$$

We will illustrate the use of these three methods by attempting to solve the initial value problem

$$y'(x) = y(x) + \frac{y(x)^2}{10 + x}, \quad y(1) = 2.$$

For each of the three methods we will take a single step for various values of the stepsize h, if we are able to do so. The reason for doubt is that methods B and C are implicit and the ability to complete a step will depend on the ability to obtain a solution to the algebraic system of equations that defines the stage values of the methods.

The results of these attempts are tabulated in Table 1, where blank entries signify that the solution method used for the implicit equations failed to converge.

The striking feature of these numerical results is that, where the data is not missing, the results found from each of the three methods are *exactly* the same. It is not hard to find the reasons for this. Method B has a completely useless second stage, because neither of the other stages, or the final output value, makes any use of the stage derivative calculated from this stage. How could they do so, when a_{12}, a_{32} and b_2 are each equal to zero? Thus, the second stage plays no other role than making the iterative solution to the system of non-linear equations which define Y_1, Y_2 and Y_3 non-convergent, for the two largest values of h attempted.

Method C gives related results for a different reason. This is that, if $|h|$ is small enough, the stage values Y_2 and Y_3 are necessarily equal. We can see this by substituting this equality and observing that the equations are still consistent with each other.

Table 1. A single numerical step with three methods

h	Method A	Method B	Method C
1.00	6.0621631333		
0.70	4.4720803146		
0.50	3.5938016529	3.5938016529	
0.30	2.8555264426	2.8555264426	
0.20	2.5373866428	2.5373866428	2.5373866428
0.10	2.2524217494	2.2524217494	2.2524217494
0.07	2.1733072392	2.1733072392	2.1733072392
0.05	2.1221829145	2.1221829145	2.1221829145
0.03	2.0723475452	2.0723475452	2.0723475452
0.02	2.0479116085	2.0479116085	2.0479116085
0.01	2.0237959759	2.0237959759	2.0237959759

The fact that the three methods give the same results (for small enough $|h|$), not only for this, but for any problem satisfying a Lipschitz condition, makes these methods very closely related. We will temporarily describe this as "computational equivalence". The fact that methods B and C can be recognised as artificial modifications of method A makes them related by what we will call "disguise equivalence".

There is a further type of relationship the three methods have and we can recognise this by working out the elementary weights for the method for a few simple trees.

The eight rooted trees with up to order four are

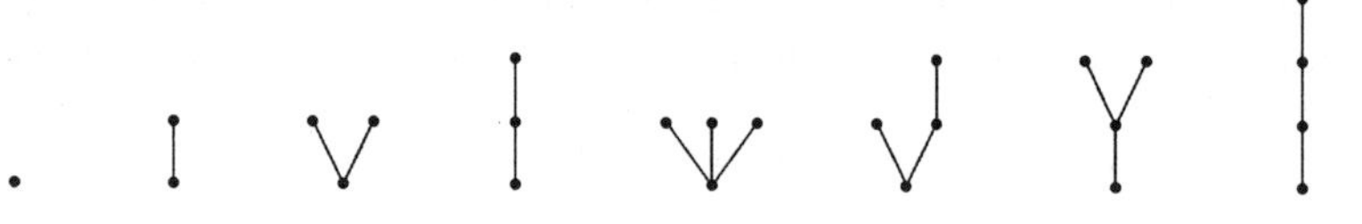

and the corresponding elementary weights are

$$b^T e \quad b^T c \quad b^T(c.c) \quad b^T Ac \quad b^T(c.c.c) \quad b^T(c.(Ac)) \quad b^T A(c.c) \quad b^T A^2 c$$

where . denotes component-by-component multiplication and e denotes the vector with every component equal to 1.

If we evaluate the expressions for each of the methods A, B and C, it is found that the eight numbers computed in this way are

$$1 \quad \frac{1}{2} \quad \frac{1}{4} \quad 0 \quad \frac{1}{8} \quad 0 \quad 0 \quad 0$$

It is easy to show that for *every* tree, these three methods will give identical values of the corresponding elementary weight. We will refer to this property as "weight equivalence".

We now have three apparently different equivalence relations on the set of all Runge-Kutta methods. However, the following statement summarizes the facts about them in a satisfactory manner:

Two Runge-Kutta methods are computationally equivalent
iff
they are disguise equivalent
iff
they are weight equivalent

3. Products of Runge-Kutta families

Let M and N denote two equivalence classes of Runge-Kutta methods. If a representative method (A, b, c), with s stages is selected from M and a second method $(\bar{A}, \bar{b}, \bar{c})$, with $\bar{s}$ stages is selected from N, then the computation of y_1 from an initial value y_0 using (A, b, c), followed by the computation of y_2 from an inital value y_1 using $(\bar{A}, \bar{b}, \bar{c})$ is equivalent to the computation of y_2 from y_0 in a single step with $s + \bar{s}$ stages using the tableau

$$\begin{array}{c|cccccccc} c_1 & a_{11} & a_{12} & \cdots & a_{1s} & 0 & 0 & \cdots & 0 \\ c_2 & a_{21} & a_{22} & \cdots & a_{2s} & 0 & 0 & \cdots & 0 \\ \vdots & \vdots & \vdots & & \vdots & \vdots & \vdots & & \vdots \\ c_s & a_{s1} & a_{s2} & \cdots & a_{ss} & 0 & 0 & \cdots & 0 \\ d+\bar{c}_1 & b_1 & b_2 & \cdots & b_s & \bar{a}_{11} & \bar{a}_{12} & \cdots & \bar{a}_{1\bar{s}} \\ d+\bar{c}_2 & b_1 & b_2 & \cdots & b_s & \bar{a}_{21} & \bar{a}_{22} & \cdots & \bar{a}_{2\bar{s}} \\ \vdots & \vdots & \vdots & & \vdots & \vdots & \vdots & & \vdots \\ d+\bar{c}_s & b_1 & b_2 & \cdots & b_s & \bar{a}_{\bar{s}1} & \bar{a}_{\bar{s}2} & \cdots & \bar{a}_{\bar{s}\bar{s}} \\ \hline & b_1 & b_2 & \cdots & b_s & \bar{b}_1 & \bar{b}_2 & \cdots & \bar{b}_{\bar{s}} \end{array}, \qquad (3)$$

where $d = b^T e$.

If members of the class M have elementary weights defined by $\alpha : T \to \mathbb{R}$, and members of N have elementary weights defined by $\beta : T \to \mathbb{R}$, then all members of the class of which (3) is a member, must share the same elementary weights. Hence, there is a mapping $T \to \mathbb{R}$ defining the elementary weights of (3), which is determined completely by α and β. Denote this by $\alpha\beta$. In the next section we will explore the nature of this product in some detail. In the meantime, we will examine some examples of multiplication of method classes.

Consider the well-known implicit Runge-Kutta method.

$$\begin{array}{c|cc} \frac{1}{2} - \frac{\sqrt{3}}{6} & \frac{1}{4} & \frac{1}{4} - \frac{\sqrt{3}}{6} \\ \frac{1}{2} + \frac{\sqrt{3}}{6} & \frac{1}{4} + \frac{\sqrt{3}}{6} & \frac{1}{4} \\ \hline & \frac{1}{2} & \frac{1}{2} \end{array}. \qquad (4)$$

To compute the result after a single step, with initial value $y(x_0) = y_0$, stage values Y_1 and Y_2 are first computed as solutions to the equations

$$Y_1 = y_0 + \frac{h}{4} f\left(x_0 + \left(\frac{1}{2} - \frac{\sqrt{3}}{6}\right)h, Y_1\right) + \left(\frac{1}{4} - \frac{\sqrt{3}}{6}\right) hf\left(x_0 + \left(\frac{1}{2} + \frac{\sqrt{3}}{6}\right)h, Y_2\right),$$

$$Y_2 = y_0 + \left(\frac{1}{4} + \frac{\sqrt{3}}{6}\right) hf\left(x_0 + \left(\frac{1}{2} - \frac{\sqrt{3}}{6}\right)h, Y_1\right) + \frac{h}{4} f\left(x_0 + \left(\frac{1}{2} + \frac{\sqrt{3}}{6}\right)h, Y_2\right),$$

and the solution at the end of the step $y_1 \approx y(x_1) = y(x_0 + h)$ is given by

$$y_1 = y_0 + \frac{h}{2} f\left(x_0 + \left(\frac{1}{2} - \frac{\sqrt{3}}{6}\right)h, Y_1\right) + \frac{h}{2} f\left(x_0 + \left(\frac{1}{2} + \frac{\sqrt{3}}{6}\right)h, Y_2\right).$$

Assuming that $|h|$ is small enough to guarantee that the various algebraic equation systems have solutions, we can rewrite the relationship between y_0, Y_1, Y_2 and y_1 so that y_1 has the role of initial information and y_0 has the role of output solution. These rewritten formulas are

$$Y_1 = y_1 - \frac{h}{4} f\left(x_1 - \left(\frac{1}{2} + \frac{\sqrt{3}}{6}\right)h, Y_1\right) - \left(\frac{1}{4} + \frac{\sqrt{3}}{6}\right) hf\left(x_1 - \left(\frac{1}{2} - \frac{\sqrt{3}}{6}\right)h, Y_2\right),$$

$$Y_2 = y_1 - \left(\frac{1}{4} - \frac{\sqrt{3}}{6}\right) hf\left(x_1 - \left(\frac{1}{2} + \frac{\sqrt{3}}{6}\right)h, Y_1\right) - \frac{h}{4} f\left(x_1 - \left(\frac{1}{2} - \frac{\sqrt{3}}{6}\right)h, Y_2\right),$$

$$y_0 = y_1 - \frac{h}{2} f\left(x_1 - \left(\frac{1}{2} + \frac{\sqrt{3}}{6}\right)h, Y_1\right) - \frac{h}{2} f\left(x_1 - \left(\frac{1}{2} - \frac{\sqrt{3}}{6}\right)h, Y_2\right),$$

corresponding to the tableau

$$\begin{array}{c|cc}
-\frac{1}{2} - \frac{\sqrt{3}}{6} & -\frac{1}{4} & -\frac{1}{4} - \frac{\sqrt{3}}{6} \\
-\frac{1}{2} + \frac{\sqrt{3}}{6} & -\frac{1}{4} + \frac{\sqrt{3}}{6} & -\frac{1}{4} \\
\hline
 & -\frac{1}{2} & -\frac{1}{2}
\end{array} \qquad (5)$$

If we carry out a step of the original implicit method (4) and then follow this by a step of (5), the combined 4 stage method becomes

$$\begin{array}{c|cccc}
\frac{1}{2} - \frac{\sqrt{3}}{6} & \frac{1}{4} & \frac{1}{4} - \frac{\sqrt{3}}{6} & 0 & 0 \\
\frac{1}{2} + \frac{\sqrt{3}}{6} & \frac{1}{4} + \frac{\sqrt{3}}{6} & \frac{1}{4} & 0 & 0 \\
\frac{1}{2} - \frac{\sqrt{3}}{6} & \frac{1}{2} & \frac{1}{2} & -\frac{1}{4} & -\frac{1}{4} - \frac{\sqrt{3}}{6} \\
\frac{1}{2} + \frac{\sqrt{3}}{6} & \frac{1}{2} & \frac{1}{2} & -\frac{1}{4} + \frac{\sqrt{3}}{6} & -\frac{1}{4} \\
\hline
 & \frac{1}{2} & \frac{1}{2} & -\frac{1}{2} & -\frac{1}{2}
\end{array}. \qquad (6)$$

This method belongs in the same equivalence class as a method with no stages at all, because stages 1 and 3 give identical values as do 2 and 4 and for each of these pairs of stages, the total final contribution in the b vector is 0.

Table 2. Elementary weights for (4), (5) and (6)

t	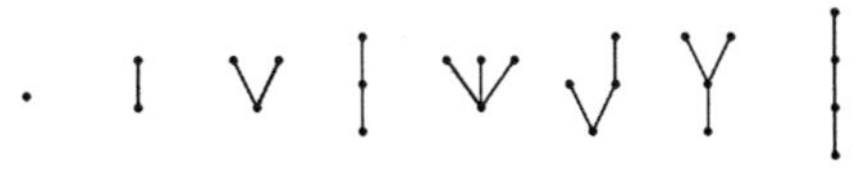 $\cdot$							
$\alpha(t)$	1	$\frac{1}{2}$	$\frac{1}{3}$	$\frac{1}{6}$	$\frac{1}{4}$	$\frac{1}{8}$	$\frac{1}{12}$	$\frac{1}{24}$
$\beta(t)$	-1	$\frac{1}{2}$	$-\frac{1}{3}$	$-\frac{1}{6}$	$\frac{1}{4}$	$\frac{1}{8}$	$\frac{1}{12}$	$\frac{1}{24}$
$(\alpha\beta)(t)$	0	0	0	0	0	0	0	0

If $\alpha(t)$, $\beta(t)$ and $(\alpha\beta)(t)$ denote the elementary weights for the three methods (4), (5) and (6) respectively, then for the first 8 trees, the values are given in Table 2.

From this example, we see why it is advantageous to use equivalent classes, rather than individual methods. The class containing (6) plays the role of the identity element of the algebraic system defined by compositions of representative methods, and the class containing (5) is the inverse of the class containing (4). Because forming compositions of individual methods is associative and because we can always construct inverses of method classes, as we have done in this example, the structure is a group. In the next section we will discuss some of the properties of this group.

4. The group of functions on trees

Denote the 8 trees with up to 4 vertices by $t_1, t_2, \ldots, t_8$ and for convenience write $\alpha_i = \alpha(t_i)$, $\beta_i = \beta(t_i)$. The expressions for $(\alpha\beta)(t_i)$ are given in Table 3.

The formula for each $(\alpha\beta)(t_i)$ can be generated recursively but it also has a simple graph-theoretic structure. For example, the terms of $(\alpha\beta)(t_6)$ can be represented schematically as in Table 4. These seven diagrams are formed from t_6 by selecting connected subgraphs, shown using black discs to represent the vertices, which, except for the first of the diagrams, where the subgraph is empty, always contain the original root. The difference graphs (formed by deleting the selected subgraphs from the original tree) consists of a number of disconnected components each of which is also a tree. These components of the difference graph have their vertices shown using empty circles. To form the corresponding term in the expression for $(\alpha\beta)(t_6)$, the subgraph becomes a β factor and the difference components become α factors.

As we have remarked, the algebraic system represented by mappings on T to $\mathbb{R}$, with the binary relation summarized in this discussion, is a group. We will denote this group by G. While it is not possible here to explore mathematical details of G, we will discuss a sequence of subgroups which have a computational significance.

For p a positive integer, denote by H_p the subset of G consisting of those mappings which take t to 0, whenever t has no more than p vertices. It can be shown that H_p is a normal subgroup of G. Denote by G_p the factor group G/H_p. We will represent the elements of G_p in the standard group-theoretic terminology as cosets of the form

Table 3. Multiplication table up to order 4

i	t_i	$\alpha(t_i)$	$\beta(t_i)$	$(\alpha\beta)(t_i)$
1		α_1	β_1	$\alpha_1+\beta_1$
2		α_2	β_2	$\alpha_2+\alpha_1\beta_1+\beta_2$
3		α_3	β_3	$\alpha_3+\alpha_1^2\beta_1+2\alpha_1\beta_2+\beta_3$
4		α_4	β_4	$\alpha_4+\alpha_2\beta_1+\alpha_1\beta_2+\beta_4$
5		α_5	β_5	$\alpha_5+\alpha_1^3\beta_1+3\alpha_1^2\beta_2+3\alpha_1\beta_3+\beta_5$
6		α_6	β_6	$\alpha_6+\alpha_1\alpha_2\beta_1+\alpha_1^2\beta_2+\alpha_2\beta_2+\alpha_1\beta_3+\alpha_1\beta_4+\beta_6$
7		α_7	β_7	$\alpha_7+\alpha_3\beta_1+\alpha_1^2\beta_2+2\alpha_1\beta_4+\beta_7$
8		α_8	β_8	$\alpha_8+\alpha_4\beta_1+\alpha_2\beta_2+\alpha_1\beta_4+\beta_8$

Table 4. Terms in $(\alpha\beta)(t_6)$

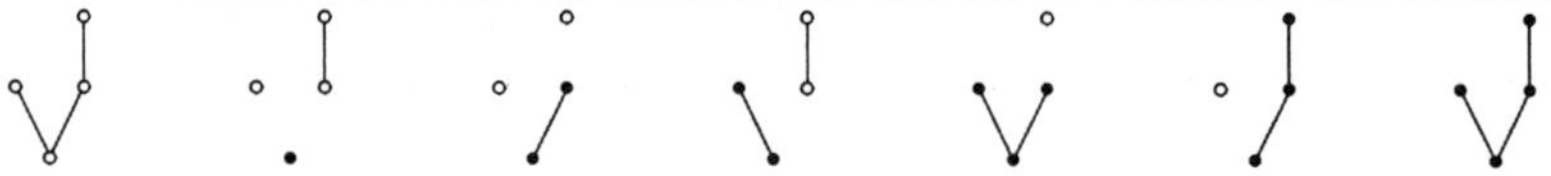

αH_p. The product of αH_p and βH_p is the coset $\alpha\beta H_p$. It is also possible, in the case of these factor groups, to represent G_p as mappings of T_p to $\mathbb{R}$, where T_p denotes the set of rooted trees with no more than p vertices. These two representations of G_p are related by the isomorphism that maps αH_p to $\alpha|_{T_p}$, the restriction of α to T_p.

The algebraic structures described here are all counterparts to natural analytic quantities. Thus, the subgroup H_p consists of representatives of Runge-Kutta method families with the properties that methods in these families, when used in the solution of a smooth problem over a single step, with initial value y_0, give a numerical result equal to $y_0+O(h^{p+1})$. This means that two Runge-Kutta families, when represented by the same members of G_p, give numerical results in a single step which differ by no more than $O(h^{p+1})$. Thus, we are able to examine questions in which order greater than h^p terms are not regarded as significant, by working in G_p, in which the values associated with trees with more than p vertices are not taken into account.

5. Generalizations of Runge-Kutta families

It would be convenient to be able to apply vector space operations to members of G. However, a difficulty arises. Suppose M and N are two families of Runge-Kutta methods and a representative member is selected from each. Since each of these Runge-Kutta methods, when applied to a smooth problem with initial value y_0, gives a result $y_0 + O(h)$. The "sum" of these methods, however, would give a result $2y_0 + O(h)$ and would therefore not be a Runge-Kutta method. We can overcome this difficulty by permitting a slight generalization in what we are able to compute with a Runge-Kutta method. In the usual formulation of an s stage method, as applied to (2), the final result takes the form $y_0 + h\sum_{i=1}^{s} b_i f(Y_i)$. We can instead consider more general computations in which the final result is given by $b_0 y_0 + h\sum_{i=1}^{s} b_i f(Y_i)$, for some constant b_0. To represent this generalization using a tableau, we can insert the additional number b_0 in the otherwise empty south-west corner of the standard Runge-Kutta tableau

$$\begin{array}{c|c} c & A \\ \hline b_0 & b^T \end{array}.$$

To carry this extra information to corresponding mappings from T to $\mathbb{R}$, we need to insert an extra "empty" tree, $t_0 = \emptyset$ and extend the function on T so that t_0 maps to b_0. It is computationally meaningless to combine two generalized methods over successive steps, unless the first of these is a genuine Runge-Kutta method (for which $b_0 = 1$). In this case, Table 3 has to be modified by inserting an extra row showing that $(\alpha\beta)(t_0) = \beta_0$ and replacing the term α_i in the formula for $(\alpha\beta)(t_i)$ by $\beta_0\alpha_i$.

The algebraic system that now results, consists of mappings from $\widetilde{T} = t_0 \cup T$ to $\mathbb{R}$. A subset of these mappings, consisting of those members which map t_0 to 1, has the role of a "multiplier set". Denote by $\widetilde{G}$ the whole set, retaining the G for the multiplier set. Members of $\widetilde{G}$ have a natural vector space structure and members of G act as linear operators over this space. Associativity holds in the sense that $(\alpha\beta)\gamma = \alpha(\beta\gamma)$ for $\alpha, \beta \in G$ and $\gamma \in \widetilde{G}$.

The 1-stage generalized method

$$\begin{array}{c|c} 0 & 0 \\ \hline 0 & 1 \end{array}, \tag{7}$$

has an interesting significance as producing, from an input value y_0, the value of $hf(y_0)$. That is, if $y_0 = y(x_0)$, then the result of applying (7) to y_0 is $hy'(x_0)$. We will denote this method, as well as its image in $\widetilde{G}$, by the symbol D. Note that $D(t) = 0$, unless $t = t_1$ and that $D(t_1) = 1$. Postmultiplication by D has an especially simple form. If a tree t consists of a root connected to subtrees $t_1, t_2, \ldots, t_k$, then

$$(\alpha D)(t) = \prod_{i=1}^{k} \alpha(t_i) \tag{8}$$

If η_1, η_2, ..., η_s denote group elements associated with the stages of a Runge-Kutta method then $\eta_i(t)$ can be calculated recursively from (8) using the formula

$$\eta_i(t) = 1 + \sum_{j=1}^{s} a_{ij}(\eta_j D)(t),$$

to give a final result represented by $1 + \sum_{i=1}^{s} b_i(\eta_i D)(t)$. Note that in these formulas, the identity group element, mapping t_0 to 1 and all other trees to 0, is denoted by 1.

To widen the type of method family further, we allow the possibility that the set of stages $\{1, 2, \ldots, s\}$ is replaced by an infinite set. In particular we consider the index set for the stages defined by $I = [0, 1]$. Define X as the set of continuous functions on I and select as A the linear operator on X defined by

$$(A\phi)(x) = \int_0^x \phi(\bar{x})d\bar{x},$$

with b^T now given by the linear functional

$$b^T\phi = \int_0^1 \phi(\bar{x})d\bar{x}.$$

If we now reinterpret the Runge-Kutta method in this setting, then the result at the end of a single step applied to an initial value $y_0 = y(x_0)$, is the *exact* solution at $x_0 + h$, to the initial value problem, as defined using the Picard integral equation.

The elementary weights can be defined recursively for this "Picard" method and for the first eight trees they are given in Table 5. The values of these elementary weights will be recognised as the numbers which appear on the right hand sides of the order conditions for Runge-Kutta methods. This is to be expected, because the order conditions require a method to agree with the exact solution, up to a certain order.

The family of methods containing the Picard method, will be denoted by E and we will use this same symbol for the corresponding member of G. $E(t)$ can be evaluated recursively by noting that if t consists of a root connected to subtrees t_1, t_2, ..., t_k, then

$$E(t) = \frac{\prod_{i=1}^{k} E(t_i)}{r(t)},$$

where $r(t)$ is the "order" (number of vertices) of t.

6. Order and effective order

A Runge-Kutta method of order p has a relationship with the exact solution, as represented by the Picard method, in that all the elementary weights of orders up to p are equal for the two methods. This means that the sequence of mappings from the numerical method and from the Picard method consisting of (i) the mapping of

Table 5. Elementary weights for Picard method

t	Elementary weight	for Picard method
	$b^T e$	$\int_0^1 dx = 1$
	$b^T c$	$\int_0^1 x dx = \frac{1}{2}$
	$b^T(c.c)$	$\int_0^1 x^2 dx = \frac{1}{3}$
	$b^T Ac$	$\int_0^1 \int_0^y x dx dy = \frac{1}{6}$
	$b^T(c.c.c)$	$\int_0^1 x^3 dx = \frac{1}{4}$
	$b^T(c.(Ac))$	$\int_0^1 y \int_0^y x dx dy = \frac{1}{8}$
	$b^T A(c.c)$	$\int_0^1 \int_0^y x^2 dx dy = \frac{1}{12}$
	$b^T A^2 c$	$\int_0^1 \int_0^z \int_0^y x dx dy dz = \frac{1}{24}$

a method into its equivalence class, (ii) the mapping of this class into G, and (iii) the canonical map from G into G_p, have a composition which takes each of the two methods into the same element of G_p. If $\alpha \in G$ corresponds to the method, then we can write this as

$$\alpha H_p = E H_p.$$

Taking this algebraic point of view, it is natural to look at the consequences of generalizing this relationship using an inner automorphism. Thus we consider the possible existence of β such that

$$\beta \alpha H_p = E \beta H_p.$$

This turns out to be a non-trivial generalization. For example, explicit Runge-Kutta methods exist with 5 stages and order 5 (as generalized in this manner), whereas it is known that order 5 in the classical sense is not possible. The computational interpretation is that there must be a "starting step" performed at the beginning (corresponding to β) and a "finishing step" at the end (corresponding to β^{-1}) to give the benefit of the order property. Details of this are given in the original paper [1] and in the author's book [3]. Recently, this question has been looked at again, but in the context of symplectic methods [5].

We now present three methods which provide an example of effective order. Method (9) is used in the preparatory starting step, and corresponds to β; method

(10) is the method of effective order 5, corresponding to α, and is used for the main sequence of steps; and (11) is the finishing method. Since this final method actually completes a further step, it corresponds (in G_5) to $\beta^{-1}E^2$, rather than to β^{-1} itself.

$$\begin{array}{c|ccccc} 0 & 0 & 0 & 0 & 0 & 0 \\ \frac{2}{5} & \frac{2}{5} & 0 & 0 & 0 & 0 \\ \frac{2}{5} & \frac{1}{5} & \frac{1}{5} & 0 & 0 & 0 \\ \frac{47}{45} & -\frac{517}{1620} & -\frac{23453}{7290} & \frac{66787}{14580} & 0 & 0 \\ 1 & -\frac{4339}{21244} & -\frac{499}{226} & \frac{44355}{13108} & \frac{4374}{154019} & 0 \\ \hline & \frac{269}{2256} & 0 & \frac{2495}{4176} & -\frac{10935}{21808} & \frac{113}{144} \end{array} \tag{9}$$

$$\begin{array}{c|ccccc} 0 & 0 & 0 & 0 & 0 & 0 \\ \frac{2}{5} & \frac{2}{5} & 0 & 0 & 0 & 0 \\ \frac{2}{5} & \frac{1}{5} & \frac{1}{5} & 0 & 0 & 0 \\ \frac{3}{5} & \frac{3}{20} & -\frac{3}{10} & \frac{3}{4} & 0 & 0 \\ 1 & \frac{9}{44} & \frac{5}{22} & -\frac{15}{44} & \frac{10}{11} & 0 \\ \hline & \frac{11}{72} & 0 & \frac{25}{72} & \frac{25}{72} & \frac{11}{72} \end{array} \tag{10}$$

$$\begin{array}{c|ccccc} 0 & 0 & 0 & 0 & 0 & 0 \\ \frac{2}{5} & \frac{2}{5} & 0 & 0 & 0 & 0 \\ \frac{2}{5} & \frac{1}{5} & \frac{1}{5} & 0 & 0 & 0 \\ \frac{3}{5} & \frac{3}{20} & -\frac{27}{110} & \frac{153}{220} & 0 & 0 \\ 1 & \frac{23}{84} & \frac{3}{14} & -\frac{15}{28} & \frac{22}{21} & 0 \\ \hline & \frac{23}{144} & 0 & \frac{5}{16} & \frac{55}{144} & \frac{7}{48} \end{array} \tag{11}$$

7. More general applications

In extending the order concept to a more general family of methods, we can be guided by the Runge-Kutta case and allow for the possibility of using a starting method. However, the fact that multivalue methods may use a variety of approximants, and certainly need information additional to the given initial value, makes the use of starting procedures even more natural than with a one-step method.

We will consider a method with s stages and r quantities passed from step to step. This type of general linear method can be characterized by a partitioned $(s+r)\times(s+r)$ matrix

$$\left[\begin{array}{c|c} A & U \\ \hline B & V \end{array}\right].$$

To explain how this method is used, we will consider just a single component of the numerical result. Let Y denotes the vector made up from the values of this component for the s stages and let F denote the corresponding components of the stage derivatives. The given differential equation provides a method of computing F once Y is known for all components. Let $y^{(n)}$ denote the vector made up from the r values of the required component at the end of step number n. The value of $y^{(n)}$ and the internal stage values are found using the equation

$$\begin{bmatrix} Y \\ y^{(n)} \end{bmatrix} = \begin{bmatrix} A & U \\ B & V \end{bmatrix} \begin{bmatrix} hF \\ y^{(n-1)} \end{bmatrix}. \tag{12}$$

To establish the order of a method of this great generality, or to derive values of the components of the coefficient matrices so that a particular order is obtained, all the quantities that appear in (12) can be replaced by corresponding members of G or of $\widetilde{G}$. If α is the vector in $\widetilde{G}^r$ representing the incoming approximations, and $\beta \in G^s$ is the vector representing the stage approximations, then the values of these quantities, together with the order conditions, are defined recursively on the trees, up to the required order, by the equations

$$\begin{bmatrix} \beta \\ E\alpha \end{bmatrix} = \begin{bmatrix} A & U \\ B & V \end{bmatrix} \begin{bmatrix} \beta D \\ \alpha \end{bmatrix}. \tag{13}$$

Using this formulation of the order conditions, a special method has been derived and is given here. This provides an alternative path past the order barrier for Runge-Kutta methods. Just as we have obtained effective order 5 with 5 explicit stages, we can now do the same with this multivalue method. However, it has been possible to choose the free coefficients in a very interesting manner. The vector α representing the starting method has its first two components equal to 1 and D respectively, indicating that the corresponding numerical components are nothing more than the exact solution and h times the derivative at the same initial point. The third component represents a crude approximation to h^2y'' at the initial point; numerically the effect of $O(h^3)$ errors in this component are extinguished in a single step of the method. The matrices defining the method, for which $c = [\frac{1}{4}, \frac{1}{2}, \frac{3}{4}, 1, 1]^T$, are

$$\left[\begin{array}{c|c} A & U \\ \hline B & V \end{array}\right] = \left[\begin{array}{ccccc|ccc} 0 & 0 & 0 & 0 & 0 & 1 & \frac{1}{4} & \frac{1}{32} \\ a & 0 & 0 & 0 & 0 & 1 & \frac{1-2a}{2} & \frac{1-2a}{8} \\ \frac{3(1-a)}{4} & \frac{9}{16} & 0 & 0 & 0 & 1 & \frac{3(4a-3)}{16} & \frac{3(a-1)}{16} \\ \frac{12a}{7} & -\frac{12}{7} & \frac{8}{7} & 0 & 0 & 1 & \frac{11-12a}{7} & \frac{7-6a}{14} \\ \frac{16}{45} & \frac{2}{15} & \frac{16}{45} & \frac{7}{90} & 0 & 1 & \frac{7}{90} & 0 \\ \hline \frac{16}{45} & \frac{2}{15} & \frac{16}{45} & \frac{7}{90} & 0 & 1 & \frac{7}{90} & 0 \\ 0 & 0 & 0 & 0 & 1 & 0 & 0 & 0 \\ b_{31} & b_{32} & b_{33} & b_{34} & b_{35} & 0 & v_{32} & 0 \end{array}\right],$$

where

$$b_{31} = \frac{128(32 - 200a + 478a^2 - 519a^3 + 216a^4)}{512 - 3712a + 10392a^2 - 13437a^3 + 6912a^4},$$

$$\begin{aligned}
b_{32} &= \frac{16(3a-2)(32-104a+87a^2)}{512-3712a+10392a^2-13437a^3+6912a^4},\\
b_{33} &= \frac{128(8a-30a^2+27a^3)}{3(512-3712a+10392a^2-13437a^3+6912a^4)},\\
b_{34} &= \frac{28a^2(3a-2)}{512-3712a+10392a^2-13437a^3+6912a^4},\\
b_{35} &= \frac{135a^3}{512-3712a+10392a^2-13437a^3+6912a^4},\\
v_{32} &= \frac{-9216+61184a-156216a^2+182655a^3-82944a^4}{3(512-3712a+10392a^2-13437a^3+6912a^4)}
\end{aligned}$$

and $a \approx 0.5015381710$ is the real solution to the equation

$$-8192+71680a-255360a^2+464400a^3-432675a^4+165888a^5=0.$$

To verify that the order of the method is 5, it is necessary to introduce the 9 additional trees of this order. We will express these in terms of the sequence of subtrees that remain when the root is removed, as in the examples $t_2 = [t_1]$, $t_3 = [t_1, t_1]$, $t_42 = [t_2]$, $t_5 = [t_1, t_1, t_1]$, $t_6 = [t_1, t_2]$, $t_7 = [t_3]$, $t_8 = [t_4]$. The new trees are $t_9 = [t_1, t_1, t_1, t_1]$, $t_{10} = [t_1, t_1, t_2]$, $t_{11} = [t_1, t_3]$, $t_{12} = [t_1, t_4]$, $t_{13} = [t_2, t_2]$, $t_{14} = [t_5]$, $t_{15} = [t_6]$, $t_{16} = [t_7]$, $t_{17} = [t_8]$.

The $\alpha(t_i)$, $i = 0, 1, 2, \ldots, 17$, are of the form

$$\alpha(t_0) = \begin{bmatrix} 1 \\ 0 \\ 0 \end{bmatrix}, \alpha(t_1) = \begin{bmatrix} 0 \\ 1 \\ 0 \end{bmatrix}, \alpha(t_2) = \begin{bmatrix} 0 \\ 0 \\ 1 \end{bmatrix}, \alpha(t_i) = \begin{bmatrix} 0 \\ 0 \\ \xi_i \end{bmatrix}, i = 3, 4, \ldots, 17,$$

where we will assume that $\xi_4 = \xi_3/2$, $\xi_6 = \xi_5/2$, $\xi_8 = \xi_7/2$. These assumptions will follow from the final form of the order conditions but are inserted at this point to simplify the details.

From this information on the input quantities, the members of $\widetilde{G}$ which represent the stage values and stage derivatives can be deduced from $\beta = A\beta D + U\alpha$, from (13). They are found in turn to be

$$(\beta D)(t_0) = \begin{bmatrix} 0 \\ 0 \\ 0 \\ 0 \\ 0 \end{bmatrix}, \quad \beta(t_0) = \begin{bmatrix} 1 \\ 1 \\ 1 \\ 1 \\ 1 \end{bmatrix}, \quad (\beta D)(t_1) = \begin{bmatrix} 1 \\ 1 \\ 1 \\ 1 \\ 1 \end{bmatrix}, \quad \beta(t_1) = \begin{bmatrix} \frac{1}{4} \\ \frac{1}{2} \\ \frac{3}{4} \\ 1 \\ 1 \end{bmatrix},$$

$$(\beta D)(t_2) = \begin{bmatrix} \frac{1}{4} \\ \frac{1}{2} \\ \frac{3}{4} \\ 1 \\ 1 \end{bmatrix}, \quad \beta(t_2) = \begin{bmatrix} \frac{1}{32} \\ \frac{1}{8} \\ \frac{9}{32} \\ \frac{1}{2} \\ \frac{1}{2} \end{bmatrix}, \quad (\beta D)(t_3) = \begin{bmatrix} \frac{1}{16} \\ \frac{1}{4} \\ \frac{9}{16} \\ 1 \\ 1 \end{bmatrix}, \quad (\beta D)(t_4) = \frac{1}{2}(\beta D)(t_3),$$

$$\beta(t_3) = \begin{bmatrix} \frac{\xi_3}{32} \\ \frac{a+2\xi_3-4a\xi_3}{16} \\ \frac{3(4-a-4\xi_3+4a\xi_3)}{64} \\ \frac{6+3a+14\xi_3-12a\xi_3}{28} \\ \frac{1}{3} \end{bmatrix}, \quad \beta(t_4) = \frac{1}{2}\beta(t_3), \quad (\beta D)(t_5) = \begin{bmatrix} \frac{1}{64} \\ \frac{1}{8} \\ \frac{27}{64} \\ 1 \\ 1 \end{bmatrix},$$

$$(\beta D)(t_6) = \frac{1}{2}(\beta D)(t_5), \; (\beta D)(t_7) = \begin{bmatrix} \frac{\xi_3}{32} \\ \frac{a+2\xi_3-4a\xi_3}{16} \\ \frac{3(4-a-4\xi_3+4a\xi_3)}{64} \\ \frac{6+3a+14\xi_3-12a\xi_3}{28} \\ \frac{1}{3} \end{bmatrix}, \; (\beta D)(t_8) = \frac{1}{2}(\beta D)(t_7),$$

$$\beta(t_5) = \begin{bmatrix} \frac{\xi_5}{32} \\ \frac{a+8\xi_5-16a\xi_5}{64} \\ \frac{3(7-a-16\xi_5+16a\xi_5)}{256} \\ \frac{30+3a+56\xi_5-48a\xi_5}{112} \\ \frac{1}{4} \end{bmatrix}, \beta(t_6) = \frac{1}{2}\beta(t_5), \beta(t_7) = \begin{bmatrix} \frac{\xi_7}{32} \\ \frac{a\xi_3+4\xi_7-8a\xi_7}{32} \\ \frac{3(3a+8\xi_3-14a\xi_3-16\xi_7+16a\xi_7)}{256} \\ \frac{12-9a-24\xi_3+39a\xi_3+28\xi_7-24a\xi_7}{56} \\ \frac{1}{12} \end{bmatrix},$$

$$\beta(t_8) = \frac{1}{2}\beta(t_7), \quad (\beta D)(t_9) = \begin{bmatrix} \frac{1}{256} \\ \frac{1}{16} \\ \frac{81}{256} \\ 1 \\ 1 \end{bmatrix}, \quad (\beta D)(t_{10}) = \frac{1}{2}(\beta D)(t_9),$$

$$(\beta D)(t_{11}) = \begin{bmatrix} \frac{\xi_3}{128} \\ \frac{a+2\xi_3-4a\xi_3}{32} \\ \frac{9(4-a-4\xi_3+4a\xi_3)}{256} \\ \frac{6+3a+14\xi_3-12a\xi_3}{28} \\ \frac{1}{3} \end{bmatrix}, \; (\beta D)(t_{12}) = \frac{1}{2}(\beta D)(t_{11}), \; (\beta D)(t_{13}) = \frac{1}{4}(\beta D)(t_9),$$

$$(\beta D)(t_{14}) = \begin{bmatrix} \frac{\xi_5}{32} \\ \frac{a+8\xi_5-16a\xi_5}{64} \\ \frac{3(7-a-16\xi_5+16a\xi_5)}{256} \\ \frac{30+3a+56\xi_5-48a\xi_5}{112} \\ \frac{1}{4} \end{bmatrix}, \quad (\beta D)(t_{15}) = \frac{1}{2}(\beta D)(t_{14}),$$

$$(\beta D)(t_{16}) = \begin{bmatrix} \frac{\xi_7}{32} \\ \frac{a\xi_3+4\xi_7-8a\xi_7}{32} \\ \frac{3(3a+8\xi_3-14a\xi_3-16\xi_7+16a\xi_7)}{256} \\ \frac{12-9a-24\xi_3+39a\xi_3+28\xi_7-24a\xi_7}{56} \\ \frac{1}{12} \end{bmatrix}, \quad (\beta D)(t_{17}) = \frac{1}{2}(\beta D)(t_{16}),$$

It is now necessary only to compute the values of $(B\beta D+V\alpha)(t_i) = \hat{\alpha}(t_i)$, say and to compare these with $(E\alpha)(t_i)$, $i = 1, 2, \ldots, 17$. The values of the $\hat{\alpha}(t_i)$ are found to be

$$\hat{\alpha}(t_1) = \begin{bmatrix} 1 \\ 1 \\ 0 \end{bmatrix}, \quad \hat{\alpha}(t_2) = \begin{bmatrix} \frac{1}{2} \\ 1 \\ 1 \end{bmatrix}, \quad \hat{\alpha}(t_3) = \begin{bmatrix} \frac{1}{3} \\ 1 \\ \frac{3(-64a+368a^2-747a^3+576a^4)}{512-3712a+10392a^2-13437a^3+6912a^4} \end{bmatrix},$$

$$\hat{\alpha}(t_4) = \frac{1}{2}\hat{\alpha}(t_3), \qquad \hat{\alpha}(t_5) = \begin{bmatrix} \frac{1}{4} \\ 1 \\ \frac{-64+352a-612a^2+189a^3+432a^4}{512-3712a+10392a^2-13437a^3+6912a^4} \end{bmatrix},$$

$$\hat{\alpha}(t_6) = \frac{1}{2}\hat{\alpha}(t_5), \qquad \hat{\alpha}(t_7) = \begin{bmatrix} \frac{1}{12} \\ \frac{1}{3} \\ \frac{9(4a^2-17a^3+24a^4)}{512-3712a+10392a^2-13437a^3+6912a^4} \end{bmatrix},$$

$$\hat{\alpha}(t_8) = \frac{1}{2}\hat{\alpha}(t_7), \qquad \hat{\alpha}(t_9) = \begin{bmatrix} \frac{1}{5} \\ 1 \\ \frac{3(-16+104a-236a^2+195a^3+36a^4)}{512-3712a+10392a^2-13437a^3+6912a^4} \end{bmatrix},$$

$$\hat{\alpha}(t_{10}) = \frac{1}{2}\hat{\alpha}(t_9), \quad \hat{\alpha}(t_{12}) = \frac{1}{2}\hat{\alpha}(t_{11}), \quad \hat{\alpha}(t_{13}) = \frac{1}{4}\hat{\alpha}(t_9),$$

$$\hat{\alpha}(t_{11}) = \begin{bmatrix} \frac{1}{15} \\ \frac{1}{3} \\ \frac{16a-52a^2+21a^3+99a^4-32\xi_3+184a\xi_3-416a^2\xi_3+438a^3\xi_3-180a^4\xi_3}{512-3712a+10392a^2-13437a^3+6912a^4} \end{bmatrix},$$

$$\hat{\alpha}(t_{14}) = \begin{bmatrix} \frac{1}{20} \\ \frac{1}{4} \\ \frac{3(16a-64a^2+57a^3+72a^4)}{4(512-3712a+10392a^2-13437a^3+6912a^4)} \end{bmatrix}, \qquad \hat{\alpha}(t_{15}) = \frac{1}{2}\hat{\alpha}(t_{14}),$$

$$\hat{\alpha}(t_{16}) = \begin{bmatrix} \frac{1}{60} \\ \frac{1}{12} \\ \frac{27(-a^3+4a^4)}{4(512-3712a+10392a^2-13437a^3+6912a^4)} \end{bmatrix}, \qquad \hat{\alpha}(t_{17}) = \frac{1}{2}\hat{\alpha}(t_{16}).$$

The first two components agree with those of the corresponding components of $E\alpha$ for each of the 17 trees up to order 5. The third component agrees for a suitable choice of $\xi_3, \xi_4, \ldots, \xi_{17}$.

Designing a method with the order attributes we have required is rather easy but there is one additional feature that makes this method live up to its claim of breaking the order barrier. This is that the stability matrix

$$M(z) = V + zB(I - zA)^{-1}U$$

has characteristic polynomial

$$\phi(w, z) = \det(wI - M(z)) = w^3 - w^2\left(1 + z + \frac{z^2}{2} + \frac{z^3}{6} + \frac{z^4}{24} + \frac{z^5}{120}\right),$$

implying a stability region exactly as for a Runge-Kutta method, if it existed, with 5 stages and order 5. This stability feature was achieved by imposing the constraints

$$\begin{array}{rclrclrcl} \mathrm{tr}(V) & = & 1, & \mathrm{tr}(BU) & = & 1, & \mathrm{tr}(BAU) & = & \frac{1}{2}, \\ \mathrm{tr}(BA^2U) & = & \frac{1}{6}, & \mathrm{tr}(BA^3U) & = & \frac{1}{24}, & \mathrm{tr}(BA^4U) & = & \frac{1}{120}, \end{array}$$

and the specific values used for the final row of B, v_{32} and a were chosen to satisfy these requirements.

Acknowledgements

The work of the author was supported by the New Zealand Foundation for Research, Science and Technology. The author expresses his gratitude to the IRISA, Rennes, France for inviting him to visit them during part of the time he was working on this paper. He particularly thanks Philippe Chartier for conversations on this and related topics.

8. References

1. J. C. Butcher, The effective order of Runge-Kutta methods, in *Conf. on the Numerical Solution of Differential Equations, Dundee, 1969* ed. J. L. Morris,(Springer, Berlin, 1972).
2. J. C. Butcher, An algebraic theory of integration methods, *Math. Comp.* **26** (1972), 79–106.
3. J. C. Butcher, *The Numerical Analysis of Ordinary Differential Equations — Runge-Kutta and General Linear Methods* (Wiley, Chichester, 1987).
4. E. Hairer and G. Wanner, On the Butcher group and and general multi-value methods, *Computing* **13** (1974), 1–15.
5. M. A. Lopez–Marcos, J. M. Sanz–Serna and R. D. Skeel, Cheap enhancement of symplectic integrators, in *Numerical Analysis 1995* (D. F. Griffiths and G. A. Watson (Eds.), Pitman Research Notes in Mathematics, Longman Scientific & Technical, 1996), pp. 107–122.

RUNGE–KUTTA METHODS ON MANIFOLDS

MARIPAZ CALVO
Departamento de Matemática Aplicada y Computación, Universidad de Valladolid
Valladolid, Spain.
E-mail: Maripaz@mac.cie.uva.es

ARIEH ISERLES
Department of Applied Mathematics and Theoretical Physics, University of Cambridge
Silver Street, Cambridge CB3 9EW, England
E-mail: A.Iserles@amtp.cam.ac.uk

and

ANTONELLA ZANNA
Newnham College, University of Cambridge
Cambridge, England
E-mail: A.Zanna@amtp.cam.ac.uk

ABSTRACT

The subject matter of this paper is the recovery of invariants and conservation laws of ordinary differential systems by numerical methods. We prove that the most likely candidates for this task, Runge–Kutta schemes, fail to stay on manifolds defined by r–tensors with $r \geq 3$. As an alternative, we suggest diffeomorphically mapping complicated manifolds to simpler ones. This procedure allows for recovery of invariants that are intractable in a classical setting and it emphasizes the crucial role of the topology of underlying manifolds.

1. Invariants and numerical methods

Let us suppose that an autonomous system of ordinary differential equations (ODEs)

$$\boldsymbol{y}' = \boldsymbol{f}(\boldsymbol{y}), \qquad t \geq 0, \tag{1}$$

is given in the Euclidean space $\mathbb{R}^d$ and that it is known that for every initial condition $\boldsymbol{y}(0) = \boldsymbol{y}_0$ its exact solution stays for all $t \geq 0$ on a manifold $\mathcal{M} = \mathcal{M}(\boldsymbol{y}_0) \subset \mathbb{R}^d$. In other words,

$$\boldsymbol{y}_0 \in \mathcal{M} \qquad \Rightarrow \qquad \boldsymbol{y}(t) \in \mathcal{M} \quad \forall\, t \geq 0. \tag{2}$$

We say that the equation (1) is $\mathcal{M}$*-invariant.*

$\mathcal{M}$-invariance often conveys an important information about the underlying differential system, for example the satisfaction of a conservation law, an integral, etc. It is usually desirable and often essential to retain this feature under discretization.

A classical approach to numerical methods for ODEs pays little attention to the issue of invariance, the emphasis being on deriving accuracy within a specified error tolerance and on the recovery of correct asymptotic behaviour of the solution. This is perfectly satisfactory in many situations and we have no intention to denigrate the impressive body of contemporary work on the numerical solution of ODEs. However, there exists a wide range of problems that require the exact retention of invariants, e.g. Hamiltonian systems and isospectral flows, whereby the classical approach falls short of our requirements. We refer the reader to (Iserles and Zanna [8]), where this issue is debated at a greater length.

Recent years have witnessed a growing amount of attention devoted to numerical retention of invariance. Much of the work has been specialized to specific manifolds: the symplectic manifold for Hamiltonian equations (Sanz-Serna and Calvo [11]), the unitary group $O(d)$ for unitary flows (Dieci, Russell and Van Vleck [5]) and the isospectral manifold for Toda lattice equations, double-bracket equations and other isospectral flows (Calvo, Iserles and Zanna [1]). However, outlines of a more general theory have started to emerge lately, mainly in the work of Peter Crouch and R. Grossman [3] and Hans Munthe-Kaas [10]. The aim of the present paper is to pursue the general theme, to understand how well existing methods can retain $\mathcal{M}$-invariance for different manifolds $\mathcal{M}$ and what can be done to improve their efficacy.

Section 2 expands upon a theme to which we have already addressed ourselves in a somewhat lesser generality in (Calvo et al. [1]), namely manifolds that are defined as a level set of a r-tensor, $r \in \mathbb{N}$. The case $r = 1$ is trivial, while $r = 2$ has been already determined by Cooper [2]. We prove that no nonconfluent Runge–Kutta method can retain $\mathcal{M}$-invariance for all manifolds described by a symmetric r-tensor whenever $r \geq 3$. In particular, no Runge–Kutta method can stay on the manifold

$$\mathcal{M} = \left\{ \boldsymbol{y} \in \mathbb{R}^d \,:\, \sum_{k=1}^{d} y_k^r = \sum_{k=1}^{d} y_k^r(0) \right\}$$

for $r \geq 3$.

In Section 3 we propose a remedy which allows us to retain $\mathcal{M}$-invariance for a large set of interesting manifolds and opens up an approach which might well have a substantially wider scope. Mapping the ODE (1) by a diffeomorphism, we obtain a differential equation which is $\mathcal{N}$-invariant and a manifold $\mathcal{N}$ which, while being topologically equivalent to $\mathcal{M}$, is tractable by Runge–Kutta methods. This technique retains $\mathcal{M}$-invariance under discretization when $\mathcal{M}$ is diffeomorphic to $\mathbb{R}^q$, $q \leq d-1$, or to the sphere $\boldsymbol{S}_q$, $q \leq d-1$. We hope to identify in the future numerical methods which retain invariance for manifolds that are diffeomorphic to torii or to manifolds of higher genera.

We refer the reader to (Guillemin and Pollack [6]) for basic concepts and terminology of differential topology, which might not be familiar to all numerical analysts.

2. Runge–Kutta methods and manifolds

We discretize the ODE system (1) by means of the ν-stage Runge–Kutta method

$$
\begin{aligned}
\boldsymbol{\varphi}_\ell &= \boldsymbol{y}_n + h \sum_{j=1}^{\nu} a_{\ell,j} \boldsymbol{k}_j, \qquad \ell = 1,2,\ldots,\nu, \\
\boldsymbol{k}_\ell &= \boldsymbol{f}(\boldsymbol{\varphi}_\ell), \\
\boldsymbol{y}_{n+1} &= \boldsymbol{y}_n + h \sum_{\ell=1}^{\nu} b_\ell \boldsymbol{k}_\ell.
\end{aligned} \tag{3}
$$

In conformity with standard terminology, we designate $A = (a_{k,\ell})$ as the *RK matrix* and call $\boldsymbol{b} = (b_\ell)$ the *RK weights* (Iserles [7]). We say that the method (3) is *nonconfluent* if the *RK nodes* $c_\ell = \sum_{j=1}^{\nu} a_{\ell,j}$, $\ell = 1,2,\ldots,\nu$, are distinct.

Assuming that the solution of (1) is $\mathcal{M}$-invariant, it is of interest to pose the question whether the solution sequence, as discretized by the Runge–Kutta method (3), shares this feature, i.e.

$$\boldsymbol{y}_0 \in \mathcal{M} \qquad \Rightarrow \qquad \boldsymbol{y}_n \in \mathcal{M} \quad \forall\, n \in \mathbb{Z}^+. \tag{4}$$

Without much abuse of terminology, we refer to the Runge–Kutta method (3) as $\mathcal{M}$*-invariant* if it obeys (4) whenever the underlying ODE system is $\mathcal{M}$-invariant. Moreover, given a family $\mathbb{M} = \{\mathcal{M}_\alpha\}$ of manifolds in $\mathbb{R}^d$, we say that a Runge–Kutta method is $\mathbb{M}$*-invariant* if it is $\mathcal{M}_\alpha$-invariant for all α.

In this section we do not attempt to address the question of $\mathcal{M}$-invariance in its full generality. Instead, we just consider a subset of all possible manifolds which, while broad enough to be of interest in numerous issues of genuine practical importance, possesses enough structure to lend itself to our analysis. Our assumption is that

$$\mathcal{M} = \{\boldsymbol{x} \in \mathbb{R}^d \,:\, S(\boldsymbol{x},\boldsymbol{x},\ldots,\boldsymbol{x}) \equiv c\} \tag{5}$$

where $c \in \mathbb{R}$ is a constant and S is a *symmetric* r*-tensor.* In other words,

$$S : \overbrace{\mathbb{R}^d \times \mathbb{R}^d \times \cdots \times \mathbb{R}^d}^{r \text{ times}} \to \mathbb{R},$$

S is linear in all its components and

$$S(\boldsymbol{x}_1,\boldsymbol{x}_2,\ldots,\boldsymbol{x}_r) = S(\boldsymbol{x}_{\pi(1)},\boldsymbol{x}_{\pi(2)},\ldots,\boldsymbol{x}_{\pi(r)}), \qquad \boldsymbol{x}_1,\boldsymbol{x}_2,\ldots,\boldsymbol{x}_r \in \mathbb{R}^d,$$

for all permutations $[\pi(1)\ \pi(2)\ \cdots\ \pi(r)]$ of the indices $[1\ 2\ \cdots\ r]$. We denote the set of all manifolds in $\mathbb{R}^d$ generated by symmetric r-tensors by $\mathcal{S}_r$.

The insistence on symmetry does not lead to loss of generality. Thus, let T be an r-tensor, which is not necessarily symmetric, and set

$$S(\boldsymbol{y}_1,\boldsymbol{y}_2,\ldots,\boldsymbol{y}_r) := \frac{1}{r!} \sum T(\boldsymbol{y}_{\pi(1)},\boldsymbol{y}_{\pi(2)},\ldots,\boldsymbol{y}_{\pi(r)}),$$

where the sum is carried out across all $r!$ permutations of the indices $[1\ 2\ \cdots\ r]$. It is easy to verify that S is a symmetric r-tensor and that

$$\{\boldsymbol{x} \in \mathbb{R}^d : S(\boldsymbol{x}, \boldsymbol{x}, \ldots, \boldsymbol{x}) \equiv c\} = \{\boldsymbol{x} \in \mathbb{R}^d : T(\boldsymbol{x}, \boldsymbol{x}, \ldots, \boldsymbol{x}) \equiv c\}.$$

Therefore we may assume without further ado that $\mathcal{M}$ is a level set of a symmetric r-tensor. This construction, although perfectly correct from the formal point of view, makes no sense when the symmetrization results in a null tensor, e.g. when T is an alternating tensor.

The case $r = 1$ corresponds to $\boldsymbol{y}(t)$ lying in an affine space: there exists $\boldsymbol{s} \in \mathbb{R}^d$ such that $\boldsymbol{s}^T\boldsymbol{y}(t) \equiv c$ for all $t \geq 0$. It is trivial to verify that every consistent numerical method is $\mathcal{S}_1$-invariant (Calvo et al. [1]).

For the quadratic case, $r = 2$, there exists a symmetric matrix $\tilde{S}$ such that $S(\boldsymbol{x}_1, \boldsymbol{x}_2) = \boldsymbol{x}_1^T\tilde{S}\boldsymbol{x}_2$. It has been proved by Graeme Cooper that the Runge–Kutta method (3) is, in our terminology, $\mathcal{S}_2$-invariant if $M = O$ (Cooper [2]), where the matrix M, ubiquitous in stability analysis of Runge–Kutta methods (Dekker and Verwer [4]; Sanz-Serna and Calvo [11]) is defined by

$$m_{k,\ell} = b_k a_{k,\ell} + b_\ell a_{\ell,k} - b_k b_\ell, \qquad k, \ell = 1, 2, \ldots, \nu.$$

Moreover, unless $M = O$, there always exists a matrix $\tilde{S}$ so that (4) is violated by the underlying nonconfluent Runge–Kutta method (Calvo et al. [1]).

Cubic manifolds have been already considered in (Calvo et al. [1]), where we have proved that for every nonconfluent Runge–Kutta method there exists a symmetric 3-tensor S such that the method is not $\mathcal{M}$-invariant with respect to the manifold $\mathcal{M}$, as defined by (5). In other words, although we cannot rule out $\mathcal{M}$-invariance for some Runge–Kutta methods and cubic manifolds, no nonconfluent Runge–Kutta method can be $\mathcal{S}_3$-invariant. In the present paper we adopt a more general approach and consider the case $r \geq 3$. Our objective, the main result of this section, is that no nonconfluent Runge–Kutta method can be $\mathcal{S}_r$-invariant for any $r \geq 3$.

We denote by $S_{k_1,k_2,\ldots,k_q}$, where $1 \leq k_1 < k_2 < \cdots < k_q \leq r$, the value of the r-tensor S when each k_jth coordinate has been replaced by $\boldsymbol{k}_{\ell_j}$, $j = 1, 2, \ldots, q$, while all the remaining coordinates are $\boldsymbol{y}_n$. By virtue of linearity of S, (3) implies that

$$\begin{aligned}
&S(\boldsymbol{y}_{n+1}, \boldsymbol{y}_{n+1}, \ldots, \boldsymbol{y}_{n+1}) \\
&= S\left(\boldsymbol{y}_n + h\sum_{\ell=1}^{\nu} b_\ell \boldsymbol{k}_\ell, \boldsymbol{y}_n + h\sum_{\ell=1}^{\nu} b_\ell \boldsymbol{k}_\ell, \ldots, \boldsymbol{y}_n + h\sum_{\ell=1}^{\nu} b_\ell \boldsymbol{k}_\ell\right) \\
&= S(\boldsymbol{y}_n, \boldsymbol{y}_n, \ldots, \boldsymbol{y}_n) + h\sum_{k_1=1}^{r}\sum_{\ell_1=1}^{\nu} b_{\ell_1} S_{k_1} + h^2 \sum_{k_1=1}^{r-1}\sum_{k_2=k_1+1}^{r}\sum_{\ell_1,\ell_2=1}^{\nu} b_{\ell_1} b_{\ell_2} S_{k_1,k_2} \\
&\quad + h^3 \sum_{k_1=1}^{r-2}\sum_{k_2=k_1+1}^{r-1}\sum_{k_3=k_2+1}^{r}\sum_{\ell_1,\ell_2,\ell_3=1}^{\nu} b_{\ell_1} b_{\ell_2} b_{\ell_3} S_{k_1,k_2,k_3} + \mathcal{O}\left(h^4\right).
\end{aligned}$$

Assuming that $\boldsymbol{y}_n \in \mathcal{M}$, we wish to check whether this is also true for $\boldsymbol{y}_{n+1}$. In other words, we examine whether

$$I := S(\boldsymbol{y}_{n+1}, \boldsymbol{y}_{n+1}, \ldots, \boldsymbol{y}_{n+1}) - S(\boldsymbol{y}_n, \boldsymbol{y}_n, \ldots, \boldsymbol{y}_n)$$

vanishes. By virtue of symmetry,

$$S_{k_1,k_2,\ldots,k_q} = S(\overbrace{\boldsymbol{y}_n, \boldsymbol{y}_n \ldots, \boldsymbol{y}_n}^{r-q \text{ times}}, \boldsymbol{k}_{\ell_1}, \boldsymbol{k}_{\ell_2}, \ldots, \boldsymbol{k}_{\ell_q}),$$

therefore

$$I = \sum_{q=1}^{r} \binom{r}{q} h^q \sum_{\ell_1,\ell_2,\ldots,\ell_q=1}^{\nu} b_{\ell_1} b_{\ell_2} \cdots b_{\ell_q} S(\boldsymbol{y}_n, \ldots, \boldsymbol{y}_n, \boldsymbol{k}_{\ell_1}, \ldots, \boldsymbol{k}_{\ell_q}). \tag{6}$$

We commence with the $q = 1$ term, noting that, according to (3), we may substitute

$$\boldsymbol{y}_n = \boldsymbol{\varphi}_\ell - h \sum_{j=1}^{\nu} a_{\ell,j} \boldsymbol{k}_j$$

for any $\ell \in \{1, 2, \ldots, \nu\}$. Therefore

$$\begin{aligned}
&\sum_{\ell=1}^{\nu} b_\ell S(\boldsymbol{y}_n, \ldots, \boldsymbol{y}_n, \boldsymbol{k}_\ell) \\
&= \sum_{\ell=1}^{\nu} b_\ell S\left(\boldsymbol{\varphi}_\ell - h \sum_{j_1=1}^{\nu} a_{\ell,j_1} \boldsymbol{k}_{j_1}, \ldots, \boldsymbol{\varphi}_\ell - h \sum_{j_{r-1}=1}^{\nu} a_{\ell,j_{r-1}} \boldsymbol{k}_{j_{r-1}}, \boldsymbol{k}_\ell\right) \\
&= \sum_{p=0}^{r-1} (-1)^p \binom{r-1}{p} h^p \sum_{\ell=1}^{\nu} \sum_{j_1,\ldots,j_p=1}^{\nu} b_\ell a_{\ell,j_1} a_{\ell,j_2} \cdots a_{\ell,j_p} \\
&\qquad \times S(\overbrace{\boldsymbol{\varphi}_\ell, \ldots, \boldsymbol{\varphi}_\ell}^{r-p-1 \text{ times}}, \boldsymbol{k}_{j_1}, \boldsymbol{k}_{j_2}, \ldots, \boldsymbol{k}_{j_p}, \boldsymbol{k}_\ell).
\end{aligned}$$

Note that, differentiating $S(\boldsymbol{y}, \boldsymbol{y}, \ldots, \boldsymbol{y}) \equiv c$ and exploiting symmetry, we deduce $S(\boldsymbol{y}, \ldots, \boldsymbol{y}, \boldsymbol{f}(\boldsymbol{y})) \equiv 0$, therefore, according to (3),

$$S(\boldsymbol{\varphi}_\ell, \ldots, \boldsymbol{\varphi}_\ell, \boldsymbol{k}_\ell) = 0.$$

Consequently,

$$\begin{aligned}
\sum_{\ell=1}^{\nu} b_\ell S(\boldsymbol{y}_n, \ldots, \boldsymbol{y}_n, \boldsymbol{k}_\ell) &= -(r-1)h \sum_{\ell,j=1}^{\nu} b_\ell a_{\ell,j} S(\boldsymbol{\varphi}_\ell, \ldots, \boldsymbol{\varphi}_\ell, \boldsymbol{k}_j, \boldsymbol{k}_\ell) \\
&+ \frac{(r-2)(r-1)}{2} h^2 \sum_{\ell,j_1,j_2=1}^{\nu} b_\ell a_{\ell,j_1} a_{\ell,j_2} S(\boldsymbol{\varphi}_\ell, \ldots, \boldsymbol{\varphi}_\ell, \boldsymbol{k}_{j_1}, \boldsymbol{k}_{j_2}, \boldsymbol{k}_\ell) \\
&+ \mathcal{O}\left(h^3\right).
\end{aligned} \tag{7}$$

Next, we proceed to the $q=2$ term. By similar substitution we readily obtain

$$\begin{aligned}
&\sum_{\ell_1,\ell_2=1}^{\nu} b_{\ell_1}b_{\ell_2}S(\boldsymbol{y}_n,\ldots,\boldsymbol{y}_n,\boldsymbol{k}_{\ell_1},\boldsymbol{k}_{\ell_2})\\
&= \sum_{\ell_1,\ell_2=1}^{\nu} b_{\ell_1}b_{\ell_2}S\left(\boldsymbol{\varphi}_{\ell_1} - h\sum_{j=1}^{\nu} a_{\ell_1,j}\boldsymbol{k}_j,\ldots,\boldsymbol{\varphi}_{\ell_1} - h\sum_{j=1}^{\nu} a_{\ell_1,j}\boldsymbol{k}_j,\boldsymbol{k}_{\ell_1},\boldsymbol{k}_{\ell_2}\right)\\
&= \sum_{\ell_1,\ell_2=1}^{\nu} b_{\ell_1}b_{\ell_2}S(\boldsymbol{\varphi}_{\ell_1},\ldots,\boldsymbol{\varphi}_{\ell_1},\boldsymbol{k}_{\ell_1},\boldsymbol{k}_{\ell_2})\\
&\quad - (r-2)h\sum_{\ell_1,\ell_2,j=1}^{\nu} b_{\ell_1}b_{\ell_2}a_{\ell_1,j}S(\boldsymbol{\varphi}_{\ell_1},\ldots,\boldsymbol{\varphi}_{\ell_1},\boldsymbol{k}_j,\boldsymbol{k}_{\ell_1},\boldsymbol{k}_{\ell_2})\\
&\quad + \mathcal{O}\left(h^2\right).
\end{aligned} \tag{8}$$

Finally, the contribution of the $q=3$ term is

$$\begin{aligned}
&\frac{(r-2)(r-1)r}{6}\sum_{\ell_1,\ell_2,\ell_3=1}^{\nu} b_{\ell_1}b_{\ell_2}b_{\ell_3}S(\boldsymbol{\varphi}_{\ell_1},\boldsymbol{\varphi}_{\ell_1},\ldots,\boldsymbol{\varphi}_{\ell_1},\boldsymbol{k}_{\ell_1},\boldsymbol{k}_{\ell_2},\boldsymbol{k}_{\ell_3})\\
&\quad + \mathcal{O}(h).
\end{aligned} \tag{9}$$

Collecting together the contributions of (7)–(9) and making a frequent, albeit trivial, use of symmetry, we expand I in powers of h,

$$\begin{aligned}
I = h^2 &\left[-(r-1)r\sum_{\ell,j=1}^{\nu} b_\ell a_{\ell,j}S(\boldsymbol{\varphi}_\ell,\ldots,\boldsymbol{\varphi}_\ell,\boldsymbol{k}_j,\boldsymbol{k}_\ell)\right.\\
&\left.\quad + \frac{(r-1)r}{2}\sum_{\ell,j=1}^{\nu} b_\ell b_j S(\boldsymbol{\varphi}_\ell,\ldots,\boldsymbol{\varphi}_\ell,\boldsymbol{k}_j,\boldsymbol{k}_\ell)\right]\\
&+ h^3\left[\frac{(r-2)(r-1)r}{2}\sum_{\ell,j,i=1}^{\nu} b_\ell a_{\ell,j}a_{\ell,i}S(\boldsymbol{\varphi}_\ell,\ldots,\boldsymbol{\varphi}_\ell,\boldsymbol{k}_j,\boldsymbol{k}_i,\boldsymbol{k}_\ell)\right.\\
&\quad - \frac{(r-2)(r-1)r}{2}\sum_{\ell,j,i=1}^{\nu} b_\ell b_j a_{\ell,i}S(\boldsymbol{\varphi}_\ell,\ldots,\boldsymbol{\varphi}_\ell,\boldsymbol{k}_j,\boldsymbol{k}_i,\boldsymbol{k}_\ell)\\
&\left.\quad + \frac{(r-2)(r-1)r}{6}\sum_{\ell,j,i=1}^{\nu} b_\ell b_j b_i S(\boldsymbol{\varphi}_\ell,\ldots,\boldsymbol{\varphi}_\ell,\boldsymbol{k}_j,\boldsymbol{k}_i,\boldsymbol{k}_\ell)\right]\\
&+\mathcal{O}\left(h^4\right).
\end{aligned}$$

In order to obtain $I=0$ we need to set to zero the coefficients of all powers of h. We start with the $\mathcal{O}(h^2)$ term. Before proceeding further, observe that

$$\begin{aligned}
&S(\boldsymbol{\varphi}_\ell,\ldots,\boldsymbol{\varphi}_\ell,\boldsymbol{k}_j,\boldsymbol{k}_\ell)\\
&= S\left(\boldsymbol{\varphi}_j + h\sum_{i=1}^{\nu}(a_{\ell,i}-a_{j,i})\boldsymbol{k}_i,\ldots,\boldsymbol{\varphi}_j + h\sum_{i=1}^{\nu}(a_{\ell,i}-a_{j,i})\boldsymbol{k}_i,\boldsymbol{k}_j,\boldsymbol{k}_\ell\right)
\end{aligned} \tag{10}$$

$$= S(\varphi_j,\ldots,\varphi_j,\boldsymbol{k}_j,\boldsymbol{k}_\ell) + h\sum_{i=1}^{\nu}(a_{\ell,i} - a_{j,i})S(\varphi_j,\ldots,\varphi_j,\boldsymbol{k}_i,\boldsymbol{k}_j,\boldsymbol{k}_\ell) + \mathcal{O}\left(h^2\right).$$

Hence, the $\mathcal{O}(h^2)$ term reduces to

$$\binom{r}{2}\sum_{\ell,j=1}^{\nu}(b_\ell b_j - b_\ell a_{\ell,j} - b_j a_{j,\ell})S(\varphi_\ell,\ldots,\varphi_\ell,\boldsymbol{k}_j,\boldsymbol{k}_\ell),$$

while adding a further contribution to the $\mathcal{O}(h^3)$ term. Consequently, $M = O$ is sufficient for setting the coefficient of h^2 to zero and later we verify that it is also necessary.

Next we turn our attention to the cubic term. We recall that, as for (10), it is true that

$$S(\varphi_\ell,\ldots,\varphi_\ell,\boldsymbol{k}_j,\boldsymbol{k}_i,\boldsymbol{k}_\ell) = S(\varphi_j,\ldots,\varphi_j,\boldsymbol{k}_j,\boldsymbol{k}_i,\boldsymbol{k}_\ell) + \mathcal{O}(h),$$

thus protracted, yet easy, algebra yields

$$\begin{aligned}\binom{r}{3}\sum_{\ell,j,i=1}^{\nu}&\{b_\ell b_j b_i - (b_\ell b_j a_{\ell,i} + b_j b_i a_{j,\ell} + b_i b_\ell a_{i,j})\\ &- [b_j a_{j,\ell}(a_{j,i} - a_{\ell,i}) + b_\ell a_{\ell,i}(a_{\ell,j} - a_{i,j}) + b_i a_{i,j}(a_{i,\ell} - a_{j,\ell})]\\ &+ (b_\ell a_{\ell,j} a_{\ell,i} + b_j a_{j,\ell} a_{j,i} + b_i a_{i,\ell} a_{i,j})\}\, S(\varphi_\ell,\ldots,\varphi_\ell,\boldsymbol{k}_j,\boldsymbol{k}_i,\boldsymbol{k}_\ell),\end{aligned}$$

which, after further simplification, reduces to

$$\begin{aligned}\binom{r}{3}\sum_{\ell,j,i=1}^{\nu}&[b_\ell b_j b_i - (b_\ell b_j a_{\ell,i} + b_j b_i a_{j,\ell} + b_i b_\ell a_{i,j})\\ &+ b_j a_{j,\ell} a_{\ell,i} + b_\ell a_{\ell,i} a_{i,j} + b_i a_{i,j} a_{j,\ell})]\, S(\varphi_\ell,\ldots,\varphi_\ell,\boldsymbol{k}_j,\boldsymbol{k}_i,\boldsymbol{k}_\ell).\end{aligned}$$

We thus deduce that the conditions for eliminating the h^2 and h^3 terms are

$$b_\ell a_{\ell,j} + b_j a_{j,\ell} - b_\ell b_j = 0, \qquad \ell, j = 1, 2, \ldots, \nu, \tag{11}$$

and

$$\begin{aligned}&(b_j a_{j,\ell} a_{\ell,i} + b_\ell a_{\ell,i} a_{\ell,j} + b_i a_{i,j} a_{i,\ell})\\ &\quad - (b_\ell b_j a_{\ell,i} + b_j b_i a_{j,\ell} + b_i b_\ell a_{i,j}) + b_\ell b_j b_i = 0,\end{aligned} \qquad \ell, j, i = 1, 2, \ldots, \nu, \tag{12}$$

respectively. Incidentally, if we impose that the numerical scheme satisfies (11), we can use the latter to express the terms which are quadratic in the components of the vector $\boldsymbol{b}$ and obtain a relation equivalent to (12), namely

$$b_i b_j b_\ell - (b_i a_{i,j} a_{i,\ell} + b_j a_{j,\ell} a_{j,i} + b_\ell a_{\ell,i} a_{\ell,j}) = 0, \qquad i, j, \ell = 1, \ldots, \nu,$$

as already derived in (Calvo et al. [1]).

Our contention is that the conditions (11) and (12) conflict for every Runge–Kutta method of nontrivial order. To this end we note that order $p \geq 1$ implies that $\mathbf{1}^T \boldsymbol{b} = 1$, therefore there exists $s \in \{1, 2, \ldots, \nu\}$ such that $b_s \neq 0$. We let $\ell = j = i = s$ in (11), (12), and this yields

$$2b_s a_{s,s} - b_s^2 = 0, \tag{13}$$

$$3b_s a_{s,s}^2 - 3b_s^2 a_{s,s} + b_s^3 = 0. \tag{14}$$

Since $b_s \neq 0$, we deduce from (13) that $a_{s,s} = \frac{1}{2} b_s$ and substitution in (14) results in $\frac{1}{4} b_s^3 = 0$, hence a contradiction.

The statement that (11) and (12) conflict does not mean, on its own, that no Runge–Kutta method can solve the underlying ODE in an $\mathcal{S}_r$-invariant manner for $r \geq 3$. To be able to argue the latter statement, we need to demonstrate that there exists $\mathcal{M} \in \mathcal{S}_r$ and an $\mathcal{M}$-invariant ODE (1) for which neither $S(\boldsymbol{\varphi}_s, \ldots, \boldsymbol{\varphi}_s, \boldsymbol{f}(\boldsymbol{\varphi}_s), \boldsymbol{f}(\boldsymbol{\varphi}_s))$ nor $S(\boldsymbol{\varphi}_s, \ldots, \boldsymbol{\varphi}_s, \boldsymbol{f}(\boldsymbol{\varphi}_s), \boldsymbol{f}(\boldsymbol{\varphi}_s), \boldsymbol{f}(\boldsymbol{\varphi}_s))$ vanish (recall that, by (3), $\boldsymbol{k}_s = \boldsymbol{f}(\boldsymbol{\varphi}_s)$). Fortunately, a relatively simple member of $\mathcal{S}_r$ furnishes us with the required example. For the sake of simplicity we thereafter assume that the underlying method (3) is nonconfluent.

Letting $d = 3$, we choose

$$\mathcal{M} = \{\boldsymbol{y} \in \mathbb{R}^3 \, : \, y_1^r + y_2^r = y_1^r(0) + y_2^r(0)\}.$$

In other words, S is a scalar multiple of the unit r-tensor (which is of course symmetric). We consider the ODE

$$\begin{aligned} y_1' &= y_2^{r-1} g(\boldsymbol{y}), \\ y_2' &= -y_1^{r-1} g(\boldsymbol{y}), \\ y_3' &\equiv 1, \end{aligned} \tag{15}$$

where $g : \mathbb{R}^3 \to \mathbb{R}$ is for the time being an arbitrary Lipschitz function. It is trivial to verify that

$$y_1^{r-1}(t) y_1'(t) + y_2^{r-1}(t) y_2'(t) \equiv 0, \qquad t \geq 0,$$

hence, as long as $\boldsymbol{y}(0) \in \mathcal{M}$, also $\boldsymbol{y}(t) \in \mathcal{M}$ for all $t \geq 0$ and the ODE (15) is $\mathcal{M}$-invariant. Moreover, simple calculation confirms that

$$\begin{aligned} S(\boldsymbol{\varphi}, \ldots, \boldsymbol{\varphi}, \boldsymbol{f}(\tilde{\boldsymbol{\varphi}}), \boldsymbol{f}(\boldsymbol{\varphi})) &= g(\boldsymbol{\varphi}) g(\tilde{\boldsymbol{\varphi}}) (\varphi_1 \varphi_2)^{r-2} \\ &\quad \times (\varphi_2 \tilde{\varphi}_2^{r-1} + \varphi_1 \tilde{\varphi}_1^{r-1}), \\ S(\boldsymbol{\varphi}, \ldots, \boldsymbol{\varphi}, \boldsymbol{f}(\tilde{\boldsymbol{\varphi}}), \boldsymbol{f}(\hat{\boldsymbol{\varphi}}), \boldsymbol{f}(\boldsymbol{\varphi})) &= g(\boldsymbol{\varphi}) g(\tilde{\boldsymbol{\varphi}}) g(\hat{\boldsymbol{\varphi}}) \\ &\quad \times (\varphi_1 \varphi_2)^{r-3} (\varphi_2^2 \tilde{\varphi}_2^{r-1} \hat{\varphi}_2^{r-1} - \varphi_1^2 \tilde{\varphi}_1^{r-1} \hat{\varphi}_1^{r-1}) \end{aligned}$$

where $\boldsymbol{\varphi}^T = [\, \varphi_1 \quad \varphi_2 \quad \varphi_3 \,]$.

We choose $\boldsymbol{y}_0 \in \mathbb{R}^3$ so that $y_{0,1}, y_{0,2}, y_{0,2}^r - y_{0,1}^r \neq 0$ and $y_{0,3} = 0$. Let $\boldsymbol{\varphi}_s$ be a real solution of the polynomial system

$$\begin{aligned}\varphi_{s,1} &= y_{0,1} + h a_{s,s} \varphi_{s,2}^{r-1},\\ \varphi_{s,2} &= y_{0,2} - h a_{s,s} \varphi_{s,1}^{r-1}.\end{aligned}$$

Note that there exists such solution as long as $h > 0$ is sufficiently small. Next, set

$$\begin{aligned}\varphi_{\ell,1} &:= y_{0,1} + h a_{\ell,s} \varphi_{s,2}^{r-1},\\ \varphi_{\ell,2} &:= y_{0,2} - h a_{\ell,s} \varphi_{s,1}^{r-1}\end{aligned}$$

for all $\ell = 1, 2, \ldots, \nu$, $\ell \neq s$ and choose a Lipschitz function g such that

$$g(\boldsymbol{\varphi}_\ell) = \begin{cases} 1, & \ell = s, \\ 0, & \ell \in \{1, 2, \ldots, \nu\},\ \ell \neq s. \end{cases}$$

Note that

$$\varphi_{\ell,3} = y_{0,3} + h \sum_{j=1}^{\nu} a_{\ell,j} = h c_\ell, \qquad \ell = 1, 2, \ldots, \nu,$$

therefore confluence implies that $\boldsymbol{\varphi}_\ell \neq \boldsymbol{\varphi}_m$, $\ell, m = 1, 2, \ldots, \nu$, $\ell \neq m$, and such function g exists.

It follows from (3) and our construction that, as long as $h > 0$ is sufficiently small, we have

$$S(\boldsymbol{\varphi}_s, \ldots, \boldsymbol{\varphi}_s, \boldsymbol{f}(\boldsymbol{\varphi}_s), \boldsymbol{f}(\boldsymbol{\varphi}_s)), S(\boldsymbol{\varphi}_s, \ldots, \boldsymbol{\varphi}_s, \boldsymbol{f}(\boldsymbol{\varphi}_s), \boldsymbol{f}(\boldsymbol{\varphi}_s), \boldsymbol{f}(\boldsymbol{\varphi}_s)) \neq 0$$

and

$$S(\boldsymbol{\varphi}_\ell, \ldots, \boldsymbol{\varphi}_\ell, \boldsymbol{f}(\boldsymbol{\varphi}_j), \boldsymbol{f}(\boldsymbol{\varphi}_\ell)), S(\boldsymbol{\varphi}_\ell, \ldots, \boldsymbol{\varphi}_\ell, \boldsymbol{f}(\boldsymbol{\varphi}_j), \boldsymbol{f}(\boldsymbol{\varphi}_i), \boldsymbol{f}(\boldsymbol{\varphi}_\ell)) = 0$$

for all $\ell \neq s$, $i, j = 1, 2, \ldots, \nu$. Consequently, both (13) and (14) must be true in order for the $\mathcal{O}(h^2)$ and $\mathcal{O}(h^3)$ terms to be eliminated. This, however, is impossible according to our analysis.

We deduce the central result of this section.

Theorem 1 *Let $r \geq 3$ be a given integer. No nonconfluent Runge–Kutta method (3) can be $\mathcal{S}_r$-invariant when $\mathcal{S}_r$ is the set of all manifolds described by symmetric r-tensors.* □

3. Invariant modification of Runge–Kutta methods

Let us suppose that it is known (typically, as a consequence of mathematical analysis) that the ODE (1) is $\mathcal{M}$-invariant for some q-dimensional manifold $\mathcal{M} \subset \mathbb{R}^d$, $q \leq d-1$. Further, we assume that $\mathcal{M}$ is diffeomorphic to another manifold, $\mathcal{N}$, say. In other words, there exists a smooth bijection

$$\boldsymbol{g} : \mathcal{M} \rightarrow \mathcal{N}$$

whose inverse is also smooth. Letting $\boldsymbol{x}(t) := \boldsymbol{g}(\boldsymbol{y}(t))$, $t \geq 0$, we readily derive from (1) a differential equation for the function $\boldsymbol{x}$,

$$\boldsymbol{x}' = \frac{\partial \boldsymbol{g}(\boldsymbol{g}^{-1}(\boldsymbol{x}))}{\partial \boldsymbol{y}} \boldsymbol{f}(\boldsymbol{g}^{-1}(\boldsymbol{x})), \quad t \geq 0, \qquad \boldsymbol{x}(0) = \boldsymbol{g}(\boldsymbol{y}_0) \in \mathcal{N}. \tag{16}$$

Since $\boldsymbol{y}(t) \in \mathcal{M}, t \geq 0$, it follows from our construction that $\boldsymbol{x}(t) \in \mathcal{N}, t \geq 0$, therefore (16) is $\mathcal{N}$-invariant. Thus, provided that the manifold $\mathcal{N}$ is sufficiently simple and we can integrate the ODE (16) with an $\mathcal{N}$-invariant numerical method, we can construct an $\mathcal{M}$-invariant solution of (1) by the inverse mapping $\boldsymbol{y}_n = \boldsymbol{g}^{-1}(\boldsymbol{x}_n)$, $n \in \mathbb{N}$.

One way of expressing our argument is that, whenever we know that an ODE is $\mathcal{M}$-invariant, it is possible to map $\mathcal{M}$ diffeomorphically into a simpler manifold, $\mathcal{N}$, for which we have our numerical analysis sorted out. An alternative point of view is that, whenever we can construct $\mathcal{N}$-invariant methods for some manifold $\mathcal{N}$, we can extend our result to all manifolds $\mathcal{M}$ that are diffeomorphic to $\mathcal{N}$.

On the face of it, we have just (in the best mathematical tradition) replaced one problem, how to construct $\mathcal{M}$-invariant methods, by two: firstly, how to find a diffeomorphism from $\mathcal{M}$ to some 'convenient' manifold $\mathcal{N}$ and, secondly, how to integrate along the second manifold in an $\mathcal{N}$-invariant manner. However, this difficulty is often illusory since many interesting manifolds lend themselves easily to a 'natural' diffeomorphism to a simple manifold. Moreover, this approach helps us to discard redundant information and focus on the real problem in numerical $\mathcal{M}$-invariance: *not the precise shape of $\mathcal{M}$ but its topology!*

We mention in passing that the idea of changing variables, to map an invariant manifold $\mathcal{M}$ into a manifold $\mathcal{N}$ with simpler topology, has been already considered in a different context and with a radically different approach in (Leimkuhler and Partick [9]).

The remainder of this section is devoted to two simple examples which illustrate its main idea. Firstly, let $\mathcal{M}$ be the unit sphere with respect to the ℓ_p norm, $1 < p < \infty$,

$$\mathcal{M} = \left\{ \boldsymbol{y} \in \mathbb{R}^d : \sum_{k=1}^{d} |y_k|^p = 1 \right\}.$$

The $(d-1)$-dimensional manifold $\mathcal{M}$ is trivially diffeomorphic to the sphere $\boldsymbol{S}_{d-1}$ (which, of course, is nothing else but the unit sphere with respect to the Euclidean norm ℓ_2) and it is easy to find a diffeomorphism. Let

$$\boldsymbol{g}(\boldsymbol{y}) = \frac{\boldsymbol{y}}{\|\boldsymbol{y}\|_2}, \qquad \boldsymbol{y} \in \mathcal{M},$$

whereby the inverse map is

$$\boldsymbol{g}^{-1}(\boldsymbol{x}) = \frac{\boldsymbol{x}}{\|\boldsymbol{x}\|_p}, \qquad \boldsymbol{x} \in \boldsymbol{S}_{d-1}.$$

We thus obtain the ODE system

$$\boldsymbol{x}' = \|\boldsymbol{x}\|_p (I - \boldsymbol{x}\boldsymbol{x}^T) \boldsymbol{f}(\boldsymbol{x}/\|\boldsymbol{x}\|_p). \tag{17}$$

Note that, although this might not be apparent at a glance, the equation (17) reduces in the special case $p = 2$ to the original ODE (1) (as it should) because $\boldsymbol{x}^T \boldsymbol{f}(\boldsymbol{x}) \equiv 0$ for equations constrained to $\boldsymbol{S}_{d-1}$.

The solution of (17) is $\boldsymbol{S}_{d-1}$-invariant. As we have already stated in Section 2, employing a different terminology, any Runge–Kutta method with $M = O$ is $\boldsymbol{S}_{d-1}$-invariant. Therefore we may solve (17) with such a method and an $\mathcal{M}$-invariant solution of the original ODE follows by letting $\boldsymbol{y}_n = \boldsymbol{g}^{-1}(\boldsymbol{x}_n)$ for all relevant values of n.

An interesting example of such a manifold $\mathcal{M}$, which harks back to the work of Section 2, is when $p \geq 4$ is an even integer. In that case $\mathcal{M}$ follows by letting S in (5) be a unit p-tensor. This is an example of a manifold which, according to Theorem 1, is intractable directly by Runge–Kutta methods in an invariant manner. Yet, subjecting it to a transformation of variables, we can solve it whilst maintaining $\mathcal{M}$-invariance.

Odd values of r in (5) can be addressed in a similar manner, except that the topology of $\mathcal{M}$ is no longer the same as of the sphere $\boldsymbol{S}_{d-1}$. Consider the manifold

$$\mathcal{M} = \left\{ \boldsymbol{y} \in \mathbb{R}^d \, : \, \sum_{k=1}^{d} y_k^r = 1 \right\},$$

where r is odd. Fixing $y_1, y_2, \ldots, y_{d-1}$, the equation

$$y_d^r = 1 - \sum_{k=1}^{d-1} y_k^r$$

has exactly one real solution. Therefore, $\mathcal{M}$ is diffeomorphic to $\mathbb{R}^{d-1}$,

$$\boldsymbol{g}(\boldsymbol{y}) = \begin{bmatrix} y_1 \\ y_2 \\ \vdots \\ y_{d-1} \end{bmatrix}$$

and

$$\boldsymbol{g}^{-1}(\boldsymbol{x}) = \begin{bmatrix} x_1 \\ \vdots \\ x_{d-1} \\ \operatorname{sgn}\left(1 - \sum_{k=1}^{d-1} x_k^r\right) \times \left|1 - \sum_{k=1}^{d-1} x_k^r\right|^{1/r} \end{bmatrix}.$$

The problem of $\mathcal{M}$-invariance is now, if at all, even easier. We map $\boldsymbol{y}$ into $\mathbb{R}^{d-1}$ by the natural projection $\boldsymbol{g}$, solve the ODE (16) by an arbitrary consistent numerical method and map back from $\{\boldsymbol{x}_n\}$ to $\{\boldsymbol{y}_n\}$.

Needless to say, an Euclidean subspace and a sphere represent the simplest possible topologies. As things stand, it is an open question how to retain invariance on torii, for example, saying nothing of more complicated topologies. Much remains to be done in elucidating numerical methods that capture qualitative attributes of the underlying ODE system. Having said this, we emphasize again our conclusion, namely that the critical factor in the retention of invariance under discretization is not the shape but the topology of the underlying manifold.

The technique of diffeomorphisms is implicit in an earlier work of the present authors. Given an arbitrary $d \times d$ matrix L_0, we consider the manifold of matrices similar to L_0, that we denote by

$$\mathcal{M}_{L_0} = \{V L_0 V^{-1} \,:\, V \in \mathrm{GL}(d, \mathbb{R})\},$$

where $\mathrm{GL}(d, \mathbb{R})$ is the group of $d \times d$ invertible matrices with real entries. Letting

$$L(t) = V(t) L_0 V(t)^{-1}, \qquad V(t) \in \mathrm{GL}(d, \mathbb{R}),$$

differentiation yields

$$L' = V'V^{-1}L - LV'V^{-1}, \tag{18}$$

thus (18) characterizes flows on $\mathcal{M}_{L_0}$. However the function V is not known a priori. An important flow on $\mathcal{M}_{L_0}$, closely allied with (18), has the form

$$L' = B(L)L - LB(L) = [B(L), L], \quad t \geq 0, \qquad L(0) = L_0. \tag{19}$$

It is known as an *isospectral flow* and its numerical solution has been debated at length in (Calvo et al. [1]).

Comparing (18) and (19) we deduce that

$$(B(L) - V'V^{-1})L - L(B(L) - V'V^{-1}) = O,$$

thus $B(L) - V'V^{-1}$ and L commute. This means that

$$B(L) - V'V^{-1} = p(L),$$

where $p(L)$ is a polynomial in L. Hence

$$V' = (B(L) - p(L))V. \tag{20}$$

Once the initial condition for (20) is fixed, the function V is uniquely determined. The initial condition $V(0) = V_0$ must commute with L_0, therefore V_0 is a polynomial in L_0. Choosing $V_0 = I$ and letting $p(L) = O$, we obtain that V is the unique solution of the differential equation

$$V' = B(V L_0 V^{-1})V, \qquad V(0) = I. \tag{21}$$

Thus,

$$\mathcal{N} = \mathrm{GL}(d, \mathbb{R})$$

and the natural diffeomorphism

$$g : \mathcal{M}_{L_0} \to \mathcal{N}$$

is $g(L) = V$, where V is the solution of (21). The inverse map is

$$g^{-1}(V) = V L_0 V^{-1}. \tag{22}$$

It follows that, solving (21) with any RK method and then mapping back to $\mathcal{M}_{L_0}$ by means of (22), we obtain a solution that is similar (an isospectral solution) to the initial condition L_0. The mapping g from $\mathcal{M}_{L_0}$ to $\mathcal{N} = \mathrm{GL}(d, \mathbb{R})$ has the advantage that classical RK methods, subject to adequate choices of parameters (for example, the stepsize h) can solve (21) whilst staying on $\mathrm{GL}(d, \mathbb{R})$, while, as proved in (Calvo et al. [1]), they cannot be isospectral, because they fail in recovering cubic conservation laws that are essential for isospectrality as soon as $d \geq 3$.

Another special case is when the matrix function $B(L)$ is skew-symmetric. In this situation the solution of (21) is orthogonal and $L(t) = V(t) L_0 V(t)^T$, therefore the most obvious diffeomorphism is

$$g(L) = V \in \mathcal{N}, \qquad \mathcal{N} = O(d),$$

where $O(d)$ is the orthogonal group of $d \times d$ matrices. Since orthogonality can be viewed as a quadratic conservation law obeyed by $V(t)$, it follows that RK schemes with $M = O$ can be used to solve (21) on $O(d)$ and, mapping back by means of $L = g^{-1}(V) = V L_0 V^T$, we obtain a solution L which is orthogonally similar to the initial condition. This is, as a matter of fact, the main idea of (Calvo et al. [1]).

The latter two examples differ from our earlier discussion, because the diffeomorphism depends in an implicit manner on the solution. Yet, as evident from (Calvo et al. [1]), this is not an unsurmountable impediment to the implementation of the main idea of this section.

Acknowledgements This paper has been written during the second author's visit to Departamento de Matemática Aplicada y Computación, Universidad de Valladolid as an IBERDROLA Visiting Professor.

4. References

1. M.P. Calvo, A. Iserles and A. Zanna (1995), "Numerical solution of isospectral flows", DAMTP Tech. Rep. 1995/NA03.
2. G.J. Cooper (1987), "Stability of Runge–Kutta methods for trajectory problems", *IMA J. Num. Anal.* **7**, 1–13.

3. P.E. Crouch and R. Grossman (1993), "Numerical integration of ordinary differential equations on manifolds", *J. Nonlinear Sci.* **3**, 1–33.
4. K. Dekker and J.G. Verwer (1984), *Stability of Runge–Kutta Methods for Stiff Nonlinear Differential Equations,* North Holland, Amsterdam.
5. L. Dieci, R.D. Russell and E.S. Van Vleck (1994), "Unitary integrators and applications to continuous orthonormalization techniques", *SIAM J. Num. Anal.* **31**, 261–281.
6. V. Guillemin and A. Pollack (1974), *Differential Topology,* Prentice–Hall, Englewood Cliffs NJ.
7. A. Iserles (1996), *A First Course in the Numerical Analysis of Differential Equations,* Cambridge University Press, Cambridge.
8. A. Iserles and A. Zanna (1995), "Qualitative numerical analysis of ordinary differential equations", DAMTP Tech. Rep. 1995/NA05, to appear in *Lectures in Applied Mathematics,* American Mathematical Society, Providence RI.
9. B. Leimkuhler and G.W. Patrick (1995), "A symplectic integrator for Riemannian manifolds", Kansas University Technical Report.
10. H. Munthe-Kaas (1995), "Lie–Butcher theory for Runge–Kutta methods", *BIT* **35**, 572–587.
11. J.M. Sanz-Serna and M.P. Calvo (1994), *Numerical Hamiltonian Problems,* Chapman & Hall, London.

NUMERICAL SOLUTIONS OF ONE AND TWO DIMENSIONAL HYPERBOLIC SYSTEMS MODELLING A FLUIDIZED BED

I. CHRISTIE, G. H. GANSER, J. W. WILDER
Department of Mathematics, West Virginia University,
PO Box 6310, Morgantown, WV 26506, USA
E-mail: na.christie@na-net.ornl.gov

ABSTRACT

Numerical solutions of fluidized bed models in one and two space dimensions are calculated. The models take the form of hyperbolic systems of conservation laws with source terms and, in the two-dimensional case, there is a coupled elliptic equation for determining the stream function. Splitting is used to separate the contribution of the source term as well as the two space dimensions. A second order formulation of Roe's method is applied to each model. Numerical results are presented to demonstrate the occurrence of slugging in the one-dimensional model. In the two-dimensional model we obtain solutions in which kidney-shaped bubbles occur.

1. Introduction

A fluidized bed consists of a vertical column containing particles. Gas is pumped through a perforated plate at the bottom of the column and flows through the spaces between the particles. When the weight of the particles is first balanced by a sufficiently strong upward flow of gas, the bed is at minimum fluidization. Increasing the gas velocity frequently leads to "bubbles" or "slugs" characterized by regions of high and low concentrations of particles moving up the bed.

Mathematical models of fluidization may or may not include a particle viscosity term. Some authors (Fanucci, Ness, and Yen[1], Needham and Merkin[2]) suggest that particle viscosity, no matter how small, is essential for the periodic behaviour corresponding to slugging to occur. In this paper we demonstrate numerically that a hyperbolic model, with no particle viscosity terms, is capable of reproducing the oscillatory slugging behaviour. The numerical methods allow the computation of large amplitude shocks. This enables us to study travelling wave solutions of the one-dimensional system, the existence of which was verified by Ganser and Lightbourne[3].

The mathematical models considered here have the form of hyperbolic systems of conservation laws with source terms. In two space dimensions there is an additional elliptic equation to be solved for the stream function. The numerical methods of solution we describe are based on the work by Christie, Ganser, and Sanz-Serna[4] for the one-dimensional problem and on the work by Christie, Ganser, and Wilder[5] for the two-dimensional problem. In both papers, Strang's second order splitting[6] is used to separate the contributions of the source term. Splitting is used to handle the space variables in the two-dimensional problem[5]. The hyperbolic systems are integrated using the second order method suggested by Roe[7]. The method uses the superbee flux limiter (Roe[8], Sweby[9]) and avoids the appearance of non-physical shocks by including the "entropy fix" of Harten and Hyman[10].

Roe's method uses an approximate Riemann solver. Methods based on exact Riemann solvers are also available. The paper by Christie and Palencia[11] applies an exact Riemann solver to the one–dimensional model considered here. Papers in which models containing particle viscosity are solved numerically include Pritchett[12] and Symalal[13].

2. One-Dimensional Equations

The scaled one–dimensional model[4] has the form

$$\begin{pmatrix} \alpha \\ m \end{pmatrix}_t + \begin{pmatrix} m \\ \alpha u^2 + F(\alpha) \end{pmatrix}_z = \begin{pmatrix} 0 \\ b(\alpha, m) \end{pmatrix} \tag{1}$$

where $z_L \leq z \leq z_R$ is the coordinate along the vertical axis and $t \geq 0$ is time. The concentration of particles by volume and the velocity of the particle phase are denoted by α and u, respectively, and the momentum is $m = \alpha u$.

This system is typical of the equations studied by Foscolo and Gibilaro[14] , Needham and Merkin[2] , Homsy, El-Kaissy, and Didwania[15], and Drew[16]. The particle phase pressure function is chosen such that its derivative is given by

$$F'(\alpha) = \left[\frac{s\alpha}{\alpha_p - \alpha} \right]^2, \tag{2}$$

where s and α_p are constants. The packing concentration $\alpha_p < 1$ sets an upper limit on α. Therefore, solutions of (1) satisfy $0 \leq \alpha < \alpha_p$.

The source term in (1) arises from the force due to gravity and the choice of a drag law linear in velocity. It is given by

$$b(\alpha, u) = -\alpha + \alpha(1 - \alpha)^{-N}(j - u). \tag{3}$$

The total volumetric flux through the bed, $j = (1 - \alpha_0)^N$, and α_0 are constants. Typical values of the positive real number N are between 3 and 4. In this paper we use a value of 3.5 in the numerical calculations. In a uniformly fluidized bed with flux $j = (1 - \alpha_0)^N$ the particle concentration would be α_0. A linearized stability analysis (Needham and Merkin[2]) shows that all of the constant solutions $\alpha = \alpha_0$, $u = 0$ are stable under the condition

$$\alpha_p > \alpha_0 > \alpha_{0u}, \tag{4}$$

where α_{0u} is a constant.

The constant s in (2) is given by

$$s = N(1 - \alpha_{0u})^{N-1}(\alpha_p - \alpha_{0u}). \tag{5}$$

Since increasing j (turning up the gas flow) implies a decrease in α_0, choosing j to violate (4) will lead to a flow where the uniform state is unstable.

The system (1) will be solved numerically subject to initial conditions for α and m, and reflecting boundary conditions

$$m = 0 \qquad \text{at} \quad z = z_L, z_R. \tag{6}$$

3. Numerical Method: 1-D Case

The model (1) can be written as

$$w_t + f(w)_z = b \tag{7}$$

where $w = [\alpha, m]^T$, $f(w) = [m, mu + F(\alpha)]^T$, $b = [0, b(\alpha, m)]^T$, and subscripts denote partial derivatives. Differentiation of f(w) in (7) gives

$$w_t + Jw_z = b. \tag{8}$$

The Jacobian matrix J is

$$J = \begin{pmatrix} 0 & 1 \\ -\lambda_-\lambda_+ & \lambda_-\lambda_+ \end{pmatrix}, \tag{9}$$

where $\lambda_\pm = u \pm c$ are the eigenvalues of J, and $c^2 = F'(\alpha)$. Strang's second order operator splitting[6] is applied to the system (7). To advance the solution over a time step Δt we solve the sequence of equations

$$w_t = b \tag{10}$$

$$w_t + Jw_z = 0 \tag{11}$$

$$w_t = b \tag{12}$$

with step lengths of $\Delta t/2$ for (10) and (12) and Δt for (11). Note that, since b is linear in m, steps (10) and (12) can be solved in closed form. For the fractional step (11), Christie, Ganser, and Sanz-Serna[4] used Roe's linearized Riemann solver[7]. Suppose we have a Riemann problem with given left and right states $[\alpha_L, m_L]^T$ and $[\alpha_R, m_R]^T$. Compute the corresponding velocities

$$u_L = m_L/\alpha_L \qquad \text{and} \qquad u_R = m_R/\alpha_R. \tag{13}$$

Although (13) requires α_L and α_R to be positive, regions where the concentration is zero can be handled by switching off those parts of the calculation.

In Roe's method an approximate Jacobian $\bar{J}$ satisfying

$$\bar{J}\begin{pmatrix} \alpha_R - \alpha_L \\ m_R - m_L \end{pmatrix} = \begin{pmatrix} m_R - m_L \\ \alpha_R u_R^2 + F(\alpha_R) - \alpha_L u_L^2 - F(\alpha_L) \end{pmatrix} \tag{14}$$

is required. One possibility is to select

$$\bar{J} = \begin{pmatrix} 0 & 1 \\ -\bar{\lambda}_-\bar{\lambda}_+ & \bar{\lambda}_-\bar{\lambda}_+ \end{pmatrix}. \tag{15}$$

where $\bar{\lambda}_\pm = \bar{u} \pm \bar{c}$ are the eigenvalues of $\bar{J}$. The averaged velocity is given, as in Roe[7] and Glaister[17], by

$$\bar{u} = \frac{\sqrt{\alpha_R}u_R + \sqrt{\alpha_L}u_L}{\sqrt{\alpha_R} + \sqrt{\alpha_L}} \tag{16}$$

and $\bar{c}$ is given by

$$\bar{c} = (F(\alpha_R) - F(\alpha_L))/(\alpha_R - \alpha_L). \tag{17}$$

Diagonalizing $\bar{J}$ using

$$P^{-1}\bar{J}P = \text{diag}(\bar{\lambda}_-, \bar{\lambda}_+) \tag{18}$$

where

$$P = \begin{pmatrix} 1 & 1 \\ \bar{\lambda}_- & \bar{\lambda}_+ \end{pmatrix} \tag{19}$$

allows (8) to be uncoupled as

$$P^{-1}w_t + P^{-1}\bar{J}PP^{-1}w_z = 0. \tag{20}$$

Each component of (20) is then equivalent to the scalar equation

$$W_t + \lambda W_z = 0 \tag{21}$$

where λ is constant and W is a component of $P^{-1}w$.

3.1. Upwind Finite Differencing

The interval $[z_L, z_R]$ is subdivided into M cells of equal length $\Delta z = (z_R - z_L)/M$ and the solution $[\alpha, m]^T$ is assumed to have a constant value $[\alpha_i, m_i]^T$ in the cell $[z_{i-1}, z_i]$, i=1, 2, ..., M, $z_i = z_L + i\Delta z$, $i = 0, 1, \ldots, M$. Simple upwind finite differencing of (21) produces a first order scheme:

$$W_i^{n+1} = W_i^n - \frac{1}{2}\nu^-\delta W_{i+\frac{1}{2}}^n - \frac{1}{2}\nu^+\delta W_{i-\frac{1}{2}}^n \tag{22}$$

$i = 2, \ldots, M-1, n = 0, 1, \ldots$, where W_i^n is the numerical approximation to W in cell $[z_{i-1}, z_i]$ at time $t = t_n$, δ is the central difference operator defined by $\delta W_{i+\frac{1}{2}}^n = W_{i+1}^n - W_i^n$, and

$$\nu^\pm = (\lambda \pm |\lambda|)\frac{\Delta t}{\Delta z}. \tag{23}$$

The time step Δt is adjusted at each time level to satisfy the CFL condition (Mitchell and Griffiths[18]). Second order accuracy is obtained following the methods of Roe[8] and Sweby[9]. The first order scheme (22) is modified by including an antidiffusive flux[9]:

$$W_i^{n+1} = W_i^n - \frac{1}{2}\nu^- \delta W_{i+\frac{1}{2}}^n - \frac{1}{2}\nu^+ \delta W_{i-\frac{1}{2}}^n + \mathrm{sgn}(\lambda)(h_{i+1/2} - h_{i-1/2}) \tag{24}$$

where

$$h_{i+1/2} = -\frac{1}{2}(1 - |\lambda|\frac{\Delta t}{\Delta z})\lambda\frac{\Delta t}{\Delta z}B(\delta W_{i+\frac{1}{2}}^n, \eta), \tag{25}$$

$$\eta = \delta W_{i-\frac{1}{2}}^n \qquad \text{if} \quad \lambda > 0 \tag{26}$$

and

$$\eta = \delta W_{i+\frac{3}{2}}^n \qquad \text{if} \quad \lambda < 0 \tag{27}$$

$i = 2, \ldots, M-1$. Mirror image cells are added at the boundaries to incorporate the reflecting boundary conditions.

The flux limiter $B(\xi, \eta)$ is chosen to preserve monotonicity in the solution. Sweby[9] describes the region in which the flux limiter should lie. As in[4] we use Roe's superbee flux limiter defined by

$$B(\xi, \eta) = \begin{cases} \mathrm{maxmod}(\xi, \eta) & \text{if } \frac{1}{2} \le \frac{\eta}{\xi} \le 2 \\ 2\,\mathrm{minmod}(\xi, \eta) & \text{if } \frac{\eta}{\xi} < \frac{1}{2} \text{ or } \frac{\eta}{\xi} > 2 \\ 0 & \text{if } \xi\eta < 0 \end{cases} \tag{28}$$

where

$$\mathrm{minmod}(\xi, \eta) = \begin{cases} \xi & \text{if } |\xi| \le |\eta| \\ \eta & \text{if } |\xi| > |\eta| \end{cases} \tag{29}$$

and maxmod is defined similarly.

Roe's method is known to allow non-physical shocks. These were encountered by Christie, Ganser, and Sanz-Serna[4] in the one–dimensional problem and avoided by using the "entropy fix" described in the paper by Harten and Hyman[10]. The same technique was used to generate the numerical results presented in this paper.

4. Numerical Results: 1–D Case

Figure 1 shows the results of solving the one–dimensional model (1) using the second order method. The packing concentration is $\alpha_p = 0.6$, and the critical state (which divides stable and unstable uniform states) is $\alpha_{0u} = 0.55$. The computational domain in space, $[z_L, z_R] = [0.0, 0.3]$, is divided into $M = 500$ cells. For $t = 0$ the bed is at uniform fluidization with $u = 0$ and $j = (1 - \alpha_{0u})^N$. Thus, the concentration in the bed is α_{0u} except in a boundary layer near the top of the bed. This initial particle concentration profile is calculated from the steady state solution of (1) and is shown as the dotted line in Figure 1. At $t = 0^+$ the flux is increased to $j = (1 - \alpha_0)^N$ where $\alpha_0 = 0.4$.

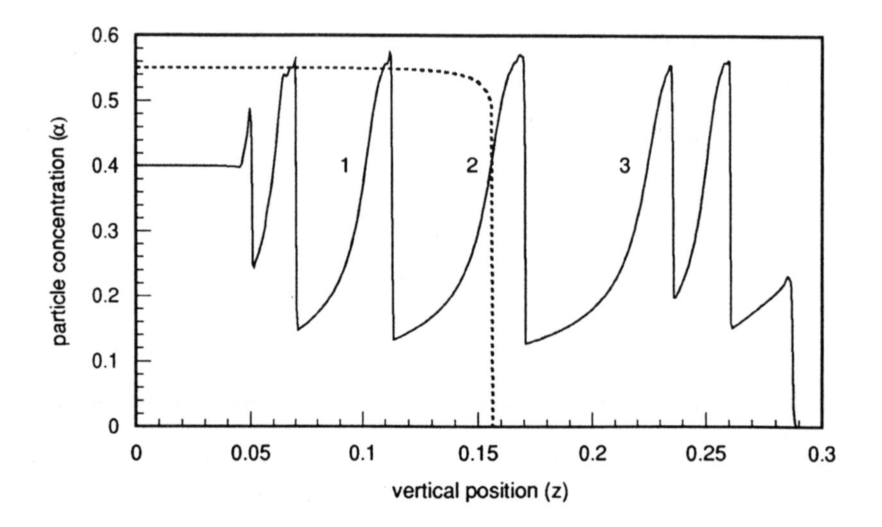

Fig. 1. Particle concentration vs. vertical position

The state of the bed at $t = 5.0$ is the solid curve in Figure 1. Clearly visable are the slugs which rise to the top of the bed. The numerical calculations show that this behaviour continues as t increases. It is easily shown[3] that the one-way travelling wave solutions of (1), $\alpha = \alpha(z - vt)$ and $m = m(z - vt)$, imply $m(z, t) = c_1 + v\alpha(z, t)$.

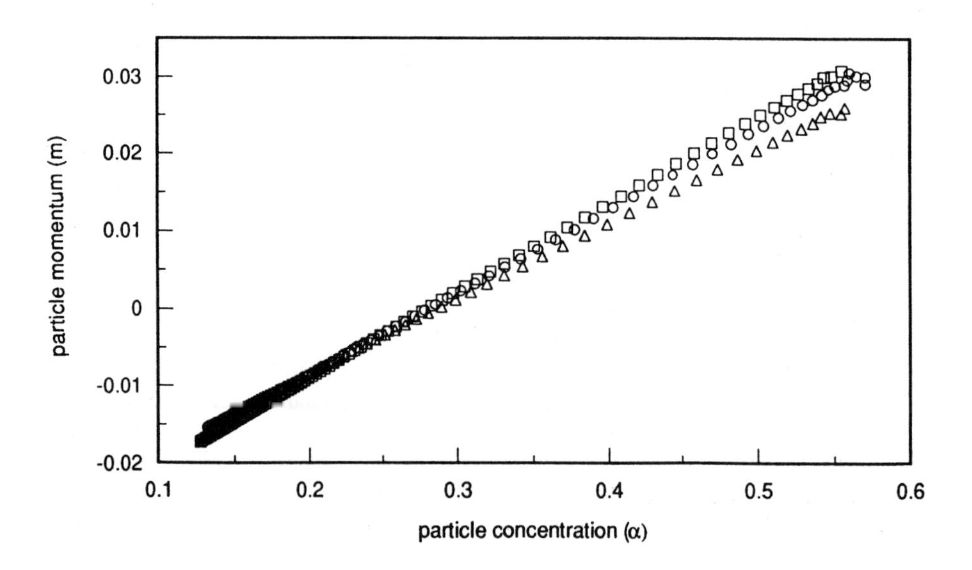

Fig. 2. Particle momentum vs. particle concentration for the travelling waves in Fig. 1.

That is, m is a linear function of α with slope v. The particle momentum, m, corresponding to the numbered slugs in Figure 1 is plotted in Figure 2 as a function of α. The data depicted as triangles, circles, and squares in Figure 2 represent the data points corresponding to waves 1, 2 and 3 labelled in Figure 1, respectively. We note that the speed of the waves nearer the bottom of the bed is smaller than that of those closer to the top.

5. Two–Dimensional Equations

The scaled two-dimensional equations[5] consist of a hyperbolic system

$$\frac{\partial \alpha}{\partial t} + \frac{\partial m}{\partial x} + \frac{\partial n}{\partial z} = 0 \tag{30}$$

$$\frac{\partial m}{\partial t} + \frac{\partial (mu + F(\alpha))}{\partial x} + \frac{\partial (nu)}{\partial z} = \frac{\alpha}{(1-\alpha)^N}\left[\frac{\partial \psi}{\partial z} - u\right] \tag{31}$$

$$\frac{\partial n}{\partial t} + \frac{\partial (mv)}{\partial x} + \frac{\partial (nv + F(\alpha))}{\partial z} = \frac{\alpha}{(1-\alpha)^N}\left[-\frac{\partial \psi}{\partial x} - v\right] - \alpha \tag{32}$$

where x, z denote the horizontal and vertical coordinates respectively, t is the time variable, $\alpha(x,z,t)$ is the particle concentration, $m(x,z,t)$ is the horizontal momentum, and $n(x,z,t)$ is the vertical momentum. The horizontal velocity $u(x,z,t)$ and the vertical velocity $v(x,z,t)$ are found from $m = \alpha u$ and $n = \alpha v$ respectively.

The stream function $\psi(x,z,t)$ which corresponds to the total volumetric flux[19] is obtained by solving the elliptic equation

$$\frac{\partial}{\partial x}\left[(\alpha\frac{\partial \psi}{\partial x} + n)(1-\alpha)^{-N}\right] + \frac{\partial}{\partial z}\left[(\alpha\frac{\partial \psi}{\partial z} - m)(1-\alpha)^{-N}\right] = 0 \tag{33}$$

if $\alpha \neq 0$, and

$$\frac{\partial^2 \psi}{\partial x^2} + \frac{\partial^2 \psi}{\partial z^2} = 0 \tag{34}$$

otherwise.

The spatial domain of the system (30)–(34) is a rectangle with $-x_R \leq x \leq x_R$ and $-z_R \leq z \leq z_R$. Reflecting boundary conditions are used to ensure zero momenta for particles travelling parallel to the coordinate axes and colliding with a wall. A diagramatic representation of the fluidized bed under consideration here is shown in Figure 3, where the white region at the top of the bed reflects an area where there are no particles initially. The boundaries $x = \pm x_R$ are streamlines with constant values for ψ. The problem studied here is a uniformly fluidized bed subjected to a centrally located jet of gas ($-x_b < x < x_b$). The background fluidizing gas entering at $z = -z_R$ is $j_M = (1-\alpha_M)^N$ where α_M is chosen between α_p and α_{0u}. At $t = 0$, the flux of gas

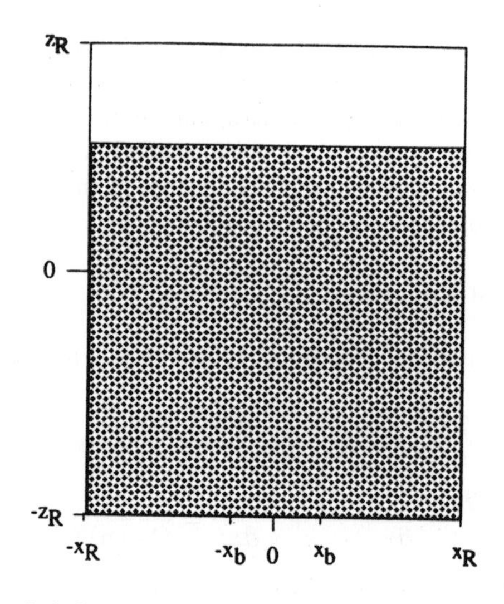

Fig. 3. A diagramatic representation of the fluidized bed

entering at $z = -z_R$ in the region $-x_b < x < x_b$ is increased to $j > j_M$. We assume that at $z = z_R$ the total volumetric flux that entered at $z = -z_R$, $\frac{x_b}{x_R}(j - j_M) + j_M$, is evenly dispersed. We choose $\psi = 0$ at $x = -x_R$ and, consequently, the boundary conditions for ψ are

$$\psi(-x_R, z, t) = 0 \tag{35}$$

$$\psi(x_R, z, t) = -2x_R j_M + 2x_b(j_M - j) \tag{36}$$

$$\psi(x, -z_R, t) = \begin{cases} -j_M(x + x_R) & -x_R \le x \le -x_b \\ -j(x + x_b) - j_M(x_R - x_b) & -x_b < x < x_b \\ -j_M(x + x_R) + 2x_b(j_M - j) & x_b \le x \le x_R \end{cases} \tag{37}$$

$$\psi(x, z_R, t) = \left[-j_M + \frac{x_b}{x_R}(j_M - j)\right](x + x_R). \tag{38}$$

6. Numerical Method: 2–D Case

The system (30)–(32) can be written

$$w_t + f(w)_x + g(w)_z = b \tag{39}$$

where $w = [\alpha, m, n]^T$, $f(w) = [m, mu + F(\alpha), mv]^T$, $g(w) = [n, nu, nv + F(\alpha)]^T$, and $b = [0, \frac{\alpha}{(1-\alpha)^N}\left(\frac{\partial\psi}{\partial z} - u\right), \frac{\alpha}{(1-\alpha)^N}\left(-\frac{\partial\psi}{\partial x} - v\right) - \alpha]^T$. The numerical method follows closely that described in Section 3 for the one–dimensional model. The non-conservation form of (39) is given by

$$w_t + J(w)w_x + K(w)w_z = b. \tag{40}$$

The Jacobian matrices J and K are defined in terms of their eigenvalues $\lambda = u$, $\lambda_\pm = u \pm \sqrt{F'}$ for J and $\mu = v$, $\mu_\pm = v \pm \sqrt{F'}$ for K.

The splitting procedure solves a sequence of problems similar to (10)-(12). Equation (11) is replaced by three hyperbolic equations obtained from dimensional splitting:

$$w_t + J(w)w_x = 0 \tag{41}$$

$$w_t + K(w)w_z = 0 \tag{42}$$

$$w_t + J(w)w_x = 0 \tag{43}$$

Equation (42) is solved with a time step of Δt and the other equations are solved with a time step of $\Delta t/2$.

The stream function equations (33), (34) are discretized using central differences with Gauss-Seidel iteration used to solve the algebraic equations. The derivatives of ψ in (31), (32) are approximated by central differences. Assuming that the first partial derivatives of ψ remain constant across a fractional step allows the equations containing the body force to be solved in closed form.

As in one space dimension, (41)–(43) are solved by applying Roe's method to the Riemann problem with left and right states $w_L = [\alpha_L, m_L, n_L]^T$ and $w_R = [\alpha_R, m_R, n_R]^T$ respectively. The Roe approximate Jacobians $\bar{J}$ and $\bar{K}$ satisfy

$$\bar{J}(w_R - w_L) = f(w_R) - f(w_L) \tag{44}$$

and

$$\bar{K}(w_R - w_L) = g(w_R) - g(w_L). \tag{45}$$

We select

$$\bar{J} = \begin{pmatrix} 0 & 1 & 0 \\ -\bar{\lambda}_+\bar{\lambda}_- & \bar{\lambda}_+ + \bar{\lambda}_- & 0 \\ -\bar{\lambda}\bar{\mu} & \bar{\mu} & \bar{\lambda} \end{pmatrix} \tag{46}$$

and

$$\bar{K} = \begin{pmatrix} 0 & 0 & 1 \\ -\bar{\lambda}\bar{\mu} & \bar{\mu} & \bar{\lambda} \\ -\bar{\mu}_+\bar{\mu}_- & 0 & \bar{\mu}_+ + \bar{\mu}_- \end{pmatrix} \tag{47}$$

where $\bar{\lambda} = \bar{u}$, $\bar{\lambda}_\pm = \bar{u} \pm \bar{c}$, $\bar{\mu} = \bar{v}$, and $\bar{\mu}_\pm = \bar{v} \pm \bar{c}$, where the averaged velocities are defined in a similar manner to (16). The equations (41)–(43) with the Jacobians replaced by the approximate Jacobians are then uncoupled and each system is solved by the second order method described in Section 3.

7. Numerical Results: 2–D Case

The dynamics of the bed resulting from calculations based on the numerical method described above vary depending on the parameters used in the simulations.

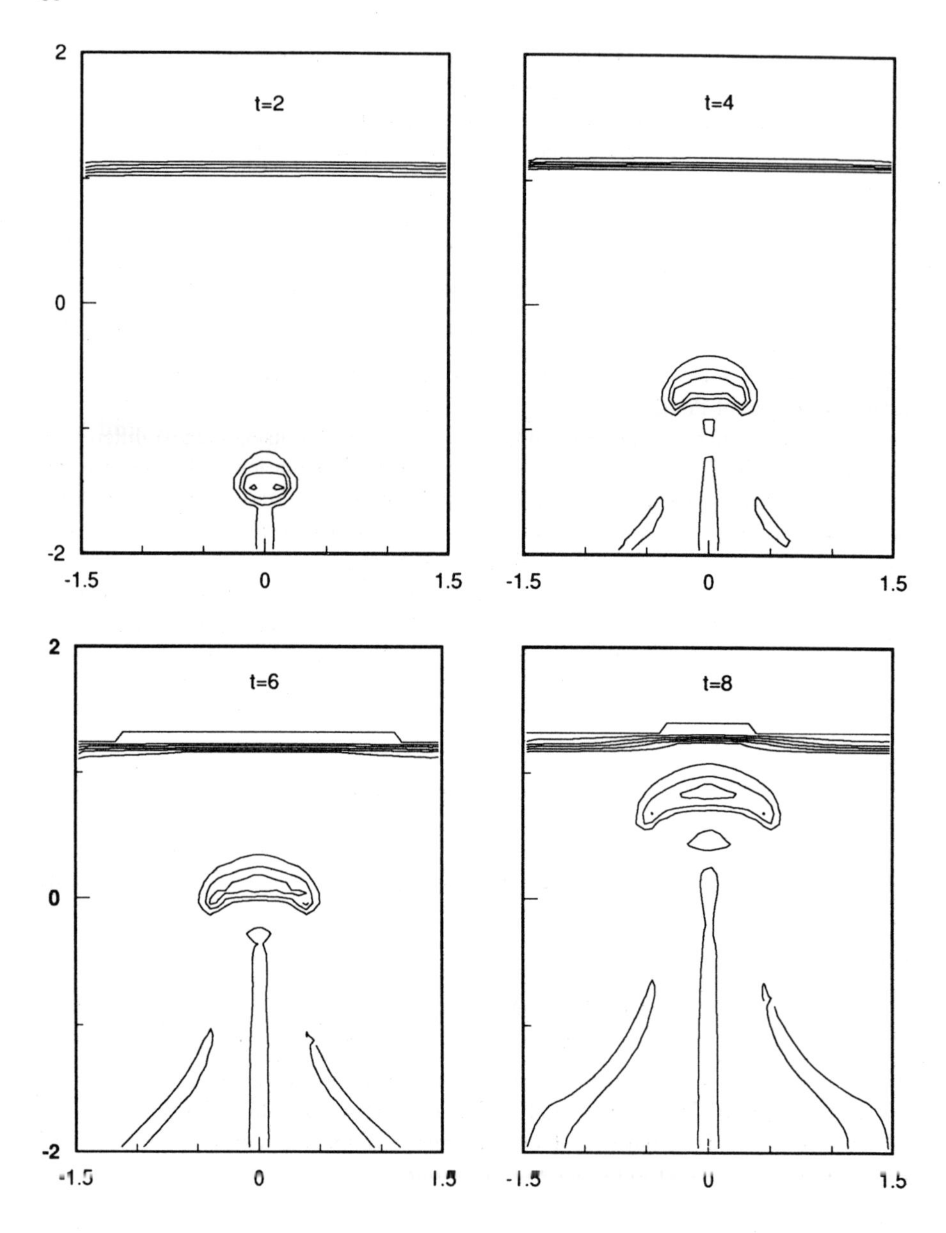

Fig. 4 Time evolution of a bubble introduced by a jet

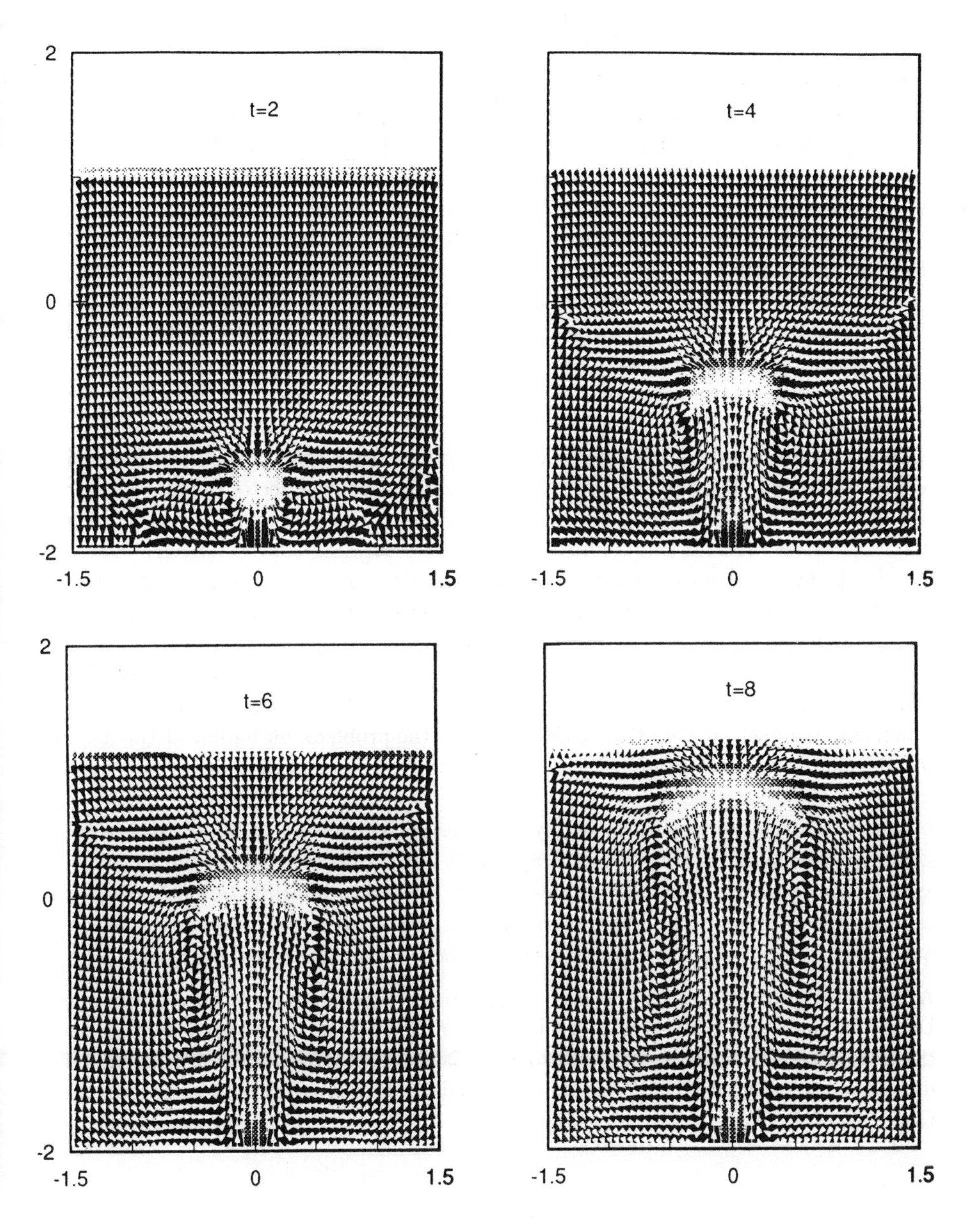

Fig. 5. Particle concentration and particle momentum for the bubbles in Fig. 4.

Under appropriate conditions, a bubble forms as a result of the jet at the bottom of the bed and then propagates upward. The total height of the bed is taken to be 4 units ($z_R = 2.0$), the width 3 units ($x_R = 1.5$), the width of the jet 0.2 units ($x_b = 0.1$), the flux at the bottom of the bed in the jet is $j = (1-\alpha_0)^N$ with $\alpha_0 = 0.2$ and the flux elsewhere at the bottom of the bed is $j_M = (1-\alpha_{0u})^N$ with $\alpha_{0u} = 0.55$. With these parameter values we obtain the particle density contours presented in Figure 4, where the particle density is shown for times of 2.0, 4.0, 6.0 and 8.0 units after the jet is turned on, with the initial state of the bed being uniform fluidization. The contours are spaced 0.1 units apart, with the outermost contour representing a particle concentration of 0.5. In performing these calculations a 50×50 grid was used to discretize the spatial domain.

As can be seen from Figure 4, the bubble is initiated by the jet, develops the characteristic kidney shape observed in experiments, and propagates away from the distributor. If one uses a gray-scale colour-map to examine the particle density and superimposes over this the particle velocities represented as a vector field, the result is Figure 5. A darker shade corresponds to a higher particle density. It can be seen that associated with the bubble is a set of counter–rotating convective rolls. These rolls are located behind the bubble and elongate with the passage of time. It is this set of rolls located behind the bubble which give rise to the characteristic kidney shape which has been noted before. This interaction/interdependence of the bubble and the set of convective rolls helps to understand the appearance of bubbles only under certain conditions. Assuming the rolls have a length scale associated with them which is dependent on the other length scales in the problem, no bubble of this type will develop if the dimensions of the bed are too small to accommodate this length scale.

Further considerations suggested by these results which need to be studied involve such things as the existence of this length scale and its dependence on the jet width and/or strength; the possibility of several jets leading to multiple bubbles; and attempts to determine if the bubble represents a formal structure propagating in the bed which obeys certain dynamical laws and the determination of these laws.

Acknowledgements I. Christie was partially supported by a grant from the Dirección General De Investigación Científica Y Técnica of Spain and acknowledges the assistance of J.M. Sanz-Serna in providing facilities at the University of Valladolid during an extended visit. J.W. Wilder was partially supported by NSF grant number OSR–9255224.

8. References

1. J. B. Fanucci, N. Ness, and R-H Yen, *Phys. Fluids* **24**(11) (1981) 1944.
2. D. J. Needham and J. H. Merkin, *J. Fluid Mech* **131** (19 83) 427.
3. G. H. Ganser and J. Lightbourne, *Chem. Eng. Sci.* **46** (5/6) (1991) 1339.
4. I. Christie, G. H. Ganser, and J. M. Sanz-Serna, *J. Comp. Phys.* **93** (1991) 297.
5. I. Christie, G. H. Ganser, and J. W. Wilder, submitted.
6. W. G. Strang, *SIAM J. Numer. Anal.* **5** (1968) 506.
7. P. L. Roe, *J. Comp. Phys.* **43** (1981) 357.
8. P. L. Roe, in Numerical Methods for Fluid Dynamics, ed. by K. W. Morton and M. J. Baines (Academic Press,1982).
9. P. K. Sweby, *SIAM J. Numer. Anal.* **21** (1984) 995.
10. A. Harten and J. M. Hyman, *J. Comput. Phys.* **50** (1983) 235.
11. I. Christie and C. Palencia, *IMA J. Num. Anal.* **11** (1991) 493.
12. J. W. Pritchett, T. R. Blake and S. K. Garg, *AIChE Symposium Series* **74** (1978) 134.
13. M Symalal, *NIMPF: A Computer Code for Nonisothermal Multiparticle Fluidization*, U.S. Department of Energy, Contract No. DE–AC21–85MC21353 Final Report (1987).
14. P. U. Foscolo and L. G. Gibilaro, *Chem. Eng. Sci.* **42** (6) (1987) 1489.
15. G. M. Homsy, M. M. El–Kaissy, and A. Didwania, *Int. J. Multiphase Flow* **6** (1980) 305.
16. D. A. Drew, *Ann. Rev. Fluid. Mech.* **15** (1983) 261.
17. P. Glaister, *J. Comp. Phys.* **77** (1988) 361.
18. A. R. Mitchell and D. F. Griffiths, *The Finite Difference Method in Partial Differential Equations*, (Wiley, 1980).
19. G. H. Ganser, X. Hu and D. Li, *SIAM J. Appl. Anal.*, in press .

NUMERICAL EXPERIMENTS FOR A NONOVERLAPPING DOMAIN DECOMPOSITION METHOD FOR PARTIAL DIFFERENTIAL EQUATIONS

JIM DOUGLAS, JR.
Department of Mathematics, Purdue University
West Lafayette, Indiana 47907-1395, U.S.A.
E-mail: douglas@math.purdue.edu

and

DAOQI YANG
Department of Mathematics, Wayne State University
Detroit, Michigan 48202, U.S.A.
E-mail: yang@math.wayne.edu

ABSTRACT

We present numerical experiments for a nonoverlapping domain decomposition method with interface relaxation for general selfadjoint and non-selfadjoint elliptic problems in two dimensions. The procedure contains two steps in each full iteration. The transmission condition on the interface is taken to be Dirichlet in the first step and Neumann in the second. However, in the presence of interior subdomains, an average mechanism is introduced at each cross-point to update the value at these points immediately after the Neumann sweep. Numerical examples show the rapid convergence of the method.

1. Introduction

Nonoverlapping domain decomposition methods have received much attention during the past few years, due to their easy implementation and computer memory savings in comparison to overlapping domain decomposition methods. For recent developments of nonoverlapping domain decomposition methods, we offer the nineteen papers in our reference list, along with numerous others in their references.

In this paper, we conduct numerical experiments for a nonoverlapping domain decomposition method for elliptic problems, selfadjoint or not, with Dirichlet boundary conditions. Our domain decomposition procedure is similar to the one considered elsewhere[19], but here we introduce underrelaxation on the interface of subdomains and an averaging mechanism at cross-points of subdomains to ensure convergence. Each full iteration of the domain decomposition procedure contains a Dirichlet sweep, in which we solve Dirichlet subdomain problems, and a Neumann sweep, in which we solve Neumann subdomain problems, except that Dirichlet boundary conditions are still imposed on the intersection of the boundaries of the original domain and the subdomains.

We will compare the convergence behavior of this method between non-selfadjoint and selfadjoint problems, between full-tensor diffusion coefficients and diagonal diffusion coefficients, and between variable coefficients and constant coefficients. In §2 and §3 we define the domain decomposition method. In §4 we give a finite-dimensional discretization for the subdomain problems based on finite differences. Finally, in §5 we report some numerical experiments and draw some conclusions in §6.

The method and the experiments are given for domains in $\mathcal{R}^2$ in this paper, but the technique clearly is applicable in $\mathcal{R}^3$.

2. The Differential Problem and Domain Decomposition

Let Ω be a smooth, bounded domain or a convex polygon in $\mathcal{R}^2$ with boundary $\partial\Omega$. Consider the following boundary value problem: find $u \in H^1(\Omega)$ such that

$$Lu = f \text{ in } \Omega, \qquad u = g \text{ on } \partial\Omega, \tag{1}$$

where $f \in L^2(\Omega)$ and $g \in H^{\frac{1}{2}}(\partial\Omega)$ are given, and the elliptic operator L is defined by

$$Lu = -\nabla \cdot \left(\begin{bmatrix} a_{11} & a_{12} \\ a_{21} & a_{22} \end{bmatrix} \nabla u \right) + [b_1(x,y), b_2(x,y)] \nabla u + a_0(x,y)u. \tag{2}$$

The coefficient matrix $\{a_{ij}(x,y)\}$ is assumed to be symmetric, uniformly positive definite, bounded, and piecewise smooth in Ω. Also, assume that b_i is smooth and bounded and that $a_0 \geq 0$.

We assume that Ω is partitioned into two nonoverlapping (not necessarily connected) subdomains Ω_1 and Ω_2 such that

$$\bar{\Omega} = \bar{\Omega}_1 \cup \bar{\Omega}_2, \qquad \Omega_1 \cap \Omega_2 = 0, \qquad \partial\Omega_1 \cap \partial\Omega \neq 0, \qquad \partial\Omega_2 \cap \partial\Omega \neq 0.$$

We denote the interface by $\Gamma = \partial\Omega_1 \cap \partial\Omega_2$. Red-black ordering allows the decomposition to contain more than two physical subdomains, such as shown in Figure 1. We should also assume that the coefficient a_0 in Eq. (2) be positive on interior subdomains in order to let the Neumann problems on interior subdomains have unique solutions.

Ω_1	Ω_2	Ω_1	Ω_2
Ω_2	Ω_1	Ω_2	Ω_1
Ω_1	Ω_2	Ω_1	Ω_2

Figure 1: A sample decomposition of the domain into two subdomains.

It is well known that, under suitable regularity conditions, the problem Eqs. (1)-(2) is equivalent to the following split problem:

$$\begin{aligned} Lu_k &= f \text{ in } \Omega_k, \qquad k=1,2, && (3)\\ u_k &= g \text{ on } \partial\Omega_k \cap \partial\Omega, \qquad k=1,2, && (4)\\ u_k &= u_m \text{ on } \Gamma, \qquad k,m=1,2, k\neq m, && (5)\\ \frac{\partial u_k}{\partial \nu_A^k} + \frac{\partial u_m}{\partial \nu_A^m} &= 0 \text{ on } \Gamma, \qquad k,m=1,2,k\neq m, && (6) \end{aligned}$$

where, for $k = 1$ or 2,

$$u_k = u|_{\Omega_k}, \qquad \frac{\partial u_k}{\partial \nu_A^k} = \sum_{i,j=1}^{d} a_{ij}\frac{\partial u_k}{\partial x_j}\nu_i^k,$$

and $\nu^k = \{\nu_1^k, \cdots, \nu_d^k\}$ is the outward normal unit vector to $\partial\Omega_k$.

3. The Domain Decomposition Method

We now define formally the following domain decomposition method: Choose $u_k^0 \in H^1(\Omega_k)$ satisfying $u_k^0|_{\partial\Omega\cap\partial\Omega_k} = g$, $k = 1,2$. For $n = 0,1,2,\ldots$, the sequence $u_k^n \in H^1(\Omega_k)$ with $u_k^n|_{\partial\Omega\cap\partial\Omega_k} = g$ is defined recursively by solving

$$\begin{aligned} &Lu_1^{n+1/2} = f \text{ in } \Omega_1,\\ &u_1^{n+1/2} = \frac{\theta}{2}\left(u_1^n + u_2^n\right) + \frac{1-\theta}{2}\left(u_1^{n-1} + u_2^{n-1}\right) \text{ on } \Gamma, \end{aligned} \qquad (7)$$

$$\begin{aligned} &Lu_2^{n+1/2} = f \text{ in } \Omega_2,\\ &u_2^{n+1/2} = \frac{\theta}{2}\left(u_1^n + u_2^n\right) + \frac{1-\theta}{2}\left(u_1^{n-1} + u_2^{n-1}\right) \text{ on } \Gamma, \end{aligned} \qquad (8)$$

$$\begin{aligned} &Lu_1^{n+1} = f \text{ in } \Omega_1,\\ &\frac{\partial u_1^{n+1}}{\partial \nu_A^1} = \frac{\theta}{2}\left(\frac{\partial u_1^{n+1/2}}{\partial \nu_A^1} + \frac{\partial u_2^{n+1/2}}{\partial \nu_A^1}\right) + \frac{1-\theta}{2}\left(\frac{\partial u_1^{n-1}}{\partial \nu_A^1} + \frac{\partial u_2^{n-1}}{\partial \nu_A^1}\right) \text{ on } \Gamma, \end{aligned} \qquad (9)$$

$$\begin{aligned} &Lu_2^{n+1} = f \text{ in } \Omega_2,\\ &\frac{\partial u_2^{n+1}}{\partial \nu_A^2} = \frac{\theta}{2}\left(\frac{\partial u_1^{n+1/2}}{\partial \nu_A^2} + \frac{\partial u_2^{n+1/2}}{\partial \nu_A^2}\right) + \frac{1-\theta}{2}\left(\frac{\partial u_1^{n-1}}{\partial \nu_A^2} + \frac{\partial u_2^{n-1}}{\partial \nu_A^2}\right) \text{ on } \Gamma, \end{aligned} \qquad (10)$$

where $\theta \in (0,1)$ is a relaxation parameter that should be determined to accelerate the convergence of the iterative procedure. In our numerical experiments reported below, we chose $\theta = \frac{1}{2}$. Eqs. (7)-(8) will be called a Dirichlet sweep, and Eqs. (9)-(10) a Neumann sweep, throughout the rest of the paper.

Due to the pathological properties of the spaces $H^{1/2}(\Gamma)$ and $H^{-1/2}(\Gamma)$, Eqs. (7)-(10) should be understood heuristically under the assumptions on the data, the domain, and the coefficients. However, its finite-dimensional discretization in variational

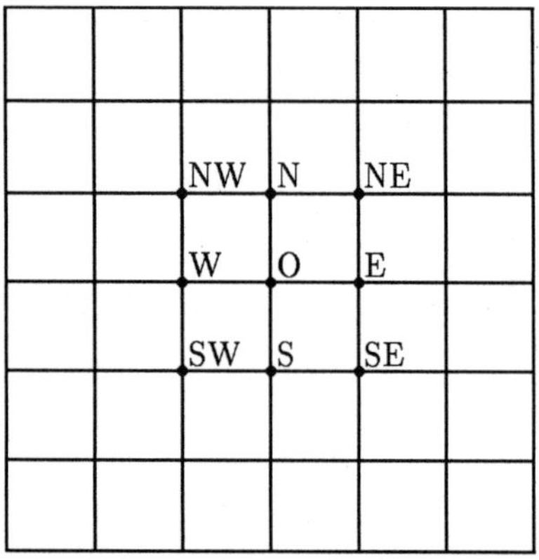

Figure 2: An interior grid point O with 8 neighboring grid points inside a subdomain.

form can be stated rigorously. For smoother data ($g \in \mathrm{H}^1(\partial\Omega)$ and $f \in \mathrm{L}^2(\Omega)$, say) and coefficients and convex, connected components of the subdomains, it is easy to see that the domain decomposition procedure is defined; what is necessary is that the trace of the conormal derivative of the solution of the subdomain problem on Ω_k lie in $L^2(\partial\Omega_k)$, so that it can be localized to portions of $\partial\Omega_k$.

The idea of the domain decomposition method is to impose continuity of Dirichlet values and Neumann values (fluxes) alternatively in the iterative process such that the limit of the solutions of Eqs. (7)-(10) converges to the solution of Eqs. (3)-(6). Note that, at the differential level, the limit of the solutions of Eqs. (7)-(10) will have continuity of Dirichlet and Neumann values simultaneously across the interface Γ if the procedure can be carried out and converges.

4. The Finite-Dimensional Discretization

In this section we describe a finite-dimensional discretization for Eqs. (7)-(10) based on a second-order finite-difference method. Assume that the domain Ω is overlaid by a rectangular grid

$$\Omega^h = \{(x_i, y_j) : x_i = a + ih_x, y_j = c + jh_y, i = 1, 2, \ldots, I, j = 1, 2, \ldots, J\}.$$

Then, at each grid point $O = (x_i, y_j)$, as in Figure 2, in the interior of a subdomain Ω_k, we have the following finite difference equation:

$$- \frac{a_{11}(EO_2)[U_k(E) - U_k(O)] - a_{11}(OW_2)[U_k(O) - U_k(W)]}{h_x^2}$$
$$- \frac{a_{22}(NO_2)[U_k(N) - U_k(O)] - a_{22}(OS_2)[U_k(O) - U_k(S)]}{h_y^2}$$

$$
\begin{aligned}
&- \frac{a_{12}(E)[U_k(NE)-U_k(SE)]-a_{12}(W)[U_k(NW)-U_k(SW)]}{4h_xh_y}\\
&- \frac{a_{21}(N)[U_k(NE)-U_k(NW)]-a_{21}(S)[U_k(SE)-U_k(SW)]}{4h_xh_y}\\
&+ b_1(O)\frac{U_k(E)-U_k(W)}{2h_x}+b_2(O)\frac{U_k(N)-U_k(S)}{2h_y}+a_0(O)U_k(O)\\
&= f(O),
\end{aligned}
$$

where $U_k(N)$, for example, denotes the approximate solution on subdomain Ω_k at point N, and $a_{11}(OW_2)$ denotes the value of a_{11} at the middle point between grid points O and W. We apply this standard nine-point finite-difference scheme for every subdomain problem at each iteration level.

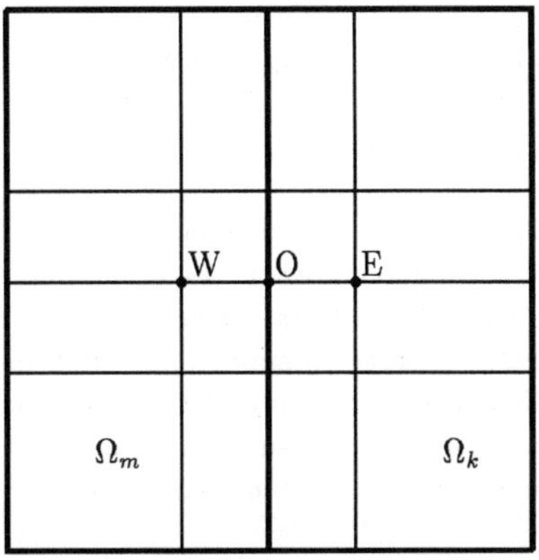

Figure 3: A grid point O at the interior of the interface of subdomains Ω_k and Ω_m.

For a grid point O (see Figure 3) at the interior of an interface of subdomains Ω_k and Ω_m, we define the Dirichlet value $D_\Gamma(O)$ and the Neumann value $N_\Gamma(O)$ at point O as follows:

$$
D_\Gamma(O)=\frac{U_k(O)+U_m(O)}{2},\quad N_\Gamma(O)=\frac{U_k(E)-U_m(W)}{2h_x}.
$$

The Dirichlet value $D_\Gamma(O)$ with an underrelaxation average will be used in the boundary condition for the Dirichlet sweep (Eqs. (7)-(8)), and $N_\Gamma(O)$ with an underrelaxation average will be used for the Neumann sweep (Eqs. (9)-(10)). This will ensure continuity of Dirichlet values and Neumann values when the iterative procedure converges. It is easy to see that the limit of the iterative solutions satisfies a nine-point finite-difference equation at all grid points, including interface points.

For a cross-point O (see Figure 4) at the interface of four subdomains $\Omega_1, \Omega_2, \Omega_3$, and Ω_4, we apply an averaging mechanism at the point, namely a second-order finite-difference equation at O:

$$
\begin{aligned}
\bar{U}(O) &= \left\{ a_0(O) + \frac{a_{11}(EO_2) + a_{11}(OW_2)}{h_x^2} + \frac{a_{22}(NO_2) + a_{22}(OS_2)}{h_y^2} \right\}^{-1} \cdot \\
&\left\{ \frac{a_{11}(EO_2)D_\Gamma(E) + a_{11}(OW_2)D_\Gamma(W)}{h_x^2} \right. \\
&+ \frac{a_{22}(NO_2)D_\Gamma(N) + a_{22}(OS_2)D_\Gamma(S)}{h_y^2} \\
&+ \frac{a_{12}(E)[U_4(NE) - U_2(SE)] - a_{12}(W)[U_3(NW) - U_1(SW)]}{4h_x h_y} \\
&+ \frac{a_{21}(N)[U_4(NE) - U_3(NW)] - a_{21}(S)[U_2(SE) - U_1(SW)]}{4h_x h_y} \\
&\left. - b_1(O)\frac{D_\Gamma(E) - D_\Gamma(W)}{2h_x} - b_2(O)\frac{D_\Gamma(N) - D_\Gamma(S)}{2h_y} + f(O) \right\}, \qquad (11) \\
& U_1(O) = U_2(O) = U_3(O) = U_4(O) = \bar{U}(O). \qquad (12)
\end{aligned}
$$

Note that each subdomain problem computes a value at point O and these values may be dramatically different from each other. Eq. (11) computes the solution to the difference equation at O using the latest values at the surrounding eight grid points, and we then assign this average to subdomain solutions at the cross-point; underrelaxation is not applied to the updating of the cross-point values in our implementation. Some such averaging mechanism at cross-points immediately after the Neumann sweep seems to be necessary to achieve convergence.

5. Numerical Examples

In this section we present some numerical experiments for the iterative procedure given by Eqs. (7)-(10); we applied the discrete scheme described above in §4 and uniform grids on each of the subdomains. The resulting linear systems of algebraic equations were approximately solved by Gauss-Seidel iteration. The initial guesses were always taken to be zero. The errors were evaluated in the L^∞-norm over all subdomains at each iteration. The following stopping criterion was used for the inner iteration (Gauss-Seidel for subdomain problems) with $\ell = -5$ and for the outer iteration (domain decomposition procedure) with $\ell = -4$:

$$\frac{\|U^{n+1} - U^n\|_\infty}{\|U^{n+1}\|_\infty} < 10^\ell, \qquad (13)$$

where $\|U^n\|_\infty$ denotes the discrete L^∞-norm over all subdomains for the discrete solution U at iteration level n.

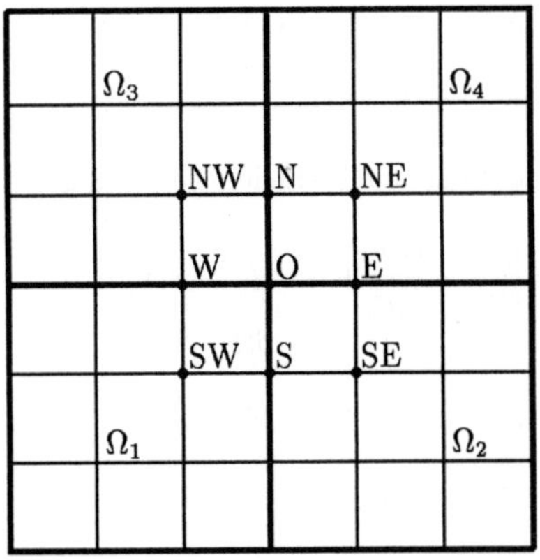

Figure 4: Cross-point O at the intersection of four subdomain boundaries.

We ran tests for the following sample problems.

Example 1. (non-symmetric with variable coefficients and full diffusion tensor)

$$-\nabla \cdot \left(\begin{bmatrix} e^x & x \\ x & e^y \end{bmatrix} \nabla u \right) + \left[\frac{1}{1+x+y}, \frac{1}{1+x+y} \right] \nabla u$$
$$+ \frac{u}{1+x+y} = f(x,y), \quad x \in \Omega,$$
$$u = g, \qquad (x,y) \in \partial\Omega,$$

where $\Omega = (0,1) \times (0,1)$. The functions f and g were chosen such that the exact solution of the differential problem is

$$u(x,y) = e^{xy}.$$

Example 2. (symmetric with variable coefficients and full diffusion tensor)

$$-\nabla \cdot \left(\begin{bmatrix} x+0.1 & \sin(xy) \\ \sin(xy) & y+0.1 \end{bmatrix} \nabla u \right)$$
$$+ (1 + \sin x + \cos y) u = f(x,y), \quad x \in \Omega,$$
$$u = g, \qquad (x,y) \in \partial\Omega,$$

where $\Omega = (0,1) \times (0,1)$, and the functions f and g were chosen such that the exact solution is

$$u(x,y) = e^{xy} \sin(10x) \cos(10y).$$

Example 3. (symmetric with variable coefficients and diagonal diffusion tensor)

$$-\frac{\partial}{\partial x}\left(e^x \frac{\partial u}{\partial x} \right) - \frac{\partial}{\partial y}\left(e^y \frac{\partial u}{\partial y} \right) + \frac{u}{1+x+y} = f, \quad (x,y) \in \Omega,$$
$$u = 0, \qquad (x,y) \in \partial\Omega,$$

where $\Omega = (0,1) \times (0,1)$. The function f was chosen such that the exact solution is

$$u(x,y) = 3e^{x+y}xy(1-x)(1-y).$$

Example 4. (symmetric with constant coefficients and diagonal diffusion tensor)

$$\begin{aligned} &-\frac{\partial^2 u}{\partial x^2} - \frac{\partial^2 u}{\partial y^2} + u = f, \quad (x,y) \in \Omega, \\ &u = 0, \qquad (x,y) \in \partial\Omega, \end{aligned}$$

where $\Omega = (0,1) \times (0,1)$. The function f was chosen such that the exact solution is

$$u(x,y) = y(1-y)\sin(\pi x/2).$$

5.1. Strip Domain Decompositions

We first decomposed the domain Ω into 10 subdomains

$$\Omega_i = (0.1(i-1), 0.1i) \times (0,1), \qquad i = 1, 2, \ldots, 10,$$

in the x-direction. (By coloring odd-numbered subdomains red and even-numbered subdomains black, we still can view this as a two-subdomain case.) The L^∞-errors between the iterative solution and the true solution over all subdomains are given in Tables 1 and 2, while in Tables 3 and 4 we show the differences in the iterative solutions between the current and previous iteration levels. For the 40×40 mesh, there are 5 grid points in the x-direction and 41 grid points in the y-direction on each subdomain. Comparing the results in Tables 3 and 4, which corresponds to an 80×80 mesh, indicates that the convergence rate is about the same for different grid sizes.

5.2. Domain Decompositions with Cross-Points

We then decomposed the domain into $N_x \cdot N_y$ subdomains with N_x subdomains in the x direction and N_y subdomains in the y direction. See Figure 1 for an example of 4×3 decomposition. At each cross-point, we applied the averaging mechanism discussed in §4. The results are shown in Tables 5 and 6. Note that in the case of 10×10 subdomains with grid $\frac{1}{60} \times \frac{1}{60}$, there are only 7 grid lines in the x or y direction on each subdomain. It should be noted that the number of iterations is getting slightly larger when the grid gets finer. The number of iterations required for the process to stop depends on the condition number of the problem and is different for different examples. Also, note that independence of the number of iterations upon the grid size was proved[19] only for strip domain decompositions.

Table 1: Numerical results with 10 subdomains in the x-direction with grid size $\frac{1}{40} \times \frac{1}{40}$. The errors are shown in the L^{∞}-norm, measured between the iterative solution and the true solution in all subdomains.

	Grid size $\frac{1}{40} \times \frac{1}{40}$			
Iteration	Example 1	Example 2	Example 3	Example 4
1	1.88	1.21	2.01	1.96
2	1.18	5.14E-1	1.16	9.10E-1
3	5.73E-1	2.81E-1	6.27E-1	4.23E-1
4	2.32E-1	1.54E-1	2.82E-1	1.85E-1
8	2.02E-2	6.01E-2	2.57E-2	1.25E-2
10	5.98E-3	2.59E-2	4.48E-3	3.54E-3
12	3.19E-3	2.61E-2	2.94E-3	3.03E-3
14	3.20E-3	1.92E-2	2.78E-3	2.74E-3

Table 2: Numerical results with 10 subdomains in the x-direction with grid size $\frac{1}{80} \times \frac{1}{80}$. The errors are shown in the L^{∞}-norm, measured between the iterative solution and the true solution in all subdomains.

	Grid size $\frac{1}{80} \times \frac{1}{80}$			
Iteration	Example 1	Example 2	Example 3	Example 4
1	5.22E-1	9.03E-1	4.93E-1	5.69E-1
2	3.87E-1	4.35E-1	3.53E-1	4.49E-1
3	2.59E-1	2.22E-1	2.37E-1	2.58E-1
4	1.37E-1	1.36E-1	1.48E-1	1.67E-1
8	2.89E-2	2.81E-2	3.29E-2	1.02E-2
10	1.36E-2	2.19E-2	7.11E-3	3.71E-3
12	2.78E-3	1.87E-2	4.66E-3	1.87E-3
14	1.51E-3	1.06E-2	4.49E-3	1.04E-3

Table 3: Numerical results with 10 subdomains in the x-direction with grid size $\frac{1}{40} \times \frac{1}{40}$. The difference between the iterative solution at the current and previous iteration levels is shown in the L^∞-norm.

	Grid size $\frac{1}{40} \times \frac{1}{40}$			
Iteration	Example 1	Example 2	Example 3	Example 4
1	1.00	1.00	1.00	1.00
2	1.36	1.06	1.07	8.68E-1
3	1.26	4.10E-1	9.10E-1	5.08E-1
4	4.94E-1	2.03E-1	6.34E-1	3.45E-1
5	1.48E-1	9.89E-2	2.77E-1	2.09E-1
10	7.90E-3	3.90E-3	7.19E-3	5.03E-3
12	3.02E-3	1.61E-3	8.59E-4	1.86E-4
14	3.73E-4	7.21E-4	2.43E-4	9.26E-5
16	3.38E-5	2.48E-4	–	–
18	–	8.93E-5	–	–

Table 4: Numerical results with 10 subdomains in the x-direction with grid size $\frac{1}{40} \times \frac{1}{40}$. The difference between the iterative solution at the current and previous iteration levels is shown in the L^∞-norm.

	Grid size $\frac{1}{80} \times \frac{1}{80}$			
Iteration	Example 1	Example 2	Example 3	Example 4
1	1.00	1.00	1.00	1.00
2	2.72E-1	4.59E-1	2.86E-1	2.21E-1
3	2.29E-1	3.15E-1	1.76E-1	1.95E-1
4	1.52E-1	1.23E-1	1.15E-1	1.54E-1
5	1.00E-1	6.89E-2	7.72E-2	8.81E-2
10	8.09E-3	6.70E-3	1.58E-2	1.69E-2
12	5.08E-3	1.87E-3	2.33E-3	1.24E-3
14	9.73E-4	9.28E-4	2.96E-4	3.68E-4
16	1.02E-4	4.81E-4	7.18E-5	1.92E-4

Table 5: Numerical results with 5×5 subdomains: the number of iterations needed to satisfy Eq. (13).

	Example 1	Example 2	Example 3	Example 4
Grid $= \frac{1}{40} \times \frac{1}{40}$	58	19	68	50
Grid $= \frac{1}{80} \times \frac{1}{80}$	70	25	82	74

Table 6: Numerical results with 5×5 subdomains: the number of iterations needed to satisfy Eq. (13).

	Example 1	Example 2	Example 3	Example 4
Grid $= \frac{1}{60} \times \frac{1}{60}$	190	42	197	126
Grid $= \frac{1}{120} \times \frac{1}{120}$	189	43	250	134

6. Concluding Remarks

We have proposed an iterative nonoverlapping domain decomposition method for elliptic partial differential problems. Each iteration in this method contains two steps, a Dirichlet sweep and a Neumann sweep.

Numerical experiments show that this method performs quite well for a wide range of problems including non-selfadjoint problems and variable coefficient problems with full diffusion tensor. Although a convergence analysis for this method can be made[14,19] when there are no cross-points, we treat here solely an experimental approach. For domain decompositions with cross-points, we provided an averaging mechanism that appears to ensure convergence; a somewhat similar treatment was also considered by Marini and Quarteroni in an unpublished note made available to the authors.

We can also consider finite-dimensional approximations to the domain decomposition method using finite element methods (with or without Lagrange multipliers) and mixed finite element methods[6,14,19].

References

1. P. Bjørstad and O. Widlund, Iterative methods for the solution of elliptic problems on regions partitioned into substructures, SIAM J. Numer. Anal., **23**(1986), pp. 1097-1120.
2. J. H. Bramble, J. E. Pasciak, and A. H. Schatz, An iterative method for elliptic problems on regions partitioned into substructures, Math. Comp., **46**(1986), pp. 361-369.

3. L. C. Cowsar and M. F. Wheeler, Parallel domain decomposition method for mixed finite elements for elliptic partial differential equations, in: R. Glowinski, Y. Kuznetsov, G. Meurant, J. Périaux, and O. B. Widlund, Eds., *Proceedings of the Fourth International Symposium on Domain Decomposition Methods for Partial Differential Equations*, SIAM, Philadelphia, PA, 1991, pp. 358-372.
4. B. Després, Domain decomposition method and the Helmholz problem, in: G. Cohen, L. Halpern, and P. Joly, Eds., *Mathematical and Numerical Aspects of Wave Propagation Phenomena*, SIAM, Philadelphia, PA, 1991, pp. 44-52.
5. J. Douglas, Jr., P. J. Paes Leme, J. E. Roberts, and J. Wang, A parallel iterative procedure applicable to the approximate solution of second order partial differential equations by mixed finite element methods, Numer. Math., **65**(1993), pp. 95-108.
6. J. Douglas, Jr., and D. Q. Yang, A domain decomposition method for elliptic problems based on mixed finite element methods (preprint).
7. D. Funaro, A. Quarteroni, and P. Zanolli, An iterative procedure with interface relaxation for domain decomposition methods, SIAM J. Numer. Anal., **25**(1988), pp. 1213-1236.
8. R. Glowinski, W. Kinton, and M. F. Wheeler, Acceleration of domain decomposition algorithms for mixed finite elements by multi-level methods, in: T. F. Chan, R. Glowinski, J. Périaux, and O. B. Widlund, Eds., *Third International Symposium on Domain Decomposition Methods for Partial Differential Equations*, SIAM, Philadelphia, PA, 1990, pp. 263–290.
9. R. Glowinski and M. F. Wheeler, Domain decomposition and mixed finite element methods for elliptic problems, in: R. Glowinski, G. Golub, G. Meurant, and J. Périaux, Eds., *Domain Decomposition Methods for Partial Differential Equations*, SIAM, Philadelphia, PA, 1988, pp. 144–172.
10. J. S. Gu and X. C. Hu, On an essential estimate in the analysis of domain decomposition methods, J. Comput. Math., **12**(1994), pp. 132-137
11. P. Le Tallec, Y. H. De Roeck, and M. Vidrascu, Domain decomposition methods for large linearly elliptic three-dimensional problems, J. Comput. Appl. Math., **34**(1991), pp. 93-117.
12. P. L. Lions, On the Schwarz alternating method III: a variant for nonoverlapping subdomains, in: T. F. Chan, R. Glowinski, J. Périaux, and O. B. Widlund, Eds., *Third International Symposium on Domain Decomposition Methods for Partial Differential Equations*, SIAM, Philadelphia, PA, 1990, pp. 202-223.
13. L. D. Marini and A. Quarteroni, An iterative procedure for domain decomposition methods: A finite element approach, in: R. Glowinski, G. H. Golub, G. A. Meurant, and J. Périaux, Eds., *Domain Decomposition Methods for Partial Differential Equations*, SIAM, Philadelphia, PA, 1988, pp. 129-143.
14. L. D. Marini and A. Quarteroni, A relaxation procedure for domain decomposition methods using finite elements, Numer. Math., **55**(1989), pp. 575-598.

15. J. R. Rice, E. A. Vavalis, and D. Q. Yang, Convergence analysis of a nonoverlapping domain decomposition method for elliptic PDEs, CSD-TR-93-048, Dept. of Computer Sciences, Purdue University, West Lafayette, IN 47907, July 1993.
16. D. Q. Yang, Different domain decompositions at different times for capturing moving local phenomena, J. Comput. Appl. Math., **59**(1995), pp. 39-48.
17. D. Q. Yang, A parallel iterative nonoverlapping domain decomposition procedure for elliptic problems, IMA J. Numer. Anal., to appear.
18. D. Q. Yang, A parallel iterative domain decomposition algorithm for elliptic problems, J. Comput. Math., to appear.
19. D. Q. Yang, A parallel iterative nonoverlapping domain decomposition method with interface relaxation for elliptic problems, in preparation.

A PIECEWISE UNIFORM ADAPTIVE GRID ALGORITHM FOR NONLINEAR DISPERSIVE WAVE EQUATIONS

ERIC S FRAGA
Department of Chemical Engineering
University of Edinburgh, Edinburgh EH9 3JL, United Kingdom
E-mail: Eric.Fraga@ed.ac.uk

and

JOHN Ll MORRIS
Department of Mathematics & Computer Science
University of Dundee, Dundee DD1 4HN, United Kingdom

ABSTRACT

An adaptive grid refinement method for one dimensional dispersive wave equations is presented. The method is intended for solutions exhibiting soliton behaviour, such as the Korteweg-de Vries (KdV) and Nonlinear Schrödinger (NLS) equations. The method has been designed to be robust, making it suitable for solving problems with non-soliton initial conditions and hence of interest for problems of more general interest. Results are presented for a variety of initial conditions for both the KdV and the NLS equations. We show that the method is stable, accurate, and insensitive to the user parameters used to define the grids. Of particular interest is the capability of preserving conservation properties of the equations, using simple numerical discretization methods adapted to non-uniform grids

1. Introduction

The numerical study of nonlinear dispersive wave equations has demonstrated the need for adaptive grid methods.[9] [7] [11] [13] [14] Modelling real phenomena requires the solution over long periods of time and for possibly large spatial dimensions. The combination of nonlinearity with long time solutions makes the use of uniform grids unfeasible.[3] [15] As these problems are evolutionary, frequently exhibiting moving waves, efficient solutions require the use of adaptive methods. These adaptive methods may be applied to both the spatial and the temporal dimensions. This paper discusses the design and implementation of an adaptive spatial grid refinement method, assuming that adaptive time stepping can be also used, as described by Sanz Serna & Christie.[13]

Numerical methods for nonlinear dispersive wave equations, such as the Korteweg-de Vries (KdV) equation:

$$\begin{aligned} \frac{\partial u}{\partial t} + u\frac{\partial u}{\partial x} + \epsilon\frac{\partial^3 u}{\partial x^3} &= 0 \\ -\infty < \; x &< \infty \\ t &> 0 \end{aligned} \qquad (1)$$

where $u(x,t)$ is a real valued function with initial condition $u(x,0) = g(x)$ and ϵ is a real valued parameter, and the Nonlinear Schrödinger (NLS) equation:

$$\begin{aligned} iu\frac{\partial u}{\partial t} + \frac{\partial^2 u}{\partial x^2} + q|u|^2 u &= 0 \\ i^2 &= -1 \\ -\infty < \; x &< \infty \\ t &> 0 \end{aligned} \tag{2}$$

where $u(x,t)$ is a complex valued function with initial condition $g(x)$ and with q a real valued parameter, must meet certain criteria. As for other types of equations, issues of accuracy, stability, and convergence are of primary importance. However, for this class of equations, there is another property that is desirable. Nonlinear dispersive wave equations have an infinite number of conservation laws which should be observed. In particular, it has been seen that numerical approximations which observe the L_2 energy

$$\frac{d}{dt}L_2 = \frac{d}{dt}\int_{-\infty}^{\infty} u^2 dx = 0. \tag{3}$$

are most effective.

To be successful, adaptive grid methods must themselves meet specific criteria. These criteria deal mostly with *ease of use* issues. If a method is to be taken up by users, it is important that it be not only easy to implement but also easy to use. Ease of use is a difficult property to quantify but for adaptive grid methods, we believe that the following criteria must be met:

- The number of user given parameters for the grid refinement method must be small. Failing that, the sensitivity of the results to the values of these parameters must be low. A user should not have to waste inordinate amounts of time attempting to determine suitable parameter values. Also, if results are insensitive to these parameter values, it should be easier to automate the selection of the values, thereby further minimizing the input from the user, input of no particular interest to that user. After all, the user is concerned with getting results, not necessarily with achieving a complete understanding of underlying techniques.

- Developing a numerical discretization for the partial differential equation should not involve significantly more work than for the equivalent discretization on a uniform mesh. As advanced numerical techniques get incorporated into *packages*, it is more difficult to rely on the user to be conversant with the finer arts of numerical method design.

- An adaptive grid method should impose a minimum amount of overhead. This ensures that even for cases where the refined grid is equivalent to the uniform grid, there is a small performance penalty to pay. This encourages users to make use of the adaptive method.

Methods described in the literature for adaptive grid refinements essentially fall into three categories: equidistribution methods, moving grid methods, and multigrid methods. The first use some local measure of difficulty based on current solution values to estimate a new grid. The second, which are often based on equidistribution methods, tie the position of grid points to the actual differential equations, creating a larger coupled system. The last use a form of error extrapolation to estimate where new grids have to be super-imposed. Although many of these methods have been successful for a variety of problem classes, none are particularly well suited to the study of nonlinear dispersive wave equations.

Multigrid methods suffer from difficulties with the artificial internal boundary conditions introduced by the subgrids. For equations such as the KdV (1), which typically require a five point wide stencil, suitable methods for dealing with these internal boundary conditions are difficult to achieve. Our experience has been that any mismatches at these points lead very quickly to nonlinear instability. Multigrid methods also violate the overhead criterion mentioned above. For problems which need to use a fine mesh over the whole interval, the cost of determining this using error extrapolation is large. Assuming a ratio of grid spacing of 2 from a coarse mesh to a finer one, the amount of work needed to refine to the desired fine mesh and perform the appropriate time stepping would be twice that needed for a uniform fine mesh.

Methods based on moving grids, as well as equidistribution class of methods, would appear to be more suitable for this class of problems (see Russell & Christiansen[12] and Hawken et al[7] for reviews of these types of methods). However, developing numerical methods that are appropriate for these methods is difficult, as described below. These methods also suffer from the limitation that they usually have a fixed number of grid points. For evolutionary problems, the number of points required at the beginning may be completely different from the number later on. For example, problems which result in dispersive wave trains from a compact initial solution will need more points as time goes on.

1.1. Energy Conservation and Convergence of the Numerical Scheme

Although numerical methods may be devised to conserve the L_2 energy on a uniform mesh, this may be difficult on a non-uniform mesh. A previous adaptive method described by the authors,[5] known as the *soliton* method, was compared to and contrasted with an adaptive scheme based on the equidistribution of the arclength of the solution (based on schemes suggested by Manoranjan[9] and Sanz-Serna and Christie[13]). The results of the comparisons,[4] led us to believe that, for the KdV equation, conservation of energy was an important factor in the success of an adaptive mesh technique. The solution of the KdV equation using an arclength equidistribution method was shown to develop phase errors and exhibited a lack of conservation of the L_2 energy. For the KdV equation, the speed of solitons in the solution is directly related to their amplitudes. As energy is lost, the solitons slow down.

However, more recent work by Garcia-Archilla and Sanz-Serna[6] has in fact shown that the numerical discretization employed by the authors for the third order term of the KdV equation is not convergent on nonuniform grids. The use of an arclength equidistribution method results in grids that are highly nonuniform. The method described by Fraga and Morris, on the other hand, consists of contiguous uniform subgrids. It appears that the use of uniform subgrids, with nonuniformities arising solely at the grid interfaces, allows the use of a method that is theoretically not convergent. In practice, the low order accuracy and lack of convergence predicted by the analysis for nonuniform grids are not exhibited when the grid consists of uniform subgrids. These problems, however, do arise with meshes that are nonuniform almost everywhere. If one wishes to use adaptive grid methods that generate such types of grids, it is necessary to use a numerical method based on a different spatial discretization of the KdV equation. This is what Garcia-Archilla and Sanz-Serna have done. However, developing exactly conservative numerical schemes for nonuniform meshes is a difficult task in general and violates the ease of use criterion mentioned above.

1.2. Mesh spacing ratio

Although the *soliton* method appeared to work well for both the KdV and NLS equations for a wide range of initial conditions, further experimentation has highlighted some difficulties. In particular, solving the KdV equation for a square pulse initial condition has proven to be difficult unless the number of points placed in the gaps between "solitons" is large. All other problems mentioned by the authors[5] were solved using small numbers of points in the gaps. Increasing the number of points in the gaps reduces the efficiency of the method although it does enable the approach to solve a wider range of problems. However, one of the criteria for successful adaptive methods is violated: the method is highly sensitive to the setting of one of the user parameters, possibly making it difficult to determine the correct value.

Russell and Christiansen[12] note that the ratio,

$$\frac{h_{max}}{h_{min}}, \tag{4}$$

for a nonuniform mesh, must be bounded. Sanz-Serna and Christie[13] also make this observation. For this reason, the latter authors introduce a parameter, β, in their implementation of the arclength equidistribution method to ensure that this ratio is kept within appropriate limits. More recently, however, Daripa[1] has studied the effect of the ratio function

$$R(x_i) \equiv \frac{x_{i+1} - x_i}{x_i - x_{i-1}} \tag{5}$$

where $\mathbf{x} = \{x_i, i = 0, ..., n\}$ is the set of n points which define the spatial grid. Daripa claims that it is important to keep this (albeit discrete) function "smooth" over the whole interval.

In light of these results, it would appear likely that the ratio in Eq. (5) exceeds some maximum when attempting to solve the square pulse initial condition problem using the *soliton* method. Although the problem was eliminated by decreasing the mesh spacing in the coarse mesh region, this is a rather ad-hoc solution and is not desirable. The *soliton* method does not provide any automatic mechanism by which the ratio can be kept small. For an adaptive grid method to be useful, it is necessary that it be able to adapt itself to a variety of circumstances.

2. The Shoulder algorithm

The difficulties noted above provided the motivation to develop a new adaptive scheme, a scheme which would ideally eliminate these problems. In particular, it was decided that the new method should preserve the property of piecewise uniformity as this allows the use of simpler numerical schemes. The numerical experiments using the *soliton* method showed that, although theoretically the numerical schemes were limited to first order accuracy on a nonuniform mesh, second order accuracy was attained in practice. Likewise, the piecewise uniformity achieved the conservation of the L_2 energy with sufficient accuracy even though the numerical scheme was not explicitly conservative on a nonuniform mesh.

No problems with stability were observed with the *soliton* method so long as the spacing ratio (Eq. 5) was kept small. With this in mind, one of the major goals of the new method was to ensure that this ratio was automatically kept to within reasonable bounds, while minimizing the number of points necessary for reasonable accuracy and energy conservation.

The combination of these goals led to the development of a new algorithm for adaptive grid generation, a method which combines the properties of piecewise uniformity for conservation and for accuracy with a gradual change in step size for stability. The new method is known as the *shoulder* method and is best described in the form of an algorithm:

1. Locate all "solitons".
2. Discretize the domain of each soliton using fine uniform spacing.
3. Fill in the gaps between the solitons with a stepwise coarser mesh.
4. Generate solution values for the new mesh.

The result is a mesh that consists of contiguous uniform submeshes, with fine discretization for regions of the solution in which solitons appear and coarse discretization elsewhere. Each of the steps in the algorithm are now described in more detail:

Locating solitons. A geometric approach is used to locate the solitons for the solution $\mathbf{v}$ at time t. First, the range of values, (v_{min}, v_{max}), in the solution vector is determined:

$$\begin{aligned} v_{min} &= \min_i |v_i| \\ v_{max} &= \max_i |v_i| \end{aligned}$$

where $\mathbf{v} = \{v_i, i = 0, ..., n\}$ are the solution values at each grid point: $v_i \approx u(x_i, t)$ for some t.

Solitons are then assumed to exist whenever the value of the solution exceeds a cut-off value

$$v_{\text{cutoff}} = v_{min} + \frac{v_{max} - v_{min}}{\text{cutoff}}. \tag{6}$$

where *cutoff* is a user defined parameter.

We define a set $\mathcal{S}$ of points

$$\mathcal{S} = \{x_i \text{ s.t. } |v_i| \geq v_{\text{cutoff}}\} \tag{7}$$

which results in $\mathcal{S}$ including all points in the current mesh at which a soliton is likely to be present. This set is then coallesced into a set of intervals $\mathcal{I} = \{I_i \text{ where } I_i = [a_i, b_i], i = 1, ..., n_s\}$, where each interval contains runs of consecutive points x_j found in $\mathcal{S}$ and n_s is the number of soliton intervals. Each interval should contain one or more actual solitons (or waves in general); no point within these intervals is below the cut-off value.

Note, this procedure can also be applied to a numerical approximation to the derivative of u and can therefore be used to locate shock fronts.

Discretization of soliton domains. The set of intervals defined above only includes points which form solitons. However, it is not guaranteed to include all such points. In particular, the leading and trailing edges of each soliton are likely to be excluded, unless the *cutoff* parameter is very small. It is therefore useful to extend each of the intervals on each side. This is also necessary because of the evolutionary nature of the problems. Therefore, each interval $I_i \in \mathcal{I}$ is redefined as

$$I_i \leftarrow (1 + \beta) I_i$$

which can be expressed by the following steps

$$\begin{aligned} l_i &= b_i - a_i \\ (a_i, b_i) &\leftarrow (a_i - \beta l_i, b_i + \beta l_i) \end{aligned}$$

where β is a user specified parameter.

A side effect of extending the intervals is that the newly defined intervals may overlap. The next step is to check each consecutive pair of intervals for a possible overlap. An overlap is indicated by the condition

$$a_i \leq b_{i-1} + rn_p h_{goal}$$

where r is the desired ratio in step size changes from one uniform region to the next, n_p is the number of points to place in each uniform region (as described in the next step of the algorithm), and h_{goal} is the desired grid spacing in the soliton intervals (a user specified parameter, typically equal to the grid spacing that would have been used for a uniform mesh). Basically, if the gap between two intervals is not enough to place a coarse uniform set of points, the two intervals are considered to overlap. If this condition occurs, the particular intervals are combined into one bigger one (and the number of intervals decreased by one).

Each final interval is now discretized uniformly using a grid spacing on the order of h_{goal}. For the ith interval, we calculate the number of grid points, n_i, needed and the actual spacing used, h, by

$$\begin{aligned} n_i &= \left\lceil \frac{b_i - a_i}{h_{goal}} \right\rceil \\ h &= \frac{b_i - a_i}{n_i} \end{aligned}$$

This guarantees a spacing no greater than the desired h_{goal}.

The values of both β and h_{goal} may be given by the user. Alternatively, the user may request that the program automatically determine their values. For some problems, the values can be estimated using the initial condition. The value of h_{goal} is the largest value which, when used to define a uniform grid, enables the accurate representation of the L_2 energy for the initial condition. Using this estimated value of h_{goal}, the value of β is defined to be the smallest value that enables the resulting non-uniform grid to also accurately represent the L_2 energy.

This step of the algorithm uses the β, h_{goal}, n_p, and r user defined parameters.

Coarse discretization of rest of interval. A novel feature of the new algorithm is how the non-soliton regions are discretized. As described above, we wish to generate nonuniform grids which enable the use of simple numerical approximation schemes while retaining their accuracy, stability, convergence, and conservation characteristics. The experience attained with the previous approach[5] indicates that we want to keep the step size ratio between adjacent discrete points to a minimum, yet we do not want the new grid to be nonuniform everywhere. This leads to a grid which is made up of uniform patches with small differences in the spacings of adjacent patches.

The set of intervals $\mathcal{I}$ describes the regions which have been discretized using the finest grid spacing h_{goal}. Conversely, the complement of $\mathcal{I}$, $\overline{\mathcal{I}} = [a,b] \backslash \mathcal{I}$, is a set of intervals which describes the regions of coarser refinement. Each of these intervals is given by (b_{i-1}, a_i), $i = 1, ..., n_s + 1$, where $b_0 \equiv a$ (a is the left end of the full problem domain) and $a_{n_s+1} \equiv b$ (b is the right end of the full problem domain).

Each element of $\overline{\mathcal{I}}$ is discretized using the following algorithm:

```
Algorithm DiscretizeInterval(a, b, h_goal, n_p, r)
          "(a, b) is the interval to discretize,"
          "h_goal is the desired spacing in fine regions,"
          "n_p is the number of points for each region, and"
          "r is the ratio between successively coarser regions."
          h ← h_goal
          "continue while there is enough room for two coarser regions"
          while b − a > 2 × n_p × r × h do
                h ← r × h                    "coarse spacing"
                place n_p points inwards from each end of the interval
                using a spacing h
                a ← a + n_p × h              "shrink the interval"
                b ← b − n_p × h
          end while
          "discretize the remaning space uniformly"
          m ← ⌈(b−a)/h⌉
          h ← (b−a)/m
          place m points using a spacing of h to fill up
          the space that is left
end Algorithm DiscretizeInterval
```

Each interval is discretized uniformly in a piecewise fashion starting from the edges and working inwards.

This algorithm will discretize the interval between any two soliton domains but is not directly applicable for the intervals that may appear between the left boundary and the first soliton domain and between the last soliton domain and the right boundary. In either of those cases, the algorithm is modified to only place points starting from the interval end that is adjacent to the soliton domain interval, working towards the boundary end.

The user parameters are required for this step are h_{goal}, n_p, and r.

Generating solution values on the new grid. Given the old grid with old solution values, we generate new solution values on the new grid using piecewise interpolation. The order of interpolation used, linear, quadratic, or cubic, is the choice of the user (via the *inter* variable).

The new *shoulder* algorithm generates a mesh that consists of contiguous uniform submeshes with the ratio of mesh spacing from one submesh to the next specified by a constant value, r. The new algorithm is a modification of the *soliton* algorithm, having changed the method by which the nonsoliton regions are discretized. In the *soliton* method, the non-soliton regions were discretized using N_{gap} points in each region. This is too simplistic. The new approach depends on two new parameters, r and n_p.

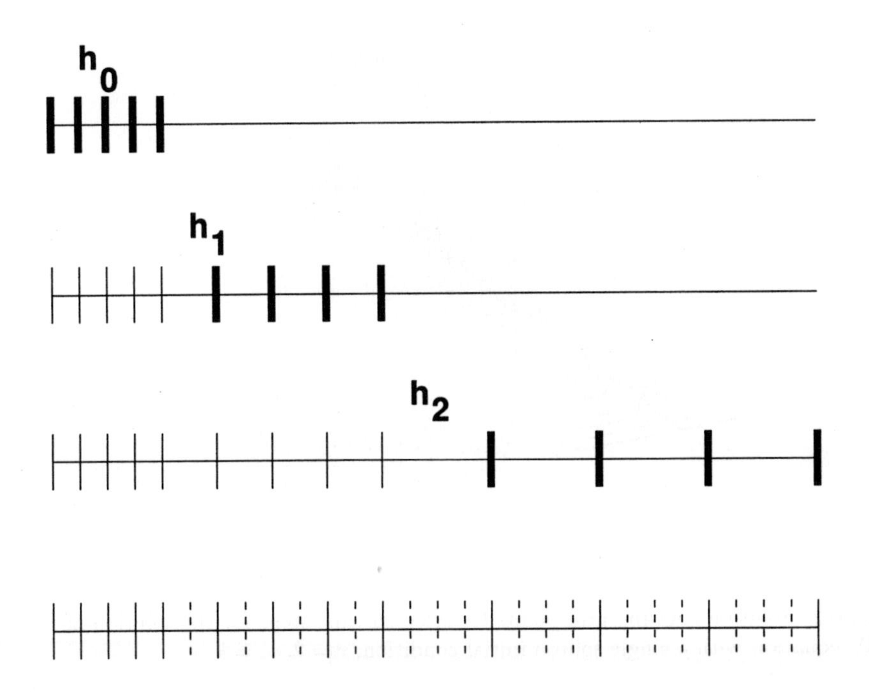

Figure 1: Step by step generation of a sample shoulder mesh.

Figure 1 shows how a mesh is generated, in this case for the interval between a soliton domain and the right problem boundary. The area of refinement is at the left of the interval. In this example, $h_0 \leftarrow h_{goal}$, $r \leftarrow 2$, and $n_p \leftarrow 4$. The top line shows the placement of the finest mesh (the one used to cover the soliton domain). The next step is to add n_p points with a spacing of $h_1 \leftarrow r \times h_0$. Then another n_p points, this time with a spacing of $h_2 \leftarrow r \times h_1 = r^2 \times h_0$, are added. The final line shows the extra points (indicated by dashed lines) that would have been placed if a uniform mesh were used.

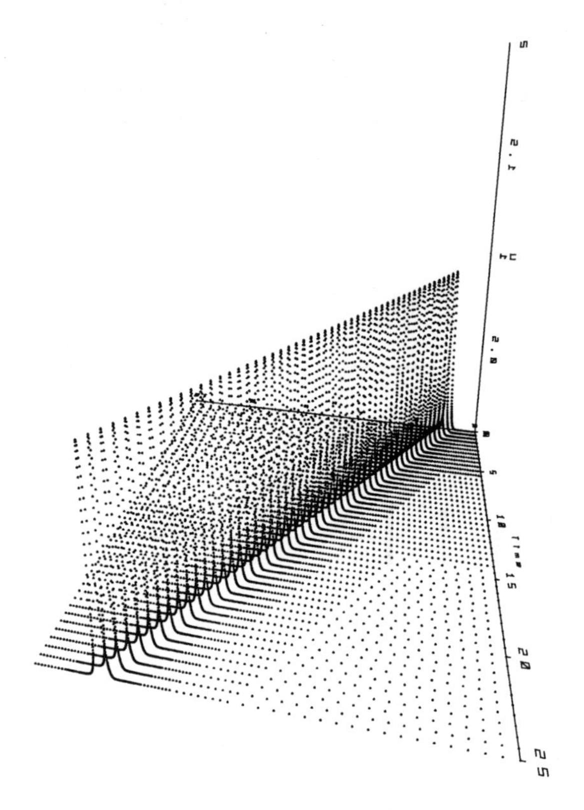

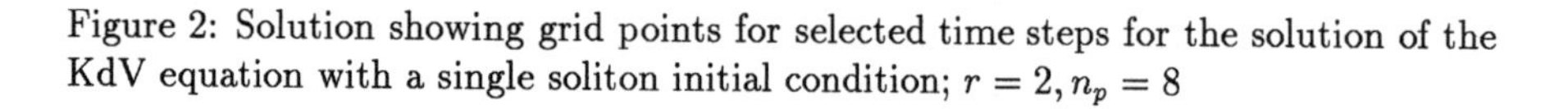

Figure 2: Solution showing grid points for selected time steps for the solution of the KdV equation with a single soliton initial condition; $r = 2, n_p = 8$

Figure 2 shows the grid points used to solve the KdV equation with a single soliton initial condition. The user parameters were set as follows: $n_n = 8$, $r = 2$, using cubic interpolation and with $h_{goal} = 0.0049$ and $\beta = 0.2$ chosen automatically by the initialization process.

3. Parameter Sensitivity and Choice

The *shoulder* algorithm uses a set of parameters, most of which must be specified by the user. The parameters are summarized here:

β: The amount to extend soliton intervals to ensure solitons are properly covered by a fine mesh.

cutoff: The cut-off ratio for locating solitons using a geometric approach.

h_{goal}: The desired grid spacing for the regions with the smallest grid spacings: the soliton covering regions. This value is typically the same as would have been used for a uniform mesh approach.

inter: The method to use for interpolating from the old grid to the new. Should be one of linear, quadratic, or cubic piecewise interpolation.

n_p: The number of points to place in each of the coarse regions (i.e. each uniform region other than those deemed to cover the domain of solitons in the solution).

r: The ratio of grid spacing between a region and the adjacent coarser region.

As described earlier, suitable values of h_{goal} and β can be automatically determined using the L_2 energy. This applies to problems where the initial condition is similar to the expected behaviour of the solution at any later time. For other problems, however, the values that would be determined will not be of use and so the user must resort to other techniques for choosing them. For h_{goal}, the user should use the techniques that would have been used for a uniform mesh. This is most likely to be achieved as a result of common practice. β can also be determined in this way.

The other user parameters are best chosen interactively. Although this might appear to be a drawback of the method proposed, in practice it is not difficult to choose these values. This is primarily because the method exhibits low sensitivity to the actual values used.

We can estimate the number of points that will be used to discretize the coarse regions of the spatial dimension. This will give us an indication of how the execution time will vary as a function of the values of r and n_p. For a non-soliton interval of length $2L$, the points will be placed symmetrically about the midpoint of the interval. We therefore estimate the number that would be placed in either half. Assuming that m intervals are placed, fitting exactly into the space required, we have

$$L = \sum_{i=1}^{m} r^i h_{goal} n_p \tag{8}$$

$$= h_{goal} n_p \sum_{i=1}^{m} r^i$$
$$= h_{goal} n_p \left(\frac{r^{m+1} - r}{r - 1} \right)$$

Eq. 8 gives us a relationship between the size of an interval, L, the number of piecewise uniform coarse intervals, m, needed to discretize it, and the parameters r, n_p, and h_{goal}. We can solve this equation for m to get

$$m = \frac{\log \left(r + \frac{(r-1)L}{n_p h_{goal}} \right)}{\log r} - 1 \tag{9}$$

The total number of points, N, in an interval of size $2L$ is $2n_p m$.

Note that Eq. 9 only gives an approximation to the number of points as it doesn't take into account the discrete nature of the intervals and the requirement that we impose that each interval have at least n_p points. It also doesn't include the points placed in the soliton intervals. Nevertheless, it gives an indication of the behaviour as n_p and r change, as illustrated in Figure 3. The number of points needed falls rapidly with increasing values of r and grows slowly with the value of n_p.

The behaviour is reflected in actual numerical experiments. Figures 4 and 5 show, respectively, the cpu time required (in seconds) and the error, e_∞, using the ∞ norm, as a function of the two parameters n_p and r. Table 1 shows a selection of the data that was used to generate these figures. The first entry in the table corresponds to the uniform mesh example, which provides a base case for comparison.

Figure 4 exhibits the same behaviour as predicted theoretically. From the table, we can see that the overhead of the grid refinement method is minimal. Each of the experiments with $r = 1$ generates grids, using the *shoulder* method, which are identical to a uniform mesh over the whole solution domain. The difference in execution times (up to 15% more for the nonuniform case) is due to the soliton location procedure and, more importantly, to interpolation from the old grid to the new. This overhead is acceptable and is unlikely to discourage the use of the adaptive method even for problems that reduce effectively to the use of a uniform mesh.

We can conclude that the ratio r and the number of points in each coarse interval, n_p, do not affect the execution time dramatically. We do see a steady but slow increase as n_p increases, as we expect: the number of points in each grid increases. Due to the type of problem solved, we do not observe any dramatic effects due to the value of r. However, past experience has indicated that this ratio can cause problems and should be as small as possible. The cpu times would indicate that a value greater than or equal to 1.5 would be reasonable.

Figure 5 shows the growth of the error as the ratio r is increased. The sensitivity of the error to the value of n_p is low, affecting only those runs with very large ratios. The combination of this with the effects on the cpu time lead us to choose small values

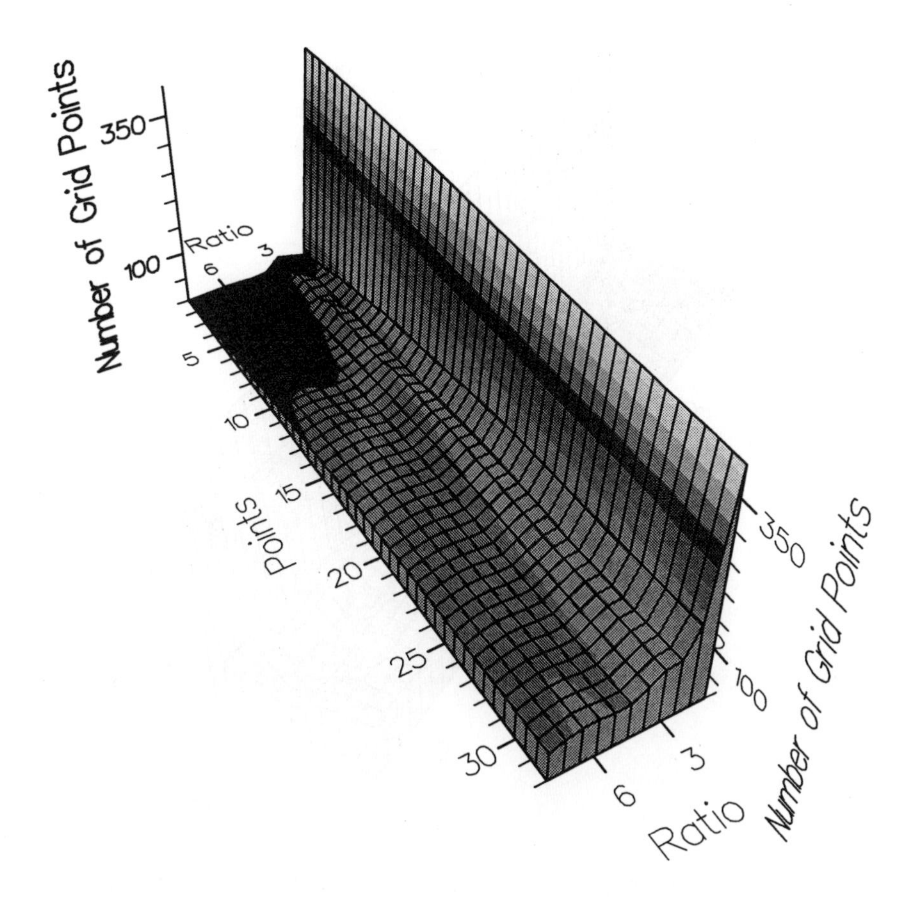

Figure 3: Number of points needed to discretize an interval for different values of r (indicated by “Ratio”) and n_p (axis labelled “Points”).

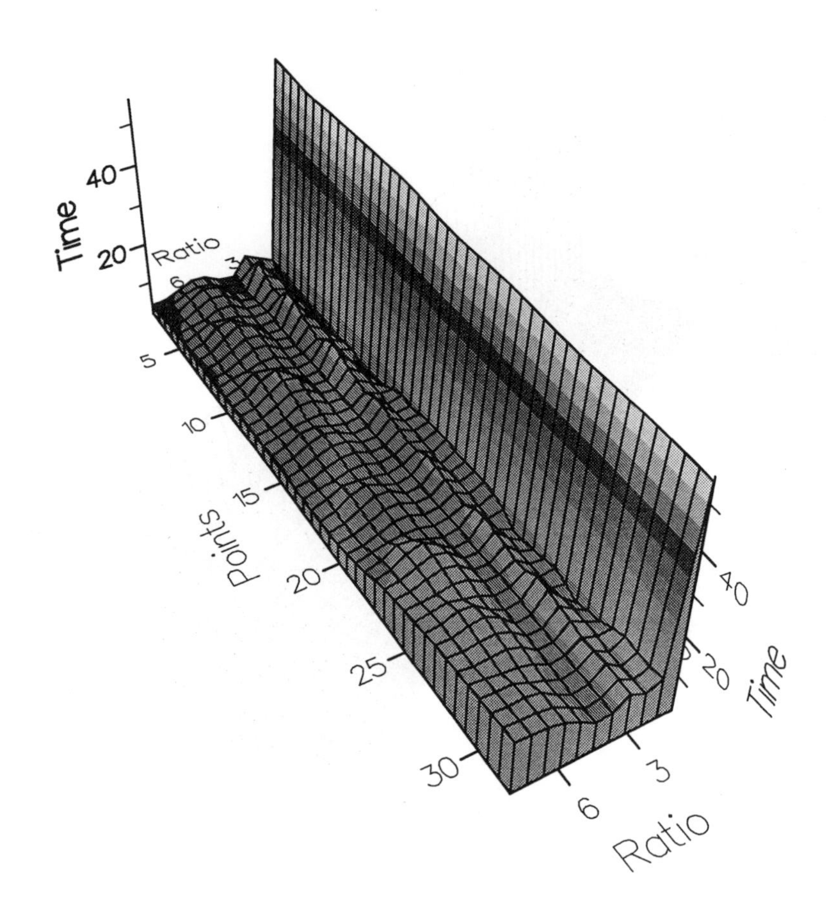

Figure 4: CPU time, in seconds, versus r and n_p for the single soliton initial condition for the KdV equation.

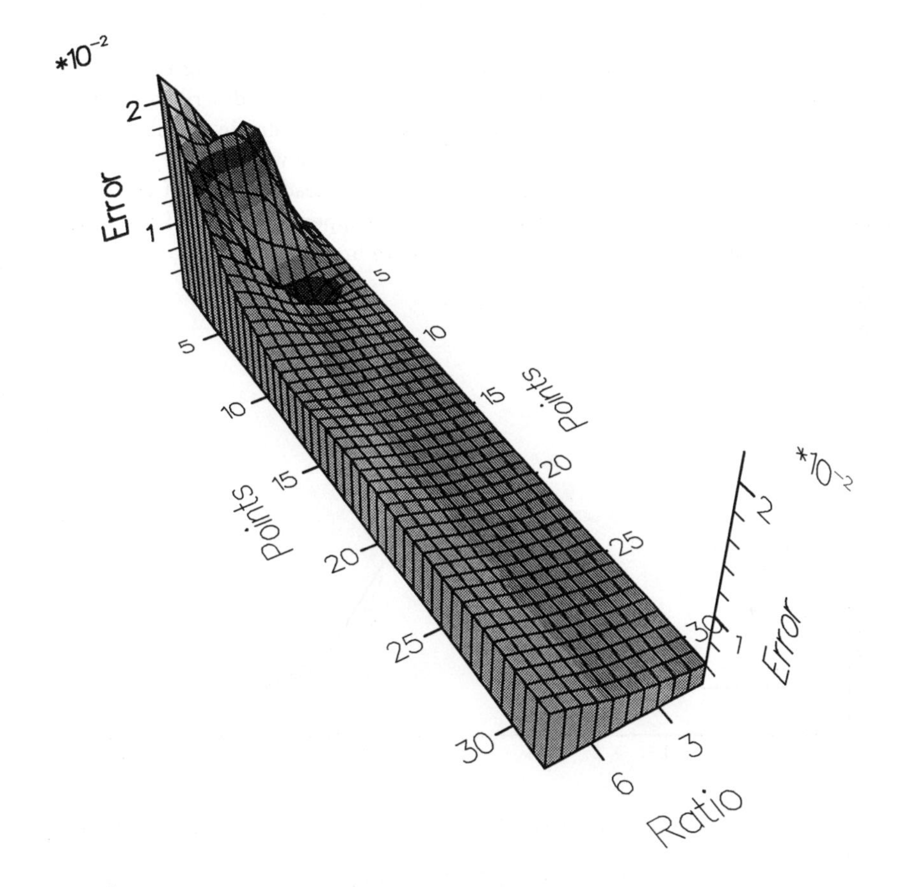

Figure 5: Error versus r and n_p for the single soliton initial condition for the KdV equation.

of r and values of n_p which are large enough to counteract any errors introduced by r while still keeping the execution times low. A lower bound on n_p is provided by the numerical approximation itself: it is desirable to keep n_p larger than the size of the difference molecule used by the approximation (eg. for the KdV equation, the difference molecule has a width of 5). In practice, we have found that a value of 8 yields good results.

The last step of the regridding algorithm involves interpolating solution values from the old to the new grid. The interpolation method used, described by the *inter* variable, is a user parameter and can be based on linear, quadratic, or cubic piecewise. In most cases, either quadratic or cubic perform equally well. Linear interpolation is not suitable as it introduces dissipative effects and leads to lack of conservation. Figure 6 shows an exaggerated view of how this happens: the top of the soliton is chopped off and the leading and trailing edges are broadened.

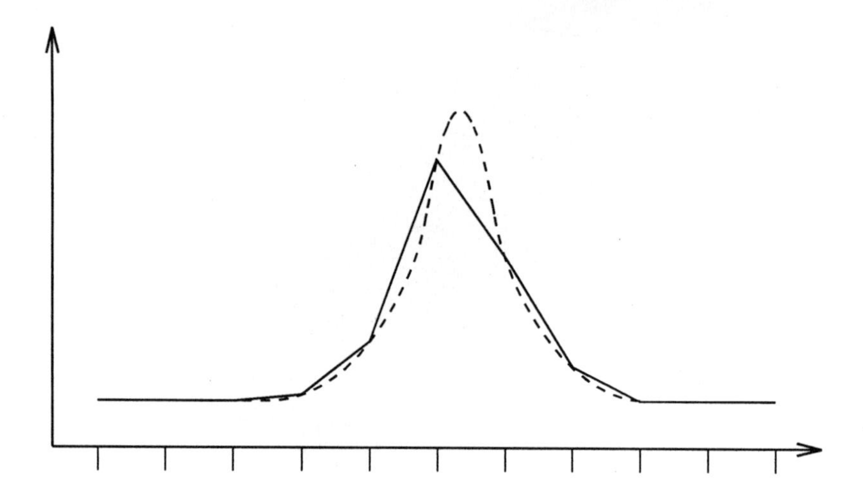

Figure 6: Illustration of dissipative effects due to the use of piecewise linear interpolation.

3.1. Summary

A new adaptive gridding algorithm, the *shoulder* method, has been introduced. By keeping changes in mesh spacing small, problems with numerical instability should be avoided. The use of contiguous uniform sub-meshes enables the use of simple numerical schemes, while still observing the conservation of energy laws. The use of uniform subgrids should also minimize the problems with the convergence and low

Table 1: Error, CPU time, and final L_2 energy for the single soliton KdV problem for a selection of different values of r and n_p.

r	n_p	e_∞	Time (s)	L_2 energy
Uniform		0.0061	48.0	0.867568E-01
1	2	0.0061	55.7	0.867563E-01
1	4	0.0061	54.8	0.867565E-01
1	8	0.0061	55.2	0.867563E-01
1	16	0.0060	54.8	0.867579E-01
1	32	0.0061	54.7	0.867569E-01
1.1	2	0.0061	7.0	0.867580E-01
1.1	4	0.0062	9.4	0.867557E-01
1.1	8	0.0061	13.3	0.867559E-01
1.1	16	0.0061	18.0	0.867561E-01
1.1	32	0.0061	23.9	0.867568E-01
1.2	2	0.0062	5.6	0.867577E-01
1.2	4	0.0061	7.2	0.867579E-01
1.2	8	0.0062	9.7	0.867550E-01
1.2	16	0.0061	13.4	0.867569E-01
1.2	32	0.0061	18.2	0.867580E-01
1.5	2	0.0066	4.5	0.867583E-01
1.5	4	0.0063	5.3	0.867586E-01
1.5	8	0.0062	6.9	0.867561E-01
1.5	16	0.0062	9.1	0.867547E-01
1.5	32	0.0062	12.5	0.867576E-01
2	2	0.0076	4.0	0.867612E-01
2	4	0.0066	4.7	0.867568E-01
2	8	0.0063	5.7	0.867559E-01
2	16	0.0063	7.5	0.867575E-01
2	32	0.0063	10.4	0.867560E-01
4	2	0.0162	3.7	0.869534E-01
4	4	0.0078	4.2	0.867651E-01
4	8	0.0070	4.9	0.867573E-01
4	16	0.0068	6.1	0.867577E-01
4	32	0.0068	8.4	0.867560E-01
8	2	0.0197	3.5	0.870753E-01
8	4	0.0155	4.6	0.867656E-01
8	8	0.0100	4.8	0.867613E-01
8	16	0.0086	5.0	0.867611E-01
8	32	0.0089	14.7	0.867565E-01

order accuracy of the numerical methods, problems that normally appear on grids with irregular spacing. Finally, the new algorithm has been designed to be robust with respect to the sensitivity of results to the choice of user parameter values.

4. Numerical experiments

The numerical methods used to solve the KdV and NLS equations are the same as were used previously.[5] The KdV equation is discretized spatially using a five point difference formula which reduces to a standard approach when applied on a uniform mesh. The NLS is approximated using a finite element method. The system of ordinary differential equations, in both cases, is fully discretized using a Crank-Nicholson scheme, which results in a system of nonlinear equations. These equations are solved using iterative schemes. The set of equations for the KdV equation are solved with a Newton type method, using the previous time step solution values as the initial guess. The NLS equation is solved in a similar way except that the initial guess is generated by applying Euler's method. Using the previous time step solution values directly as an initial guess did not enable the Newton method to converge.

The problems considered are described below:

K-1 : The KdV equation with a single soliton initial condition using the theoretical solution

$$u(x,t) = 3c\text{sech}^2(Ax - Bt + d)$$

where $A = \frac{1}{2}\{bc/\epsilon\}$, and $B = \frac{1}{2}c\sqrt{bc/\epsilon}$. This problem was solved with

$$x \in [0,10]$$

$$t \in [0,25]$$

and

$$b = 1.0, c = 0.3, d = -12.0, \epsilon = 4.84 \times 10^{-4}.$$

The initial L_2 energy is 0.0867593.

K-2 : The KdV equation with a double soliton initial condition where the theoretical solution, due to Hirota,[8] is

$$u(x,t) = 12(\ln F)_{xx}$$

$$F = 1 + f_1 + f_2 + \left(\frac{\alpha_2 - \alpha_1}{\alpha_2 + \alpha_1}\right)^2 f_1 f_2$$

where

$$f_i = e^{[-\alpha_i(x-x_i)+\alpha_i^3 t]}, i = 1,2.$$

The parameters x_1, x_2, α_1, and α_2 represent the initial displacements and amplitudes of the two solitons. In the numerical experiments conducted, the following values were used:

$$x \in [-250, 1000]$$

$$t \in [0, 7000]$$

$$x_1 = -1.0, \alpha_1 = 0.2$$

$$x_2 = -150.0, \alpha_2 = 0.3$$

The initial energy is 0.84.

K-P : The KdV equation with a square pulse initial condition for which the theoretical solution is not known. The initial condition used was

$$u(x,0) = \begin{cases} h, & \text{if } -\frac{l}{2} \leq x \leq \frac{l}{2}, \\ 0, & \text{o.w.} \end{cases}$$

Mitchell[10] shows that the number of solitons, N, generated by a rectangular well (width l and depth h) initial condition is given by

$$N = \left\lfloor \frac{A}{\pi} + 1 \right\rfloor \tag{10}$$

$$A = h^{\frac{1}{2}} l$$

However, this formula was derived for the equation

$$\frac{\partial u}{\partial t} - 6u\frac{\partial u}{\partial x} + \frac{\partial^3 u}{\partial x^3} = 0 \tag{11}$$

which can be transformed to equation 1 (with $\epsilon = 1$) by replacing u with $-u/6$. With this transformation, a rectangular well of depth h becomes a square pulse with height $h/6$ so the number of solitons generated by equation (1) is given by Eq. 10 with

$$A = \left(\frac{h}{6}\right)^{\frac{1}{2}} l \tag{12}$$

where h and l are the height and width, respectively, of the square pulse.

For the runs mentioned in Table 2, $h = 0.9$ and $l = 20$ were used which, according to eqs. 12 and 10, should result in three solitons with an initial L_2 energy of 16.403. Figure 7 shows that this is the case. We see three solitons emerge and proceed to the left with a dispersive wave train moving to the right. Figure 8 shows the intervals of refinement as identified by the *shoulder* algorithm.

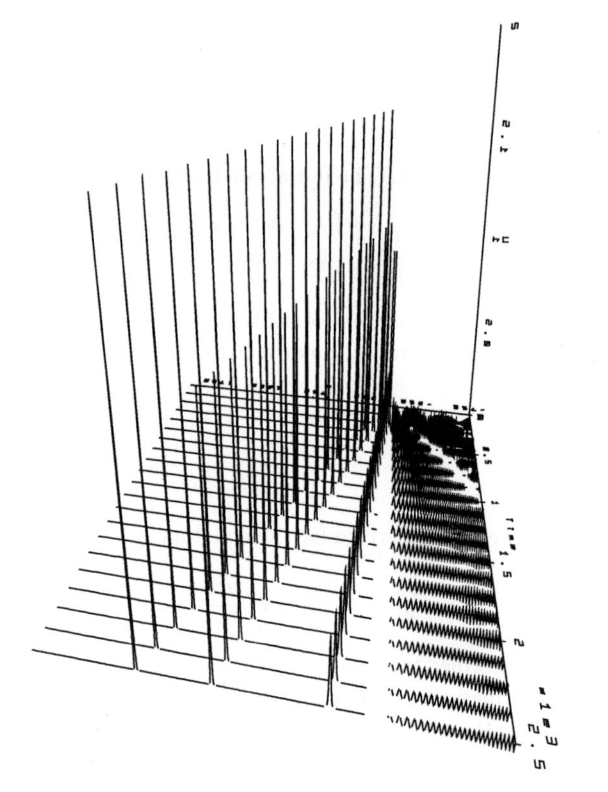

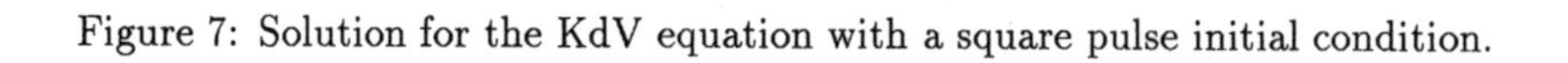

Figure 7: Solution for the KdV equation with a square pulse initial condition.

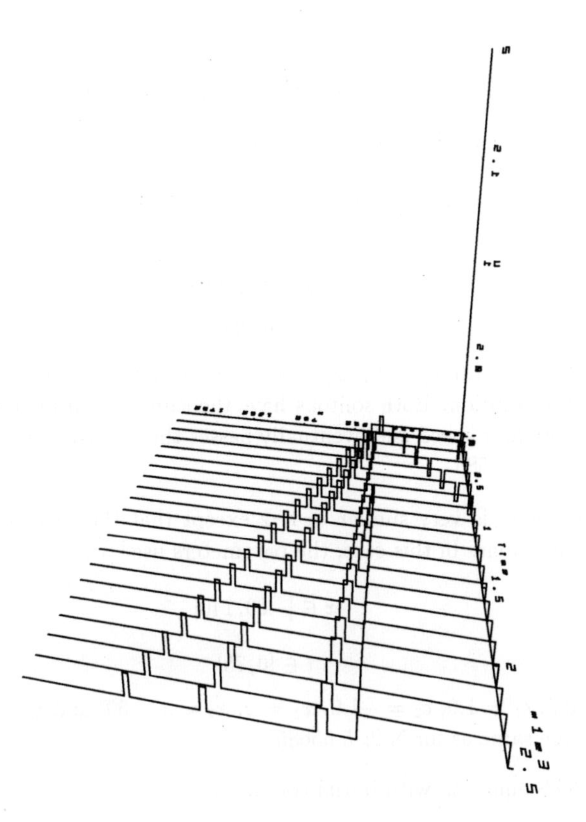

Figure 8: Intervals determined by the system as needing refinement for the KdV equation with a square pulse initial condition.

N-1 : The NLS equation with a single soliton initial condition:

$$u(x,0) = \sqrt{2\alpha}\text{sech}(\sqrt{\alpha}x)e^{0.5icx}$$

with

$$x \in [-20, 80]$$
$$t \in [0, 48.75]$$

$\alpha = 0.5$, $c = 1$, and $q = 1$. The L_2 energy is 2.82843 initially.

N-2 : The NLS equation with a double soliton initial condition:

$$u(x,0) = \sqrt{2\alpha}\left[\text{sech}(\sqrt{\alpha}x_1)e^{0.5ic_1x_1} + \text{sech}(\sqrt{\alpha}x_2)e^{0.5ic_2x_2}\right]$$

with

$$x \in [-20, 300]$$
$$t \in [0, 250]$$

and $\alpha = 0.5$, $c_1 = 1.0$, $c_2 = 0.2$, $x_1 = x$, $x_2 = x - 100$, and $q = 1$. This results in a two soliton solution. Both solitons have the same amplitude but the left one moves at a greater speed and eventually passes through the right one. Initially, the energy is 5.65685.

N-2x : This problem is very similar to `N-2` except that the two solitons move in opposite directions. In this case, the parameters used are

$$x \in [-60, 140]$$

$$t \in [0, 75]$$

and $\alpha = 0.5$, $c_1 = 1.0$, $c_2 = -1.0$, $x_1 = x$, $x_2 = x - 50$, and $q = 1$. The initial energy is the same as for N-2: 5.65685.

N-q18 : The NLS equation with initial condition

$$u(x,0) = \text{sech}(x)$$

and with $q = 18$. A value of $q = 2 * N^2, N = 1, \ldots$ generates bound states of N solitons. High values of q result in problems that are hard to solve, and thereby pose a good test problem for any adaptive scheme. Steep spatial and temporal gradients develop. The L_2 energy is initially 2.0.

N-P : The NLS equation with a square pulse initial condition

$$u(x,0) = \begin{cases} 0.52, & \text{if } -2.6 \le x \le 2.6 \\ 0, & \text{o.w.} \end{cases}$$

with $q = 1$. This should generate a single soliton after sufficiently long time has elapsed.[2] Although this is a good test for stability and accuracy, the *shoulder* method is not expected to do particularly well (in terms of computer time needed versus a uniform grid) as dispersive wave trains are sent off in both directions, requiring refinement over most of the spatial domain. The energy is 1.3858.

All results were generated on a Sun SPARCstation 5 workstation, with the programs written in FORTRAN-77, and are summarized in Table 2.

Table 2: Summary of results for all KdV and NLS problems.

Problem	h	τ	cpu Time (min)		L_2 Energy	
			Uniform	Shoulder	Uniform	Shoulder
K-1	0.0049	0.005	10.37	1.11	0.0868	0.0867
K-2	0.3052	0.5	32.90	4.50	0.8400	0.8398
K-P	0.25	0.125	198.84	80.67	16.4022	15.3114
N-1	0.195	0.125	0.68	0.22	2.8284	2.8261
N-2	0.195	0.125	10.38	2.12	5.6561	5.6402
N-2x	0.195	0.125	1.99	0.64	5.6567	5.6502
N-P	0.125	0.02	0.73	0.20	1.3858	1.3963
N-q18	0.025	0.025	17.32	4.70	2.0003	1.9988

The *shoulder* method runs were done with $n_p = 8$ and $r = 2$. For all the problems except those with square pulse initial conditions, the value of h_{goal} and, for the adaptive methods, the value of β were chosen automatically, using the procedure described by the authors [5]. Cubic interpolation was used for generating the solution on the new grids for the *shoulder* method. A cut-off value of 50 was used for all runs except for the KdV square pulse problem where a value of 1000 was used to capture the dispersive waves properly.

The new method appears to be successful in achieving the design goals described earlier. In particular, we see that both the accuracy and energy conservation targets are met. In particular, the new method does successfully tackle the KdV equation with a square pulse initial condition, which the *soliton* method was unable to do.

5. Conclusions

The *shoulder* method was developed with the goal of creating a mesh refinement technique that took the best features of previous methods and combined them successfully. The goal was also to reduce problems with stability and convergence that previous methods may have had. The result was the *shoulder* method with the following features:

- The mesh generated is finest in the most critical areas of the solution.

- The mesh generated consists of contiguous uniform submeshes, thereby minimizing problems with the convergence of simple numerical schemes, and also reducing the loss of energy due to the spatial approximation on a nonuniform grid, especially for the KdV equation.

- The ratio of successive mesh spacings is bounded which reduces the problems associated with stability of the numerical method and the interpolation methods used.

- The method has been shown to be insensitive to the values of most of the user parameters, making it easy to use for a wide range of problems.

The new method is still simple to implement but, with these new features, is much more robust and can be applied to more problems in general.

6. Acknowledgements

Both authors were supported in part by the Natural Sciences and Engineering Research Council, Grant Number A3597; the second author was also supported in part by a NATO grant, Number RG0358/82; both authors also wish to acknowledge the support of the University of Dundee in providing research facilities; and, the first author would also like to gratefully acknowledge the use of computer facilities in the Department of Chemical Engineering at the University of Edinburgh.

7. References

1. P. Daripa, *J. Comput. Phys.* **100** (1992) p. 284.
2. M. Delfour, M. Fortin, and G. Payre, *J. Comput. Phys.* **44** (1981) p. 277.
3. R. K. Dodd, J. C. Eilbeck, J. D. Gibbon, and H. C. Morris, *Solitons and Nonlinear Wave Equations* (Academic Press, London, 1982).
4. E. S. Fraga and J. Ll. Morris, *Department of Mathematics Technical Report* **NA/102** (University of Dundee, 1987).
5. E. S. Fraga and J. Ll. Morris, *J. Comput. Phys.*, **101** (1992) p. 94.
6. B. Garcia-Archilla and J. M. Sanz-Serna, *Math. Comp.* **57** (1991) p. 239.
7. D. F. Hawken, J. J. Gottlieb, and J. S. Hansen, *J. Comput. Phys.* **95** (1991) p. 254.
8. R. Hirota, *Phys. Rev. Lett.* **27** (1971) p. 1192.
9. V. S. Manoranjan, *Department of Mathematics Technical Report*, **NA/76** (University of Dundee, 1984).
10. A. R. Mitchell, in *Proceedings of the First International Conference on Mathematics in the Gulf Area*, ed. Y. Al-Khamees.
11. M. A. Revilla, *Int. J. Numer. Methods*, **23** (1986) p. 2263.

12. R. D. Russell and J. Christiansen, *SIAM J. Numer. Anal.* **15** (1980) p. 59.
13. J. M. Sanz-Serna and I. Christie, *J. Comput. Phys.* **67** (1986) p. 348.
14. J. F. Thompson, *Applied Numerical Mathematics* **1** (1985) p. 3.
15. G. B. Whitham, *Linear and Nonlinear Waves* (John Wiley & Sons, New York, 1974).

DIAGONAL DOMINANCE AND POSITIVE DEFINITENESS OF UPWIND APPROXIMATIONS FOR ADVECTION DIFFUSION PROBLEMS

GENE GOLUB
Department of Computer Science, Stanford University
GATES 2B MC 9025, Stanford, CA 94305-9025
E-mail: golub@sscm.stanford.edu

DAVID SILVESTER
Department of Mathematics, UMIST
Manchester, M60 1QD
E-mail: na.silvester@na-net.ornl.gov

and

ANDY WATHEN
Oxford University Computing Laboratory
Wolfson Building, Parks Road, Oxford, OX1 3JD
E-mail: wathen@comlab.ox.ac.uk

ABSTRACT

We examine whether three different upwind schemes for the positive definite advection diffusion problem can yield indefinite coefficient matrices. In particular, in line with the increasing use of adaptive meshes, we are concerned with discretisations on arbitrary grids but only in one dimension. We show that indefinite coefficient matrices can arise from certain approaches: this can present difficulties with efficient iterative solution techniques which might be required for corresponding approximations in higher dimensions.

1. Introduction

The numerical approximation of advection-diffusion operators continues to provide a significant challenge for Numerical Analysts. Many schemes have been proposed in the framework of finite differences , finite volumes and finite elements (see for example Morton[6]). A common theme is the use of upwinding: approximations which are biased to the direction of advection. A common goal is to achieve accurate (and possibly also monotone) solutions over a range of Peclet or mesh Peclet numbers. In this paper we examine another aspect of the problem: the structure of the discrete (matrix) equations arising from alternative approximation techniques.

The simplest model problem is in one dimension:

$$-u'' + \sigma u' = 0 \quad , \qquad x \in [0,1], \qquad u(0) = 1,\ u(1) = 0, \tag{1}$$

where σ is a positive constant. It is well known that for large σ the solution of this equation develops a boundary layer of thickness $O(1/\sigma)$.

The issue we address is of the qualitative approximation of the advection-diffusion operator by a discrete approximation. In particular we are concerned with whether or not the positive definiteness of the continuous differential operator is mirrored by various discretisation schemes. This is a significant issue when iterative solution methods are employed for the resulting linear systems, since certain methods are convergent only for systems which are positive definite:

$$x^T A x > 0 \qquad \text{for nonzero real vectors } x$$

(see for example Elman[1]). The related but different property of diagonal dominance:

$$|a_{i,i}| \geq \sum_j |a_{i,j}| \qquad \text{for each } i$$

with strict inequality in at least one row is a sufficient condition for the convergence of most simpler fixed point iterations such as the Jacobi and Gauss-Seidel iterations. The issue of reducibility which rarely arises in differential equation applications is also a general consideration—see Varga[12]. Tridiagonal matrices with non-zero sub- and super-diagonal entries are certainly irreducible and we will only deal with such matrices here without further mention of this issue. The close connection between diagonally dominant and positive definite matrices (through scaling) was established by Tartar[11]. In some situations the related M-matrix property is also useful (for example for preconditioning: see Meijerink and van der Vorst[5]).

In the one-dimensional situation many solution methods are applicable, but in particular in three-dimensional problems the applicability and rapid convergence of iterative methods is the central practical issue. In general a discretisation which is not positive definite in one-dimension will also not be in higher dimensions. We will thus consider only the model problem (1) for simplicity with the understanding that corresponding discretisations in higher dimensions will share similar qualitative features.

2. Symmetry and skew-symmetry

Multiplying the differential equation (1) by an appropriate test function v which vanishes at the end points of the domain and integrating employing integration by parts on the diffusion (second derivative) term yields

$$\langle u', v' \rangle + \langle \sigma u', v \rangle = 0$$

where $\langle \cdot, \cdot \rangle$ is the L_2 inner product. Now it is apparent that the first term is symmetric and positive definite whereas the second term is skew-symmetric at least if σ' is zero (the corresponding property in higher dimensions would be $\nabla . \sigma = 0$ which for example in incompressible fluid dynamics corresponds to conservation of mass).

Many approximations respect this structure: for example if the Galerkin method is employed using a conforming approximation space $V_h = \text{span}\{\phi_1, \phi_2, \ldots, \phi_n\}$ then discrete equations result of the form

$$A\mathbf{u} + C\mathbf{u} = \mathbf{f}$$

where $A = \{a_{i,j}\}, a_{i,j} = \langle \phi_j', \phi_i' \rangle$ is a symmetric and positive definite matrix, $C = \{c_{i,j}\}, c_{i,j} = \langle \sigma\phi_j', \phi_i \rangle$ is skew-symmetric and $\mathbf{f}$ arises from the inhomogeneous boundary term. Thus, for example, for a Galerkin spectral approximation or a Galerkin finite element approximation on any grid, the association of the symmetric part of the discretised matrix with the self-adjoint part of the continuous problem and correspondingly the skew-symmetric part with the skew-adjoint part of the differential operator holds true. With appropriate scaling, the elementary central finite difference approximation also shares this property—see below.

Much emphasis has however been put on preserving diagonal dominance of discretisations rather than ensuring positive definiteness of the symmetric part. In many situations, diagonal dominance corresponds to the existence of a discrete maximum principle (which may be useful in the suppression of oscillations in the discrete solution—but see Gresho and Lee[2]). In all cases diagonal dominance implies that all of the eigenvalues lie in the right half plane by simple application of the Gershgorin theorem, but it does not follow that the symmetric part and thus the matrix itself is positive definite. A simple example is illustrative:

Using the simplest finite differences on a grid of variable spacing

$$\ldots, h = x_j - x_{j-1}, k = x_{j+1} - x_j, \ell = x_{j+2} - x_{j+1}, \ldots$$

the second derivative term $-u''$ might be replaced by

$$-\left(\frac{u_{j+1} - u_j}{k} - \frac{u_j - u_{j-1}}{h}\right) \Big/ \left(\frac{1}{2}(h+k)\right)$$

and the first derivative could either be approximated by a central difference

$$\sigma\frac{(u_j + u_{j+1})/2 - (u_j + u_{j-1})/2}{\frac{1}{2}(h+k)} = \sigma\frac{u_{j+1} - u_{j-1}}{h+k}$$

or an upwind difference

$$\sigma\frac{u_j - u_{j-1}}{h}.$$

Thus, after division by the common factor $1/(h+k)$, the coefficient matrix is of the form

$$A = \text{tri}\left(-\sigma - \frac{2}{h}, \frac{2}{k} + \frac{2}{h}, \sigma - \frac{2}{k}\right)$$

in the central difference case. The condition for diagonal dominance (which is the same as the condition that A be an M-matrix) is the well-known mesh Peclet number condition $\sigma h/2 \leq 1$. (One would have to take the maximal h to satisfy dominance in every row of the matrix). However, the symmetric part of the matrix comes only from the second derivative and is diagonally dominant and hence positive definite because of the Gershgorin theorems (see Varga[12]).

Turning to the upwind difference, we obtain the matrix

$$A = \mathrm{tri}\,(-\sigma(h+k)/h - 2/h, 2/k + 2/h + \sigma(h+k)/h, -2/k)$$

which is diagonally dominant for every σ, h and k (i.e. for any mesh). But considering the symmetric part we have

$$(A+A^T)/2 = \mathrm{tri}\,(-\sigma(h+k)/2h - 2/h, 2/k + 2/h + \sigma(h+k)/h, -\sigma(k+\ell)/2k - 2/k)$$

which is diagonally dominant only in rows for which $\ell h \leq k^2$. Many practical meshes that might be employed for this simple problem could satisfy this condition: for example a smoothly graded mesh with $k = \alpha h, \ell = \alpha^2 h$ would yield a diagonally dominant matrix (with the strict inequalities holding in the first and last row). However it is clear that there are reasonable meshes (such as one with a regular mesh spacing in one part of the domain and a different regular spacing elsewhere) which do not satisfy this condition in every row. This still does not imply indefiniteness or otherwise. However a specific case shows that indefiniteness is possible: for the simple mesh $0, 0.5, 0.51, 0.52, 1$ with $\sigma = 10$ it is readily checked that

$$H := (A+A^T)/2 = \begin{pmatrix} 214.2 & -210 & 0 \\ -210 & 420 & -445 \\ 0 & -445 & 694.166^r \end{pmatrix}.$$

This matrix has determinant equal to -10579695 and thus has a negative eigenvalue. This is a coarse and not very suitable mesh for the given problem, however for meshes with many more points but similarly clustered around 0.5 we have similarly observed indefiniteness of the symmetric part. For related but more complicated problems an interior layer at 0.5 might be expected: a mesh of the given form might be reasonable in such a situation. Here it is the non-monotonic change in mesh size which is necessary to give indefiniteness: for a monotonically graded mesh this could not occur (see Morton[6], page 76). In higher dimensions and for example with mesh adaptivity such monotonicity may not be so easy to guarantee.

Of relevance to the matrix theory for this problem, we note that H certainly has positive diagonal entries and non-positive off-diagonal entries, thus by a theorem of Tartar[11] there will be a diagonal scaling matrix D for which DA is positive definite if and only if H^{-1} has positive entries. It is apparent by considering the inner product of the second row of H with the second column of H^{-1} that the second column of H^{-1} must have at least one negative entry, thus the difficulty here is not one which can be overcome by simple scaling.

Further, we comment that without the $h+k$ scaling it is a much simpler matter to demonstrate that indefinite matrices arise even with meshes which grade smoothly into the right hand boundary. The scaling that we have employed (which is unique in preserving symmetry of the approximation when $\sigma = 0$) is therefore apparently the most sensible.

The basic point is that the upwind approximation of the first order derivative contributes to the symmetric part of the coefficient matrix. This is well known on regular meshes where it strengthens the positive definiteness. The example given here shows that weakening is also possible to the extent that the discretisation of the underlying positive definite problem becomes indefinite.

3. Finite Element and Finite Volume methods

The simple example above demonstrates the issue we wish to highlight. In this section we consider two different and popular upwind strategies which are usually described in the frameworks of finite element and finite volume methods respectively.

There are a large number of finite element approaches including the use of upwind test functions in a so called Petrov-Galerkin setting and bubble functions in the context of Galerkin least squares (see Griffiths and Lorenz[3], Quarteroni and Valli[9]). However, we shall consider only one of the simpler and earlier upwind approaches namely that due to Heinrich et al[4]. In this approach standard piecewise linear finite element trial (expansion) functions ϕ_i are used together with test functions of the form $\phi_i + \psi_i$ where

$$\psi_i(x) = \begin{cases} 3\alpha_i(x - x_{i-1})(x_i - x)/(x_i - x_{i-1})^2 & x_{i-1} \leq x \leq x_i \\ -3\alpha_{i+1}(x - x_i)(x_{i+1} - x)/(x_{i+1} - x_i)^2 & x_i \leq x \leq x_{i+1} \end{cases}$$

and α_i is related to the local mesh Peclet number: here we use the popular choice $\alpha_i = \coth(\beta_i) - 1/\beta_i$ where $\beta_i = \sigma(x_i - x_{i-1})/2$ is the local mesh Peclet number (but see Simo, Armero and Taylor[10] for a more recent approach). The form of the (i,j) entry of the coefficient matrix is then

$$\int_0^1 \phi_j'(\phi_i' + \psi_i') + \sigma\phi_j'(\phi_i + \psi_i)\mathrm{dx}$$

which is integrated to give $A = K + \sigma C$ where K is the standard 'stiffness' matrix

$$K = \mathrm{tri}\,(-1/(x_i - x_{i-1}), 1/(x_i - x_{i-1}) + 1/(x_{i+1} - x_i), -1/(x_{i+1} - x_i))$$

and

$$C = \mathrm{tri}\,(-1/2 - \alpha_i/2, \alpha_i/2 + \alpha_{i+1}/2, 1/2 - \alpha_{i+1}/2)\,.$$

Thus here the first derivative (advection) term does give rise to a symmetric as well as a skew symmetric part: the symmetric part being due solely to the augmentation ψ_i of the trial function. However, since $\alpha_j \geq 0$ for any positive β_j, the symmetric part of C is diagonally dominant and it follows that the symmetric part of A has positive diagonal entries and negative sub- and super-diagonal entries and is diagonally dominant. Hence for any mesh, this discretisation yields a positive definite coefficient matrix.

The second method we consider is the four-point cell-vertex finite volume scheme due to Morton, Rudgyard and Shaw[8]. In the notation employed above this has a typical row of the form

$$\ldots, 0, \frac{-1}{\ell(k+\ell)}, \frac{1}{\ell(k+\ell)} - \frac{\ell}{k^2(k+\ell)} + \frac{h}{k^2(k+h)} + \frac{\sigma}{k},$$

$$\frac{1}{h(k+h)} - \frac{h}{k^2(k+h)} + \frac{\ell}{k^2(k+\ell)} - \frac{\sigma}{k}, \frac{-1}{h(k+h)}, 0, \ldots .$$

It is apparent that this method is cell-based and so there is little chance of diagonal dominance for the coefficient matrix in this linear system to determine the nodal unknowns. Indeed, it is not immediately apparent which are the diagonal entries. It is possible to consider the matrix as the sum of the two tridiagonal matrices (with different diagonals)

$$\text{tri}\left(\frac{-1}{\ell(k+\ell)}, \frac{1}{\ell(k+\ell)} - \frac{\ell}{k^2(k+\ell)} + \frac{\sigma}{k}, \frac{\ell}{k^2(k+\ell)}\right)$$

and

$$\text{tri}\left(\frac{h}{k^2(k+h)}, \frac{1}{h(k+h)} - \frac{h}{k^2(k+h)} - \frac{\sigma}{k}, \frac{-1}{h(k+h)}\right)$$

at least away from the boundary, however for large enough σ the second of these has a large negative diagonal and must be indefinite or even negative definite.

One approach to construct equations for nodal (rather than cell) residuals is through 'distribution matrices' and artificial viscosity: the analysis of such schemes is beyond the scope of this short note.

Though the analysis of coercivity and stability for cell-vertex finite volume methods is achieved without recourse to matrix theory (see Morton[7]), for advection-diffusion equations it remains a challenging problem to find iterative solution techniques for these methods (but see Wash[13]).

4. Conclusions

We have here only considered three of the great many upwind schemes for the advection diffusion equation. Our concern has been to show that some upwind schemes on certain meshes can give indefinite coefficient matrices even though the partial differential equations problem is positive definite.

One consequence is that certain iterative solution techniques can be expected not to perform well (or quite possibly fail) for discrete systems of equations derived from such schemes. This is a serious practical issue for large three-dimensional problems.

Dedication: We dedicate this paper to Professor Ron Mitchell who has been an inspiration to many generations of numerical analysts. In addition his generous spirit has infused our discipline with a sense of positive endeavour.

5. References

1. H. C. Elman, Iterative methods for linear systems, in *Advances in Numerical Analysis, Vol III: Large Scale Matrix Problems and the Numerical Solution of Partial Differential Equations*, J. Gilbert and D. Kershaw, eds., Cambridge University Press, 1994.
2. P. Gresho and R. L. Lee, Don't suppress the wiggles—they're telling you something, *Computers and Fluids*, **9**, 1981, 223–253.
3. D. F. Griffiths and J. Lorenz, An analysis of the Petrov-Galerkin finite element method, *Comput. Meths. Appl. Mech. Engrg.* **14**, 1978, 39–64.
4. J. C. Heinrich, P. S. Huyakorn, A. R. Mitchell and O. C. Zienkiewicz, An upwind finite element scheme for two-dimensional convective transport equations, *Int. J. Numer. Meths. Engrg.* **11**, 1977, 131–143.
5. J. A. Meijerink and H. A. van der Vorst, An iterative solution method for linear systems of which the coefficient matrix is a symmetric M-matrix, *Math. Comput.* **31**, 1977, 148–162.
6. K. W. Morton, *Numerical Solution of Convection-Diffusion Problems,* Chapman Hall, 1996.
7. K. W. Morton, Coercivity for one-dimensional cell vertex approximations, this volume.
8. K. W. Morton, M. A. Rudgyard and G. J. Shaw, Upwind iteration methods for the cell-vertex scheme in one-dimension, *J. Comput. Phys.* **114**(2), 1994, 209–226.
9. A. Quarteroni and A. Valli, *Numerical Approximation of Partial Differential Equations,* Springer-Verlag, 1994.
10. J. Simo, F. Armero and C. Taylor, Stable and time-dissipative finite element methods for the incompressible Navier-Stokes equations in advection dominated flows, *Int. J. Numer. Meths. Engrg.* **38**, 1995, 1475–1506.
11. L. Tartar, Une nouvelle caracterisation des M-matrices, *Revue Francaise d'Informatique et de Recherche operationnelle* **R-3**, 1971, 127–128.
12. R. S. Varga, *Matrix Iterative Analysis,* Prentice-Hall, 1962.
13. N. D. Wash, *Upwind Iteration Techniques for Compressible Flow Computations,* D. Phil. Thesis, Oxford University, 1995.

CHAOS IN NUMERICS

B. M. HERBST, G. J. Le ROUX
Department of Applied Mathematics, University of the Orange Free State, P.O. Box 339, Bloemfontein 9300, South Africa
E-mail: herbst@ibis.uovs.ac.za

and

M. J. ABLOWITZ
Program in Applied Mathematics, University of Colorado, Boulder, CO 80309, USA

ABSTRACT

Three numerical experiments changed the course of nonlinear science during the second half of this century. The soliton was rediscovered numerically by Zabusky and Kruskal [26] in their attempt to explain the results of the numerical experiment of Fermi, Pasta and Ulam [10]. Lorenz [19] discovered the strange attractor and was largely responsible for the modern interest in dynamical systems. One can only admire the skills and vision of these early investigators. It was very much in the spirit of these investigations that Ron Mitchell encouraged so many of us to investigate nonlinearity from a numerical point of view; he realized very clearly how much numerical analysis stood to be gained from its interaction with nonlinear phenomena. This paper is a tribute to Ron Mitchell. We wish to indicate, by means of two simple examples, how numerical studies of nonlinearity almost inevitably lead to unexpected results, and how we may understand these results in terms of nonlinear dynamical systems theory.

1. Chaos

Let us start with a very simple example and apply Newton's method to the function,

$$f(x) = 1 + x^2. \tag{1}$$

Undeterred by the fact that it does not have any real zeros (what about infinity?), we go ahead and apply Newton's method in any case to obtain,

$$x_{n+1} = g(x_n) \tag{2}$$

with $g(x) = \frac{1}{2}(x - 1/x)$, x real.

Since $f(x)$ does not have any real zeros, it means that $g(x)$ does not have any real fixed points, $g(p) = p$. However, $g(x)$ does have *periodic* points, i.e. points for which one can find an integer, N, such that $x_{n+N} = x_n$ for all integer n. The simplest one is the 2-periodic orbit, i.e. $g(g(x)) = x$ and it is easy to find the two different values of x required by an orbit of minimal period two, $x^2 = 1/3$. Let us calculate this orbit numerically by specifying $x_0 = 1/\sqrt{3}$. The result is shown in Figure 1.

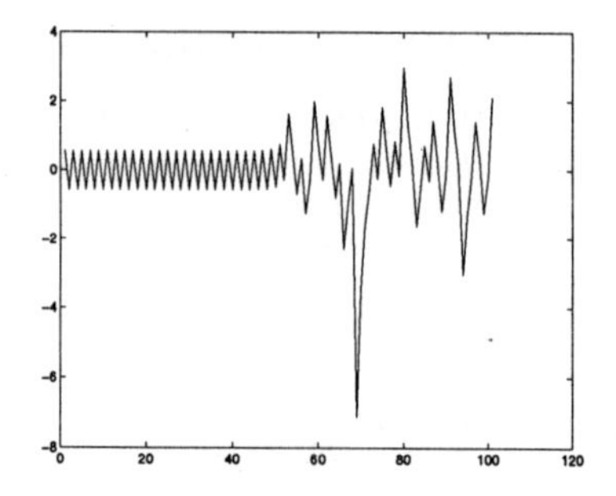

Figure 1: All accuracy is lost after a while.

Everything goes well for about 50 iterations, then, rather surprisingly, the periodicity is lost and subsequent iterations have a distinct random character.

Apart from the periodic orbits another set of orbits is clearly of interest—we may wish to avoid those points that will end up at infinity. These orbits consist of zero and all its pre-images, and there are lots of them. The question is how to identify them. The answer is to transform the problem into one whose dynamics are transparent.

1.1. Topological conjugacies

Given a sequence generated by

$$x_{n+1} = g(x_n),$$

the basic idea is to construct a new sequence, generated by

$$y_{n+1} = G(y_n)$$

which, ideally, is much simpler than the original. In order for this procedure to have any value, one has to be able to go back and forth between the two sequences which means that they have to be related by a continuous invertible map, $y_n = \phi(x_n)$. It follows that

$$y_{n+1} = \phi(x_{n+1}) = \phi \circ g(x_n) = \phi \circ g \circ \phi^{-1}(y_n)$$

and the relationship between the two maps defines a *topological conjugacy*

$$G = \phi \circ g \circ \phi^{-1}. \tag{3}$$

In practical terms it means that the two maps are completely equivalent with respect to their dynamics. For example, if p is a fixed point of g, then $\phi(p)$ is a fixed point of G. Similarly, periodic orbits of g are mapped to periodic orbits of G, etc.

Let us now return to the Newton iteration (2). Randy Bullock showed* that a good choice for the homeomorphism, $\phi : \mathbb{R} \to I$, with $I := (0,1)$, is

$$\phi(x) = \tfrac{1}{2} + \frac{1}{\pi}\tan^{-1} x.$$

As explained above, let $G : I \to I$ be given by $G = \phi \circ g \circ \phi^{-1}$. After some algebra where various trigonometric identities are used, one arrives at the so-called baker map [5],

$$G(y) = \begin{cases} 2y & 0 < y < \frac{1}{2} \\ 2y - 1 & \frac{1}{2} < y < 1. \end{cases} \tag{4}$$

Before we study this map in a little more detail, note that the baker map is equivalent to a continuous map on the unit circle, $\mathbf{S}^1$. Let $\theta = 2\pi y$ be an angular coordinate on $\mathbf{S}^1$, then the equivalent circle map is given by,

$$\theta_{n+1} = 2\theta_n. \tag{5}$$

The circle map (5) is suggestive—simple multiplication by 2 or, in binary notation, a shift of the decimal point. This leads us straight to *symbolic dynamics*.

1.2. Symbolic dynamics

Returning to the baker map, (4) we express y in binary notation. For instance, we may choose an initial value of the form,

$$y = 0.10011011\cdots$$

Iterating the baker map now amounts to shifting the decimal point one position to the right and discarding everything to the left. Thus, for the example above, we find,

$$G(y) = 0.0011011\cdots, \quad G(G(y)) = 0.011011\cdots, \text{ etc.}$$

Note that there is a simple interpretation of the binary sequence in terms of the original Newton iteration (2): the original iteration is reduced to a shift on a *symbolic* sequence where only the signs of the entries in the original sequence $\{x_n\}_{n=0}^{\infty}$ are kept. For instance for $x_j > 0$ a 1 might be entered, or a zero if $x_j < 0$. However, there is no reason why one should use 0, 1; $\pm$ for instance, work equally well, hence the terminology, *symbolic dynamics*. For simplicity we stay with binary sequences.

The space of symbolic sequences is easily made into a metric space [5]. The distance between two sequences $s = (s_0 s_1 \cdots)$ and $t = (t_0 t_1 \cdots)$ is defined by,

*While studying this problem as part of a homework assignment!

$$d(s,t) = \sum_{i=0}^{\infty} \frac{|s_i - t_i|}{2^i}. \tag{6}$$

Since $|s_i - t_i|$ is either 0 or 1, this series is dominated by the geometric series,

$$\sum_{i=0}^{\infty} \frac{1}{2^i}$$

and therefore converges.

Everything is now in place to answer the questions raised in the beginning. We start with periodic orbits.

1.3. Periodic orbits

One can ask the following questions:

- Which initial conditions will lead to orbits of given period n?
- How many orbits of minimal period n do we have?

The first question is easily answered using the circle map. An orbit of period n is obtained using an initial condition, θ_0, satisfying,

$$2^n \theta_0 = \theta_0 + 2\pi m,$$

for any integer m. Choosing $\theta_0 = 2\pi p/q$, (p, q integer), it follows that,

$$\frac{p}{q} = \frac{m}{M}, \text{ with } M = 2^n - 1.$$

Thus orbits of period n are generated from initial conditions given by,

$$\frac{p}{q} = \frac{n}{M}, \quad m = 1, \ldots, M-1. \tag{7}$$

However, not all orbits generated by (7) are different. For instance, choosing $n = 4$, the following orbits are obtained (written modulo $2^4 - 1$),

$$\begin{array}{rrrrr} 1 & 2 & 4 & 8 & 1 \\ 3 & 6 & 12 & 9 & 3 \\ 5 & 10 & 5 & 10 & 5 \\ 7 & 14 & 13 & 11 & 7 \end{array}$$

Thus we find 3 different orbits of minimal period 4 and one orbit of period 2.

This brings us to the second of the two questions raised above. Define $Q(d)$ to be the number of different orbits of minimal period d, i.e. $Q(n)$ is the number we are looking for. Since there are 2^n orbits (not all different) of period n—the number of different arrangements of a binary sequence of n entries—we have,

$$2^n = \sum_{d|n} Q(d)d,$$

where the sum, $d|n$, is over all the divisors of n. Our next task is to invert this formula to find $Q(n)$. Fortunately this is well-known from number theory and is given in terms of the Möbius function, $\mu(n)$, defined by (see, for example, Hardy and Wright [12]),

(i) $\mu(1) = 1$,

(ii) $\mu(n) = 0$, if n has a squared factor,

(iii) $\mu(p_1 p_2 \dots p_k) = (-1)^k$ if all the primes $p_1, p_2, \dots, p_k$ are different.

Since (see, for example, Hardy and Wright [12]),

$$g(n) = \sum_{d|n} f(d),$$

is inverted as,

$$f(n) = \sum_{d|n} \mu(d) g\left(\frac{n}{d}\right),$$

it follows immediately that

$$Q(n) = \frac{1}{n} \sum_{d|n} \mu(d) 2^{n/d}. \tag{8}$$

For example, if $n = 4$ it follows that

$$Q(4) = \frac{1}{4}[2^4 - 2^2] = 3,$$

as it should, according to the example above.

In summary, starting from rational initial conditions, the baker map generates,

(a) Periodic orbits

(b) Eventually periodic orbits

(c) Fixed points, if the initial conditions are chosen as, $m/2^n$, $m = 1, \dots, 2^n - 1$, any (integer) n.

A few remarks are in order.

- The fixed points correspond to unbounded solutions of the Newton iteration. Thus, all orbits that start from initial conditions that have a finite binary representation in the baker map, will eventually become unbounded in the Newton iteration. These orbits form a dense set of measure zero in the space of all orbits.

- The periodic orbits also form a dense set of measure zero.

- The set formed by the union of all the sets mentioned above is of measure zero. These orbits are all unstable.

The last remark implies that periodic, eventually periodic and unbounded orbits are rather untypical behavior. The typical behavior turns out to be chaotic.

1.4. Chaotic solutions

Suppose we restrict the number of consecutive 1's to two and also suppose that we do not allow any consecutive 0's. Is it possible to form an interesting orbit in this way? Consider the following,

$$0.0101010110101101010101101010101010101101010101\ldots$$

It should be obvious that it is impossible to predict the next occurrence of two consecutive 1's. In fact, it is quite impossible to predict whether the next entry should be 1 or 0; either one is admissible and corresponds to a different orbit. It means that two orbits starting arbitrarily close (made precise by means of the metric, (6)), may eventually diverge completely. This sensitivity is an essential ingredient of chaotic behavior, and easily verified using a symbolic representation. In fact, it is possible to define a map as being chaotic if the iterates in the space of all bounded orbits is topologically equivalent to a shift map on the space of all symbolic sequences. In practice this definition is often relaxed and one often finds that orbits are called chaotic if it remains bounded and is *sufficiently unpredictable.* Normally this means that one cannot predict future behavior from observing the system for any (finite) amount of time.

The results of the numerical experiment shown in Figure 1 should now be clear: Since real numbers are represented by a finite number of bits, all accuracy is lost as soon as the iteration runs out of bits.

2. Hamiltonian systems.

Let us now address the question how chaos may occur in differential equations. Instead of addressing this huge subject, let us restrict ourselves to a very special and important class of differential equations, namely Hamiltonian systems. Even among this class of problems we shall only consider the following special situation. Let $\mathbb{R}^2$ be an two dimensional oriented Euclidean space with coordinates (p, q) and let $H(p, q)$ be a function defined on a (open, connected) domain, $\Omega \subset \mathbb{R}^2$. Then a Hamiltonian system with one degrees of freedom can be defined by (see, for instance, Arnold [3] for a discussion of the general situation),

$$\frac{dp}{dt} = -\frac{\partial H}{\partial q}, \quad \frac{dq}{dt} = \frac{\partial H}{\partial p} \tag{9}$$

The *flow*, ϕ_t, generated by the Hamiltonian system (9) is a transformation of Ω into itself, such that

$$(p(t), q(t)) = \phi_t(p^0, q^0), \tag{10}$$

is the solution of (9) at time t with initial conditions $(p(0), q(0)) = (p^0, q^0)$, i.e. ϕ_0 is the identity transformation. If (p^0, q^0) in (10) is kept fixed and t is varied, then the solution of the Hamiltonian system is recovered. On the other hand one can fix t and vary (p^0, q^0) in which case ϕ_t defines a map of Ω into itself, assuming of course, that the solutions using initial values (p^0, q^0) exist at time t. Since this is not necessarily the case, the domain of definition of ϕ_t for given t may be smaller than Ω.

2.1. Area preservation and integrability.

It follows directly from (9) that the Hamiltonian vector field is divergence free,

$$\frac{\partial}{\partial p}\left(-\frac{\partial H}{\partial q}\right) + \frac{\partial}{\partial q}\left(\frac{\partial H}{\partial p}\right) = 0.$$

An immediate consequence is Liouville's theorem which states that for each fixed t the flow ϕ_t is an *area-preserving* transformation in Ω in the sense that for each bounded $\Sigma \subset \Omega$ for which $\phi_t(\Sigma)$ is defined, Σ and $\phi_t(\Sigma)$ have the same oriented areas. In fact this is the fundamental property of planar Hamiltonian systems—it holds *only* for Hamiltonian systems, see, for example, Arnold [3].

It is quite straightforward to verify that a map is area preserving in practice. Let $(P, Q) = \phi(p, q)$ define a smooth transformation in a domain Ω. According to the standard rules for changing variables in an integral, this transformation preserves area if and only if its Jacobian is identically 1, i.e.

$$\frac{\partial P}{\partial p}\frac{\partial Q}{\partial q} - \frac{\partial P}{\partial q}\frac{\partial Q}{\partial p} = 1, \quad \forall (p, q) \in \Omega. \tag{11}$$

Thus, suppose we are dealing with a one parameter family of transformations (the parameter may, for instance, denote time), such that (11) holds. This defines an area preserving or Hamiltonian flow.

Autonomous planar Hamiltonian systems have another remarkable property—they are *integrable*. It follows easily from (9) that

$$\frac{dH}{dt} = 0,$$

i.e. $H(p(t), q(t)) = H(p^0, q^0)$ along the solution curves. Away from fixed points where

$$\frac{\partial H}{\partial p} = 0 \text{ and } \frac{\partial H}{\partial q} = 0, \tag{12}$$

the implicit function theorem ensures that we can either solve for $p = p(q; p^0, q^0)$ or $q = q(p; p^0, q^0)$. This can be used to eliminate one of the variables in the planar system, reducing it to a single quadrature.

2.2. Consequences of area-preservation.

Area preservation has a marked impact on the long time dynamics of (planar) Hamiltonian systems. For instance, asymptotically stable equilibrium (fixed) points or limit cycles are not allowed—in their vicinity area has to shrink. This has an important consequence.

Recall the Poincaré–Bendixon theorem for general planar dynamical systems: An orbit which remains bounded for all time ($t > 0$) and does not have a fixed point, approaches a closed orbit (limit cycle). We have just excluded limit cycles for planar Hamiltonian systems, therefore, the only possibility for bounded planar Hamiltonian orbits not containing any fixed points, is periodicity.

Another important consequence of the Poincaré–Bendixon theorem is that planar dynamical systems do not exhibit chaos. In order to find chaos we need to go to higher dimensions.

3. Homoclinic tangles and chaos.

Let us now assume that our Hamiltonian system (9) has a homoclinic orbit to the fixed point, $\bar{X} = (\bar{p}, \bar{q})$, i.e. assume that there is an orbit $(p(t), q(t))$ connecting $\bar{X}$ with itself, $\lim_{t\to\pm\infty}(p(t), q(t)) = (\bar{p}, \bar{q})$. Another way of putting it is to say that the stable and unstable manifolds emanating from the fixed point are identical. However, a moment's reflection will convince the reader that a homoclinic orbit is a very unstable structure in the sense that almost any small perturbation will break the homoclinic connection and separate the stable and unstable manifolds.

In order to ensure that the system remains Hamiltonian, we add such a perturbation to the Hamiltonian itself, $H = H_0 + \epsilon H_1$. It is clearly not sufficient to add a autonomous perturbation. If H_1 does not depend on time the resulting system is still integrable—one may observe quantitative but not qualitative changes. In any case, as we observed before, the Poincaré-Bendixon theorem excludes the possibility of chaos for planar systems. Therefore, the more interesting perturbations are those that introduce another dimension. Accordingly, consider a perturbed system with Hamiltonian,

$$H(p,q,t) = H_0(p,q) + \epsilon H_1(p,q,t)$$

where $|\epsilon| << 1$ and H_1 is assumed to be periodic in t, $H(p,q,t) = H(p,q,t+T)$. The periodicity allows us to reduce the three dimensional Hamiltonian flow to a two dimensional area preserving map, P, defined by $P : (p(t_0), q(t_0)) \to (p(t_0+T), q(t_0+T))$, i.e. P maps the solution at time t_0 to the solution at one period later, $t_0 + T$. Note that if we change t_0 over one period then the map reproduces a particular orbit of the Hamiltonian flow, given as a projection of (p,q,t) space onto (p,q).

Recall that the homoclinic connection is (almost always) broken under arbitrary perturbations. This can happen in basically two different ways, the stable and unstable manifolds either separate completely without any intersections, as shown in Figure 2(a) or they are broken in such a way that they intersect in a point, X_0 at time t_0 say, as shown in Figure 2(b). Since the figures show the time T map, P, defined above and the intersection is actually a result of the projection of 3-dimensional extended phase space onto 2-dimensional space, the intersection does *not* indicate a loss of uniqueness. Let us now map X_0 to $X_{n+1} = PX_0$. Since X_0 lies on the (perturbed) stable and unstable manifolds, X_{n+1} is the solution on the unstable manifold at the time $t_0 + T$ as well as the solution on the stable manifold, but at time $t_o - T$. Therefore, X_{n+1} is another intersection point of the stable and unstable manifolds, and so is $P^i X_0$ for all integer values i, as shown in Figure 2(b). Note that orientation preservation maps skip an intersection point going from X_0 to PX_0.

Figure 2: (a) Separation but no intersection. (b) Homoclinic intersections

This is the generic situation—nonintersecting stable and unstable manifolds are not compatible with area preserving maps.

Finally we note that the homoclinic tangles as shown in Figure 2(b) lead to the formation of horseshoes, which can be described in terms of symbolic dynamics and ultimately, to chaos (see, for instance, Devaney [5]). Thus, conservative perturbations of planar Hamiltonian systems generically lead to chaos in the vicinity of homoclinic orbits.

What has all of this to do with the numerical solution of Hamiltonian problems? Everything! A numerical discretization of a differential equation is a map taking the solution from one time step to the next. This map can be viewed as a perturbations of the continuous system, with the perturbation intimately connected with the truncation error. Moreover, in the case of planar Hamiltonian systems we shall take care that the numerical map is area preserving, i.e. *symplectic*. This implies (see for example, Meyer and Hall [21], p117), that the numerical map is a period map of a periodical ly perturbed Hamiltonian system, of the form considered above. Thus, a symplectic (area preserving) discretization of a planar Hamiltonian system, is generically chaotic, for all values of the discretization parameter. One can indeed think of this as a severe unconditional nonlinear instability—almost every symplectic discretization of an integrable Hamiltonian system, leads to a chaotic numerical solution. It will be our task in the rest of this study to show that this instability is not as bad as it may sound.

4. Measuring the splitting distance

In the previous section we pointed out that nonintegrability and chaos is characterized by a splitting of the stable and unstable manifolds. It is therefore of considerable interest to be able to estimate the magnitude of the splitting distance, hence the deviation from integrability. Only for specific cases—the standard map for instance—are rigorous results available, see Gelfreich *et. al* [11] and Lazutkin [15,16,17]. However, if one is prepared to sacrifice rigor, it is easy to estimate the splitting dist ance, using Mel'nikov's method.

A second, not entirely unrelated approach is provided by the Mackay-Meiss-Percival (MMP) *action principle*[20]. Unlike the Mel'nikov function which provides a first order estimate of the splitting distance between the stable and unstable manifolds, the MMP principle gives an exact expression for the lobe area which we shall use to calculate the lobe area numerically.

4.1. The Mel'nikov function

Consider a planar symplectic mapping given by

$$Q_\epsilon : \boldsymbol{x} \to \boldsymbol{F}(\boldsymbol{x}) + \epsilon \boldsymbol{G}(\boldsymbol{x}, \epsilon),\ \boldsymbol{x} \in \mathbb{R}^2, \tag{13}$$

and $\boldsymbol{F}, \boldsymbol{G} : \mathbb{R}^2 \to \mathbb{R}^2$. Assume that the unperturbed map ($\epsilon = 0$) admits a constant of motion, $H(\boldsymbol{x}_n) = \text{const}$, for all n where $\boldsymbol{x}_{n+1} = \boldsymbol{F}(\boldsymbol{x}_n)$, i.e.

$$H(\boldsymbol{F}(\boldsymbol{x}_n)) = H(\boldsymbol{x}_n). \tag{14}$$

Also assume that the homoclinic orbit of the unperturbed map is obtained from $H(\hat{\boldsymbol{x}}_n(\xi)) = 0$, where ξ is a phase factor indicating the position of the initial condition on the homoclinic orbit. For this map the splitting distance between the stable and unstable manifolds at the phase point, ξ, is given to first order in ϵ by the Mel'nikov function (see, for example, Easton [9]),

$$M(\xi;\epsilon) = \sum_{n=-\infty}^{\infty} \boldsymbol{G}(\hat{\boldsymbol{x}}_{n-1}(\xi),\epsilon) \wedge \hat{\boldsymbol{v}}_n(\xi) \tag{15}$$

where $\hat{\boldsymbol{v}}_n(\xi)$ denotes the n-th iterate of the tangent vector to the unperturbed homoclinic orbit from the phase point ξ. Writing the vectors in component form as $\boldsymbol{u} := (u^{(1)}, u^{(2)})^T$, the wedge product is defined by $\boldsymbol{u} \wedge \boldsymbol{v} = u^{(1)}v^{(2)} - u^{(2)}v^{(1)}$. Note that $\hat{\boldsymbol{x}}_n(\xi)$ is evaluated on the homoclinic orbit of the unperturbed problem.

It will be useful to give another interpretation of the Mel'nikov function. The change in the constant of motion $H(\boldsymbol{x}_n)$ under the perturbed system (13) is calculated from,

$$\begin{aligned} H(\boldsymbol{x}_{n+1}) &= H(\boldsymbol{F}(\boldsymbol{x}_n) + \epsilon \boldsymbol{G}(\boldsymbol{x}_n)) \\ &= H(\boldsymbol{F}(\boldsymbol{x}_n)) + \epsilon DH(\boldsymbol{F}(\boldsymbol{x}_n)) \cdot \boldsymbol{G}(\boldsymbol{x}_n) + O(\epsilon^2) \\ &= H(\boldsymbol{x}_n) + \epsilon DH(\boldsymbol{F}(\boldsymbol{x}_n)) \cdot \boldsymbol{G}(\boldsymbol{x}_n) + O(\epsilon^2), \end{aligned} \tag{16}$$

where we have made use of (14) and DH denotes the gradient of H. Rewriting (16) as

$$\begin{aligned} \Delta H_n &= \epsilon DH(\boldsymbol{F}(\boldsymbol{x}_n)) \cdot \boldsymbol{G}(\boldsymbol{x}_n) + O(\epsilon^2) \\ &= \epsilon \boldsymbol{G}(\hat{\boldsymbol{x}}_n(\xi),\epsilon) \wedge \hat{\boldsymbol{v}}_{n+1}(\xi) + O(\epsilon^2) \end{aligned} \tag{17}$$

(recall that $\boldsymbol{x}_{n+1} = \boldsymbol{F}(\boldsymbol{x}_n)$ on the homoclinic orbit), it follows that the Mel'nikov function is a first order estimate of total change in the constant of motion, $H(\hat{\boldsymbol{x}}_n)$, over the homoclinic orbit.

Let us for the moment assume that the Mel'nikov function given by (15) can be written as

$$M(\xi,\tau) = \sum_{n=-\infty}^{\infty} m(n\tau - \xi), \tag{18}$$

where τ is related to the discretization parameter. Since we sum over all values of n, it follows that M is periodic of period τ, $M(\xi+\tau,\tau) = M(\xi,\tau)$. Since $m(s)$ is evaluated on the unperturbed homoclinic orbit, it decays exponentially fast as $s \to \pm\infty$. The Mel'nikov sum therefore converges uniformly for small but finite τ.

The Fourier coefficients of M are given by

$$\hat{M}_n(\tau) = \frac{1}{\tau}\int_0^\tau M(\xi,\tau)\exp(-i\mu_n\xi)d\xi, \tag{19}$$

where $\mu_n = 2\pi n/\tau$. Inserting expression (18) for $M(\xi,\tau)$ into (19), and interchanging the summation and integration, we find

$$\hat{M}_n(\tau) = \int_{-\infty}^{\infty} m(s)\exp(-i\mu_n s)ds. \tag{20}$$

If the perturbation and homoclinic orbit are analytic in a strip around the real axis, $m(s)$ is analytic in a strip surrounding the real axis. This allows one to conclude (see, for instance, Murray [22]) that the leading order behavior of the integral is determined by the distance of the nearest singularities of $m(s)$ to the real axis. More specifically, if the nearest singularity is located at a distance $i\rho$ from the real axis, then

$$\hat{M}_n(\tau) \propto \exp(-2\pi\rho n/\tau). \tag{21}$$

Thus we have established that all modes with $n \neq 0$ decay exponentially fast in τ. In fact, we know that the stable and unstable manifolds either intersect or are identical in the case of planar Hamiltonian systems. Therefore, the average splitting distance, $\hat{M}_0(\tau)$, decays at least as fast as the higher order coefficients, proving the exponential decay for the $n = 0$ mode. A simple geometric argument shows that $\hat{M}_0$ is, in fact, zero and an analytical proof will be given towards the end of the next section.

4.2. The Mackay-Meiss-Percival (MMP) action principle

Instead of estimating the splitting distance one can also measure the *lobe area*, i.e. the area enclosed by pieces of the stable and unstable manifolds between two consecutive homoclinic intersection points, see Figure 3. A particularly elegant derivation involving differential forms, is given by Easton [8].

Let $f : \mathbb{R}^2 \to \mathbb{R}^2$ be a planar mapping with constant Jacobian $J = 1$, written as

$$Q = Q(q,p), \quad P = P(q,p).$$

The first step is to define the so-called action function, F, obtained from integrating the expression,

$$p\,dq - f^* p\,dq = dF \tag{22}$$

where f^* denotes the *pull-back* of f. A specific example will be given in section 7.3.2.

The next step is to determine two neighboring intersection points, say a_0 and b_0, cf. Figure 3. The lobe area, i.e. the area of the region between the stable and unstable manifolds between a_0 and b_0 is then given by

$$\int_D dp \wedge dq = \sum_{j=-\infty}^{\infty} [F(b_j) - F(a_j)], \tag{23}$$

where $a_{j+1} = f(a_j)$ and $b_{j+1} = f(b_j)$, $j \in \mathbb{Z}$

An immediate consequence of (23) is that all of the areas enclosed between the stable and unstable manifolds are equal, as one would expect from an area preserving map. In addition, let us consider two neighboring lobes,

$$D_1 = \sum_{-\infty}^{\infty} [F(b_j) - F(a_j)]$$

and

$$D_2 = \sum_{-\infty}^{\infty} [F(a_{j+1}) - F(b_j)],$$

showing that

$$D_1 = -D_2.$$

This shows that the average splitting distance is zero, as pointed out in the previous section.

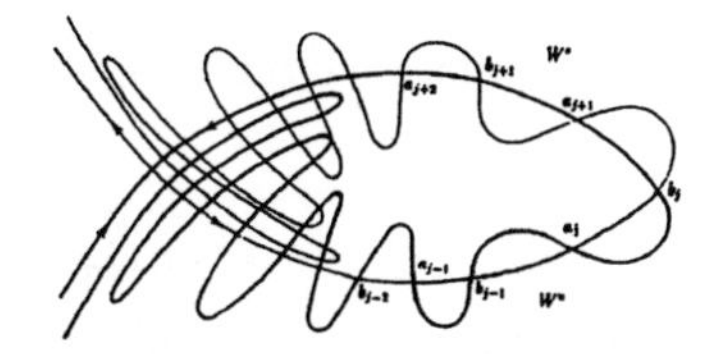

Figure 3: Equality of areas of different lobes.

5. Examples

5.1. The pendulum equation.

In order to illustrate some of the ideas discussed above, let us consider the well-known the pendulum equation,

$$q'' + \sin q = 0. \tag{24}$$

Recall that (24) can also be written as the Hamiltonian system,

$$\begin{aligned} q' &= p \\ p' &= -\sin q, \end{aligned} \tag{25}$$

with the Hamiltonian function given by

$$H(q,p) = \tfrac{1}{2}p^2 - \cos q. \tag{26}$$

If we identify all the points $(p, \pm\pi)$, it is straightforward to show that there is a homoclinic orbit to $(p,q) = (0,\pi)$, given by

$$\tfrac{1}{2}p^2 - \cos q = 1,$$

or,

$$q(t) = \pi + 4\arctan[\exp(t+\gamma)]. \tag{27}$$

Note that the nearest singularities in the complex plane are located at

$$t_0 = \pm\tfrac{1}{2}i\pi. \tag{28}$$

5.2. Discretizations

The first discretization to be considered is the straightforward second order scheme given by,

$$Q_{n+1} - 2Q_n + Q_{n-1} + k^2 \sin Q_n = 0, \tag{29}$$

which can be rewritten as a first order system of the form

$$\begin{aligned} Q_{n+1} &= Q_n + kP_n \\ P_{n+1} &= P_n - k\sin Q_{n+1}. \end{aligned} \tag{30}$$

It is important to note that this scheme is indeed area preserving, i.e. it is symplectic. Thus, from our discussion in the previous sections we expect to see some unusual, nonintegrable behavior, in the numerical solution. The solution obtained from this discretization for $k = 0.2$ is shown in Figure 4(a). Apparently all is well; no abnormal behavior is observed. However, if one takes a close up look as in Figure 4(b) (note the change in scale from Figure 4(a)), the situation changes dramatically. Now the familiar KAM features—resonant islands, chaos and invariant curves—become visible.

In order to better understand the reasons for this, we turn to a different discretization,

$$\begin{aligned} Q_{n+1} &= Q_n + kP_n \\ P_{n+1} &= P_n + \frac{2}{k} i \ln\left[\frac{1+\frac{1}{3}k^2\exp(iQ_{n+1})}{1+\frac{1}{3}k^2\exp(-iQ_{n+1})}\right]. \end{aligned} \tag{31}$$

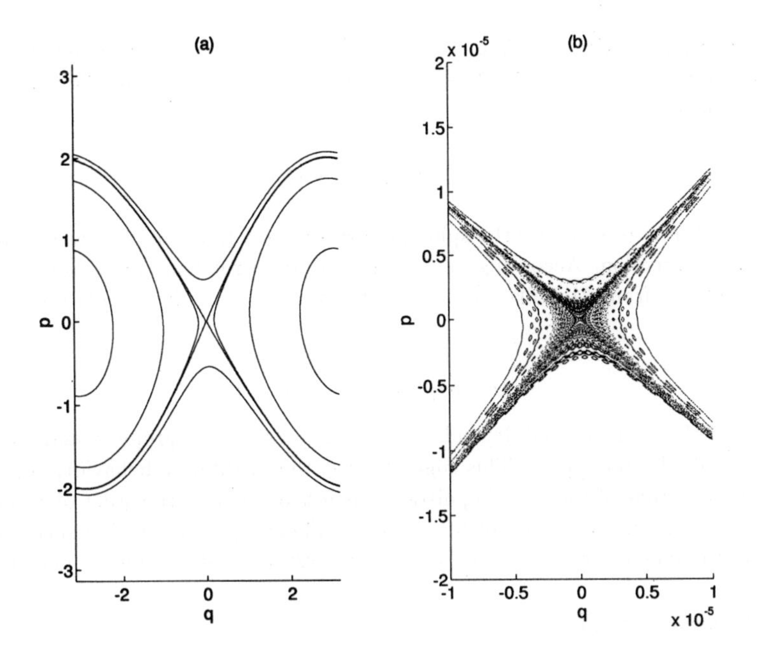

Figure 4: The solution of the pendulum equation. (b) is a close up view of (a)

Again it is not difficult to check that this scheme is area preserving. But even more importantly, it is *integrable* with a constant of motion given by (see Bobenko *et al.*[4]),

$$H := \tfrac{1}{2}(\cos Q_n + \cos Q_{n-1}) + (k^2/16)\cos(Q_n + Q_{n-1}) + \frac{1}{k^2}[\cos(Q_n - Q_{n-1}) - 1]. \quad (32)$$

It is also not hard to find an expression for the homoclinic associated with (31) (see Ablowitz and Herbst [1]),

$$Q_n = \pi + 4\arctan(\exp(pnk + \gamma)), \quad (33)$$

where

$$\cosh pk = \frac{1 + \frac{1}{4}k^2}{1 - \frac{1}{4}k^2}. \quad (34)$$

A comparison of (27) and (34) shows that the discrete homoclinic orbit is exactly the same as the continuous one, apart from a phase error.

One can now write the symplectic discretization (30) as a perturbation of the integrable discretization (31). A standard Taylor expansion shows that the perturbation is of the form,

$$\frac{1}{8}k^3\sin(2Q_{n+1}) + O(k^5)$$

This enables us to estimate the splitting distance as a function of k, as explained in the previous section. According to (33) the nearest singularities are located at $\pm i\pi/2$ and it follows from Mel'nikov's method that the splitting distance is given to first order by,

$$D \propto \exp(-\pi^2/k), \text{ as } k \to 0. \tag{35}$$

Note that the estimate (35) depends on the area preservation property and not on the order of the discretization. This suggests that the qualitative behavior, determined by the magnitude of the splitting distance, is independent of the order of accuracy of the discretization. We do not expect to see significant qualitative differences between symplectic schemes of different orders. Let us test some of these ideas numerically.

5.3. Numerical Results

In this section we compare a number of symplectic discretizations of Duffing's equation numerically. More precisely, we compare a first, fourth and eighth order symplectic discretization, referred to as SI1, SI4 and SI8, respectively. We are particularly interested to learn how well the integrability of the pendulum equation is preserved in practice by the different schemes. In order to quantify the deviation from integrability of the different schemes, two quantities are measured numerically namely, the splitting distance and the lobe area, as discussed in sections 5.1 and 5.2.

5.3.1. Measuring the splitting distance.

In the first series of experiments the distance to the first invariant curve is measured, as one moves away from the fixed point at π (homoclinic orbit)[†] In practice it is not hard to identify the invariant curve and the distance is measured as a function of k. The measured distance is then compared with the estimate obtained from the Mel'nikov function.

The relationship between the measured quantity and the Mel'nikov function requires a little explanation. Let us identify a particular invariant curve by its energy (e.g. the value of its Hamiltonian function). It follows from (32) that the homoclinic orbit is given by $H_h = -1 + k^2/16$, and a particular invariant curve by, $H_i = \cos Q_i + (k^2/16)\cos 2Q_i$. The latter is obtained from the intersection of the

[†]This corresponds to what is usually referred to in dynamical systems as the *last* invariant curve.

invariant curve with the line $P = 0$ (or equivalently, $Q_{i+1} = Q_i$). Assuming Q_i to be close to the homoclinic orbit, it follows that $Q_i = \pi - D_i$ where D_i is the distance that we measure numerically. Assuming D_i to be small, it is given by,

$$D_i \propto \sqrt{\Delta H}, \tag{36}$$

where $\Delta H := H_h - H_i$, is the corresponding change in the Hamiltonian.

Recalling from section 5.1 that the Mel'nikov function estimates the change in the Hamiltonian function under the perturbation, it follows from (35) that,

$$D_i \propto \exp(-\pi^2/2k). \tag{37}$$

Since this estimate suggests behavior of the form

$$D \propto \exp(-\alpha/k),$$

we measure α from the following numerical experiment:

- Measure the distance between the first invariant curves for k ranging between $k = 0.34$ and $k = 0.4$. Doing this for first–, fourth– and eighth–order SI's, the plots of $\ln D$ against $-1/k$ are shown in Figure 5.

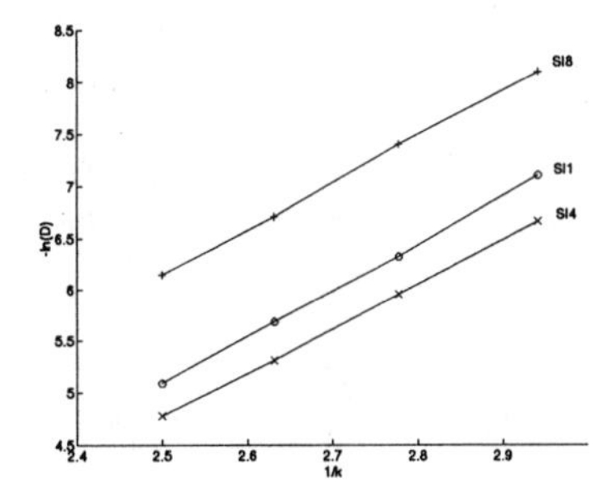

Figure 5: Exponential decay of the chaotic region as k is decreased, for SI 1, SI 4 and SI 8.

The graphs verify the expected exponential decay of the chaotic region around the separatrix, as a function of k. Note that there is very little qualitative difference between the results obtained from symplectic integrators of different orders.

Table 6.1 give the values of α for our numerical experiments.

Table 6.1 Decay-rate obtained from varying k.

Integrator	α
SI 1	4.6
SI 4	4.3
SI 8	4.4

Table 6.1 shows that the splitting distances decay at more or less the same exponential rate for each SI. This is of course expected—the exponential smallness dominates the polynomial rates of convergence. Moreover, the values of the rate of decay agree with the theoretical value predicted by the Mel'nikov function, at least within the order of numerical accuracy.

5.3.2. Measuring the Lobe Area

In this second set of experiments the lobe area as given by (23) is measured instead of the distance between the first invariant curves. Let us return to the symplectic scheme (30). We shall need its inverse in a moment, given by,

$$\begin{aligned} Q_{n+1} &= Q_n - kP_{n+1} \\ P_{n+1} &= P_n + k\sin Q_n. \end{aligned} \tag{38}$$

A linearized analysis of (30) yields the directions of the stable and unstable manifolds at the fixed point. With the linearized system given by

$$\begin{pmatrix} \epsilon_{n+1} \\ \mu_{n+1} \end{pmatrix} = \begin{pmatrix} 1 & k \\ k & 1+k^2 \end{pmatrix} \begin{pmatrix} \epsilon_n \\ \mu_n \end{pmatrix}.$$

the eigenvalues become

$$\lambda_\pm = 1 + \frac{k^2}{2} \pm \sqrt{k^2 + \frac{1}{4}k^4},$$

and the eigenvectors are given by

$$\begin{pmatrix} \epsilon \\ \mu \end{pmatrix}_\pm = \begin{pmatrix} -k \\ 1-\lambda_\pm \end{pmatrix}.$$

Iterating a small section of the *linear* stable and unstable manifolds at the origin (in the positive direction using (30), and in the negative direction using its inverse (38)), a good approximation of the stable and unstable manifolds of the symplectic integrator is obtained, as shown in Figure 6.

In order to apply the MMP action principle, two consecutive intersection points, a_0 and b_0, are required as well as the action function. The intersection points are determined by brute force and for the mapping (30), we have

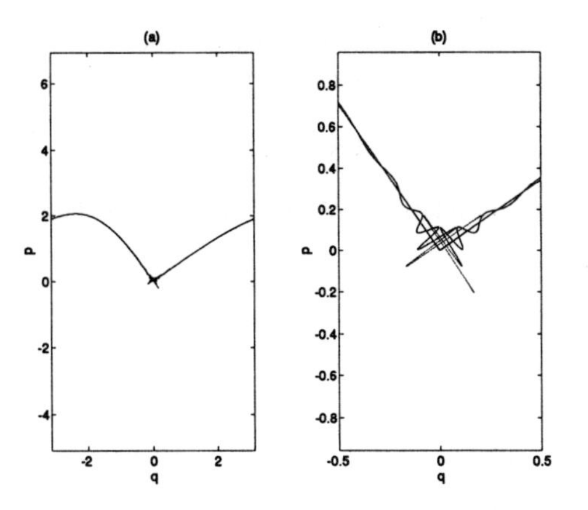

Figure 6: Homoclinic tangles in the numerical solution of Duffing's equation obtained from the first-order SI.

$$\begin{aligned} f^*p &= p - k\sin(q+kp) \\ f^*dq &= dq + kdp \end{aligned}$$

and it follows that

$$\begin{aligned} dF &= p\,dq - f^*p\,dq \\ &= k\sin(q+kp)dq - (kp - k^2\sin(q+kp))dp \end{aligned} \tag{39}$$

This can be integrated to obtain,

$$F = -\tfrac{1}{2}kp^2 - k\cos(q+kp). \tag{40}$$

Making use of the MMP action principle, the lobe area is now computed experimentally and the results are summarized in Table 6.2.

Table 6.2 Areas enclosed between the stable and unstable manifolds.

k	Area ($\times 10^{-4}$)
0.70	5.8
0.72	8.3
0.74	11.7
0.76	16.0

The Mel'nikov function (35) suggests that he lobe area, A (see Kaper *et al.*[14] for the relationship between the Mel'nikov function and the lobe area), is given by,

$$A \propto \exp(-\alpha/k),$$

where $\alpha = \pi^2$. The graph relating $\ln A$ and $-1/k$ for the data of Tables 6.2 is shown in Figure 7.

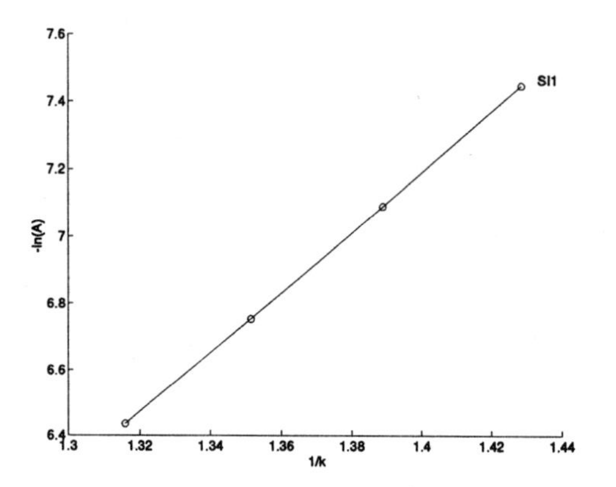

Figure 7: Exponential decay in the size of the area enclosed between the stable and unstable manifolds, as a function of k.

The average value of α is given by 8.95. Bearing in mind the relatively large values of k necessitated by the exponential smallness in k, these results again confirm the theoretical estimates.

For a more detailed discussion, also including higher dimensional systems the reader should consult le Roux [18].

6. Discussion.

Our discussion points towards an intimate relationship between numerics and dynamics. For instance, although it is easily seen that symplectic discretizations of planar Hamiltonian systems generically lead to chaotic solutions, a more detailed analysis showed that the chaos is confined to regions that are exponentially small in the discretization parameter. (Note that this result is also implied by the backward error analysis interpretation of modified Hamiltonian introduced by Yoshida[25], see also Sanz–Serna and Calvo [23]). Thus, the *integrability* of planar Hamiltonian systems is preserve with exponential accuracy in the discretization parameter by symplectic discretizations and this result does not depend on the order of the discretization—from this qualitative point of view all symplectic discretizations are the same.

It is natural to ask what happens with higher, including infinite dimensional systems. Although a detailed discussion will take us beyond the scope of this introduction, one might anticipate a much more complicated situation. Not only does one find very complicated homoclinic *structures* (see Ercolani *et al.* [6]) in the infinite dimensional Hamiltonian systems of soliton theory, there is also evidence that the performance of symplectic integrators is not clearly understood. For example, Sanz-Serna and Calvo [23] give numerical examples demonstrating that the performance of different symplectic schemes may be very different. *It is not only the symplectic property that is important.* Ablowitz *et al.* [2] find very little difference between the qualitative properties (precisely defined) of symplectic and nonsymplectic schemes as applied to spatial discretizations of the sine-Gordon equation. One can therefore ask, when are symplectic schemes good, and in what sense are they good?

These questions raised above cannot be answered without a detailed study of the dynamics of the discretizations. Numerical analysis has entered exciting new territory; Ron Mitchell realized this a long time ago.

Acknowledgements: It is a pleasure to thank Willem Fouche and Johan Meyer who introduced us to the Möbius function and Bob Easton, Jim Meiss and Harvey Segur for valuable discussions.

7. References

1. M.J. Ablowitz, B.M. Herbst and Constance Schober. On the Numerical Solution of the sine-Gordon equation. I. Integrable Discretizations and Homoclinic Manifolds. Submitted to J. Comput. Phys. (1995).
2. M.J. Ablowitz, B.M. Herbst and Constance Schober. On the Numerical Solution of the sine-Gordon equation. II. Numerical Schemes. Preprint (1995).
3. V.I. Arnold. *Mathematical Methods of Classical Mechanics.* Springer-Verlag, New York (1978).
4. A. Bobenko, N. Kutz and U. Pinkall. The discrete quantum pendulum. SFB 288 Preprint 42, Berlin (1992).
5. R.L. Devaney. *An Introduction to Chaotic Dynamical Systems* (second edition). Addison Wesley, Redwood City (1989).
6. N. Ercolani, M.G. Forest and D.W. McLaughlin. Geometry of the Modulational Instability Part III: Homoclinic Orbits for the Periodic Sine-Gordon Equation. Physica D, **43**, pp349-384 (1990).
7. A. Erdélyi. *Asymptotic Expansions.* Dover, New York (1956).
8. R.W. Easton. Transport through chaos. Nonlinearity, **4**, pp583-590 (1991).
9. R.W. Easton. Computing the Dependence on a Parameter of a Family of Unstable Manifolds: Generalized Mel'nikov Formulas. Nonlinear Analysis, Theory , Methods and Appl., **8**, pp1–4 (1984).

10. E. Fermi, J. Pasta and S. Ulam. Studies of nonlinear problems. I. Los Alamos Report LA1940 (1955). Reprinted in, Nonlinear wave motion. A. Newell (ed.). AMS, Rhode Island (1974).
11. V.G. Gelfreich, V.F. Lazutkin and M.B. Tabanov. Exponentially small splitting distances in Hamiltonian systems. Chaos, **1**, pp137–142 (1991).
12. G.H. Hardy and E.M. Wright. *An Introduction to the Theory of Numbers* (fifth edition). Clarendon Press, Oxford (1979).
13. B.M. Herbst, M.J. Ablowitz. Numerical chaos, symplectic integrators and exponentially small splitting distances. J. Comput. Phys., **105**, pp122–132 (1993).
14. T.J. Kaper, G. Kovacic and S. Wiggins. Melnikov functions, action and lobe area in Hamiltonian systems. To appear Dynamics and Differential equations (1990).
15. V.F. Lazutkin. Analytic integrals of the semistandard map, and splitting of separatrices. Leningrad Math. J., **1**, pp427–445 (1990).
16. V.F. Lazutkin. On the width of the instability zone near the separatrix of a standard map. Soviet Math. Dokl., **42**, pp5–9 (1991).
17. V.F. Lazutkin, I.G. Schachmannski and M.B. Tabanov. Splitting of separatrices for standard and semistandard mappings. Physica D, **40**, pp235–248 (1989).
18. G. le Roux. *Numerical Homoclinic Instabilities.* M.Sc. Thesis, Department of Applied Mathematics, University of the Orange Free State, Bloemfontein (1993).
19. E.N. Lorenz. Deterministic nonperiodic flow. J. Atmos. Science, **20**, pp130–141 (1963).
20. R.S. MacKay, J. Meiss and I. Percival. Transport in Hamiltonian systems. Physica D, **13**, pp55–81 (1984).
21. K.R. Meyer and G.R. Hall. *Introduction to Hamiltonian dynamical systems and the N–body problem.* Springer–Verlag, New York (1992).
22. J.D. Murray. *Asymptotic Analysis.* Springer–Verlag, New York (1984).
23. J.M. Sanz-Serna and M.P. Calvo. *Numerical Hamiltonian Problems.* Chapman and Hall, London (1994).
24. G. Strang. A chaotic search for i. Preprint. (1990).
25. H. Yoshida. Construction of higher order symplectic integrators. Phys. A., **150**, pp262-268 (1990).
26. N.J. Zabusky and M.D. Kruskal. Interaction of "solitons" in a collisionless plasma and the recurrence of initial states. Phys. Rev. Lett., **15**, pp240-243 (1965).

EXACT DIFFERENCE FORMULAS FOR LINEAR DIFFERENTIAL OPERATORS

DIRK P. LAURIE
Department of Mathematics, Potchefstroom University for C. H. E.
P. O. Box 1174, Vanderbijlpark, South Africa
E-mail: dirk@calvyn.puk.ac.za

and

ALAN CRAIG
Department of Mathematical Sciences, University of Durham
South Road, Durham DH1 3LE, U.K.
E-mail: alan.craig@durham.ac.uk

ABSTRACT

A difference approximation to a linear ordinary differential operator can in many cases be represented as an integral. If the integral is evaluated exactly, one obtains an exact method for obtaining a table of values of the solution to an ordinary differential equation. Many well-known methods for ODE's can be derived by approximating the integral by a quadrature formula. Such exact representations can in principle be obtained for any n-th order linear differential operator and any set of $n+1$ nodes, under mild technical conditions. Our construction involves a generalized B-spline that instead of being piecewise polynomial, is a piecewise solution of the homogeneous adjoint equation.

1. Introduction

Traditionally difference methods for a differential equation of the form

$$Ly(x) = f(x, y(x)), \tag{1}$$

where L is a linear n-th order differential operator, have been derived by approximating L by a linear difference operator L_h and solving the difference equation

$$L_h y_h(x) = f(x, y_h(x)) \tag{2}$$

for the approximate solution y_h. Much of the literature on the topic of numerical solution of differential equations is devoted to estimating the error $y - y_h$.

It has been noticed that better results are often obtained by solving

$$L_h y_h(x) = Qf(x, y_h(x)) \tag{3}$$

where Q is a suitable smoothing operator: e.g. in the well-known 'royal road' method for the case when $Ly = y''$, the order of the method is raised from 2 to 4 by taking $Qf(x) = \frac{1}{12}(f(x-h) + 10f(x) + f(x+h))$. The 'Mehrstellenverfahren' of Collatz[5] is a systematic method for obtaining such compact high order approximations.

The main point of this contribution is that it is often possible to represent the difference formula exactly in the form

$$L_h y = Q(Ly) \tag{4}$$

by appropriately matching L_h and Q. In other words, equation (4) is an *identity* satisfied by all sufficiently smooth functions, not only solutions to some differential equation. We show how to obtain L_h and Q when L and the set of discretization points is given, and also examine some cases where L_h has been obtained by some other method, and one has to work backwards to identify a suitable L and thence Q.

When (4) is available, its application to the solution of ordinary differential equations becomes straightforward. The differential equation gives an expression for Ly that can be substituted into (4) to obtain an algebraic equation that is exactly satisfied by the solution to the ODE. An approximate difference method can then be derived in which the only approximation required is the replacement of the smoothing operator by a quadrature formula. This point of view is developed for the case $Ly = y''$, by Butcher et al.[3]

The form (3) also arises when the discretization is obtained by a Galerkin method, although in such cases the order of approximation is not necessarily higher than that obtained from (2). The typical Galerkin approximation arises from a weak form of (1), namely

$$\langle Ly, v\rangle = \langle f, v\rangle, \tag{5}$$

where $\langle\cdot,\cdot\rangle$ is an inner product, and v ranges over a suitably chosen space V of test functions. The usual procedure (after possibly performing one or more integrations of (5) by parts, in order to reduce the smoothness required of y) is to discretize (5) by approximating y by a member of a finite-dimensional space U_h and considering only a finite-dimensional subspace V_h of V : when U_h is not the same as V_h the method is commonly called the Petrov-Galerkin method.[4]

In this paper we make the following points:

- It is possible to choose the test functions in the Petrov-Galerkin method in such a way that the evaluation of $\langle Ly, v_h\rangle$ reduces to forming a linear combination of values of y at discrete points.

- In the case of a linear differential equation, this choice leads to an exact difference equation

  $$L_h y(x) = \langle f, v_h\rangle \tag{6}$$

  satisfied by the solution y of the original equation (1). The solution to a system of such equations reproduces y exactly at the chosen points. When the correct test functions have been chosen, it does not matter what the trial functions are: we get precisely the same difference equation.

- When the equation to be solved involves some other differential operator than the one for which the exact difference equation holds, or the right-hand side is not independent of the solution, the approach is still useful in the derivation of approximate difference equations.

- When L_h has been determined by some other method, it is in principle possible to work back to an operator L with constant coefficients that would have given that L_h, and thus also to find the proper test function that corresponds to it.

The present paper is an exposition of the basic idea, and a sampling of various application possibilities. There are connections with Green's function methods[6] (indeed, for second order equations v reduces to Green's function), defect correction,[2] Peano's kernel theorem,[7], B-splines,[8] (for the differential operator $Ly = y^{(n)}$ the proper test function simply is a B-spline) and the theory of Ghizzetti and Ossicini for quadrature formulas.[9]

2. Choosing the test functions

We make the following assumptions:

1. $Ly = y^{(n)} + [\text{ lower order terms}]$.

2. v has compact support and is n times continuously differentiable, except for $n+1$ jump discontinuities in $v^{(n-1)}$ at points $x_0 < x_1 < \cdots < x_n$.

3. $\langle u, v\rangle = \int_{-\infty}^{\infty} u(x)v(x)\,\mathrm{d}x$. In practice the interval only runs over a finite interval because of the compact support of v.

Integrate the left-hand side of (5) n times by parts to obtain

$$\langle y, L^*v\rangle + (-1)^{n-1}y(x)v^{(n-1)}(x) + [\text{ lower order terms }] = \langle f, v\rangle, \tag{7}$$

where L^* is the formal adjoint differential operator to L. If we now choose v to satisfy $L^*v = 0$ inside each subinterval (x_j, x_{j+1}), the integral is zero and the lower order terms cancel, so that we are left with

$$\sum_{j=0}^{n}(-1)^n(v^{(n-1)}(x_j+) - v^{(n-1)}(x_j-))y(x_j) = \langle f, v\rangle. \tag{8}$$

To obtain the magnitude of the jumps, we substitute n linearly independent solutions y_i of $Ly = 0$ into (8) to obtain

$$\sum_{j=0}^{n} y_i(x_j)c_j = 0,\ i = 1, 2, \ldots, n, \tag{9}$$

where $c_j = v^{(n-1)}(x_j+) - v^{(n-1)}(x_j-)$. This is a homogeneous system of $n+1$ equations in n unknowns and therefore has a solution. In fact, if the system of solutions satisfies the Haar condition on an interval containing all the x_j, this solution is unique modulo a constant multiplier. Once the jumps are known, the construction of v on each subinterval is straightforward.

The function v can be thought of as a generalized B-spline. In fact, in the case where $Ly = y(n)$, it *is* simply a B-spline.

The above construction of v is possible whenever the system of homogeneous equations for the coefficients c_j can be solved, and for any function $f = Ly$ such that $\langle f, v\rangle$ makes sense. We therefore have the following theorem:

Let L be a linear n-th order differential operator such that there exists a system of n solutions to the homogeneous equation $Ly = 0$ which satisfies the Haar condition on some open interval I. For any $n+1$ real numbers $x_0 < x_1 < \cdots < x_n$ in I, there exist coefficients c_j, $j = 0, 1, \ldots, n$ and a function $v \in C^{n-2}(-\infty, \infty)$ which is supported on the interval (x_0, x_n), such that the identity

$$c_0 y(x_0) + c_1 y(x_1) + \cdots + c_n y(x_n) = \int_{x_0}^{x_n} v(x) Ly(x)\,\mathrm{d}x \tag{10}$$

holds for any function y such that Ly is piecewise integrable.

3. Applications

For the practical application of (10), there are two difficulties: the integration in (10) is likely to require numerical techniques, and the differential equation to be solved may not be of the simple form $Ly = f$, i.e. Ly might not be available directly. The impact of these difficulties is that one is forced to make approximations, which will then lead to a difference formula that looks very much like the standard difference formulas obtained by the usual methods. In fact, we shall see that the 'royal road' method is obtained in a straightforward way by using a quadrature formula in (10).

3.1. Second-order equations

In the constant coefficient case

$$Ly := y'' + 2\mu y' + (\mu^2 - \sigma^2) y, \tag{11}$$

two linearly independent solutions are

$$\begin{aligned} y_1(x) &= e^{-\mu x}\cosh\sigma x, && (12)\\ y_2(x) &= \tfrac{1}{\sigma} e^{-\mu x}\sinh\sigma x; && (13) \end{aligned}$$

which are to be replaced by their limiting values when $\sigma = 0$. Note that y_1 and y_2 remain real even when σ is imaginary.

When $y = y_i$, $i = 1, 2$, we have

$$[\ y(x-h_1)\quad y(x)\quad y(x+h_2)\ ]\begin{bmatrix} e^{-h_1\mu}\sinh(\sigma h_2) \\ -\sinh(\sigma(h_1+h_2)) \\ e^{h_2\mu}\sinh(\sigma h_1) \end{bmatrix} = 0, \tag{14}$$

where $h_j = x_j - x_{j-1}$. In the constant step case $h_1 = h_2 = h$ this equation reduces to

$$e^{-h\mu}y(x-h) - 2\cosh h\sigma y(x) + e^{h\mu}y(x+h) = 0. \tag{15}$$

The test function v corresponding to (14) is given by

$$v(x) = \begin{cases} \frac{1}{\sigma}e^{\mu(x-x_1)}\sinh\sigma h_2\, \sinh\sigma(x-x_0), & x_0 \le x \le x_1; \\ \frac{1}{\sigma}e^{\mu(x-x_1)}\sinh\sigma h_1\, \sinh\sigma(x-x_2), & x_1 \le x \le x_2. \end{cases} \tag{16}$$

The obvious thing to do is to apply an interpolatory quadrature formula

$$\int_{x_0}^{x_2} v(x)f(x)\,\mathrm{d}x = w_0 f(x_0) + w_1 f(x_1) + w_2 f(x_2) + E(f)$$

such that $E(f)$ vanishes when $f(x) = x^j$, $j = 0, 1, 2$. Rather than perform any integrations, it is much easier to solve the equations

$$c_0 y(x_0) + c_y(x_1) + c_2 y(x_2) = w_0 Ly(x_0) + w_1 Ly(x_1) + w_2 Ly(x_2) \tag{17}$$

with y replaced by $y_j(x)x^j$, $j = 0, 1, 2$ as Collatz[5] does. This does require $\mu^2 \neq \sigma^2$, since otherwise Ly_j does not span the quadratic polynomials.

In the case where $Ly = L^*y = y''$ and $h_1 = h_2$, the royal road method is obtained by using the interpolatory quadrature formula (scaled to the interval (x_0, x_2))

$$\int_{-1}^{1}(1-|x|)f(x)\,\mathrm{d}x \approx \tfrac{1}{12}f(x-h) + \tfrac{5}{6}f(x) + \tfrac{1}{12}f(x+h),$$

which is exact when f is a cubic polynomial.

If f does not depend on y, it is possible to improve the approximation by using a more accurate quadrature formula. For example, the three-point Gaussian formula

$$\int_{-1}^{1}(1-|x|)f(x)\,\mathrm{d}x \approx \tfrac{5}{24}f(x-\sqrt{0.4}h) + \tfrac{7}{12}f(x) + \tfrac{5}{24}f(x+\sqrt{0.4}h)$$

for the weight function $1 - |x|$ is exact when f is a quintic polynomial. If f does depend on y, the method of the next section is applicable.

3.2. Applying a formula to the wrong equation

In practice, the problem we are solving is seldom simply $Ly = f$, where L is the operator in the theorem. Nevertheless, since (10) is an identity, we can in some cases obtain a useful formula even when a 'wrong' equation is being solved.

As an example we take the equation

$$y''(x) + q(x)y(x) = f(x). \tag{18}$$

Even when q is linear and $f \equiv 0$, the theoretical solution of this equation involves special functions. Therefore the exact difference formula for this operator may be constructible in principle but not in practice. Instead, we compute the exact formula (10) using $Ly := y'' + q(x_1)y$. The differential equation (18) reduces to $Ly(x) = f(x)+(q(x_1)-q(x))y(x)$. Substituting into (10), we obtain the exact difference formula

$$c_0y(x_0) + c_1y(x_1) + c_2y(x_2) = \int_{x_0}^{x_2} v(x)(f(x) + [q(x_1) - q(x)]y(x))\,\mathrm{d}x. \tag{19}$$

Example. Consider the differential equation

$$y''(x) - (\tfrac{1}{4}x^2 + a)y(x) = 0 \tag{20}$$

satisfied by the parabolic cylinder functions (see Chapter 19 of the Handbook of Mathematical Functions[1]) for the case $x_j = j$, $j = 0, 1, 2$ and $a = -\frac{3}{2}$. Using $Ly = y'' + 1.25y$ (note that since σ^2 in (11) is negative, the hyperbolic functions in (16) become trigonometric functions) we obtain the exact difference formula (coefficients given to five digits)

$$1.11101(y(0) + y(2)) - 0.97203y(1) = \int_0^2 \tfrac{1}{4}(x^2 - 1)y(x)\,v(x)\,\mathrm{d}x. \tag{21}$$

This formula has been normalized so that the three-point interpolatory quadrature formula exact for polynomials of degree 2 for approximating the right-hand side, namely

$$\int_0^2 v(x)f(x)\,\mathrm{d}x \approx 0.08881(f(0) + f(2)) + 0.82238f(1), \tag{22}$$

has coefficients that sum to 1. Using the quadrature formula, we obtain the approximate difference formula

$$1.13322y_0 - 0.97203y_1 + 1.04441y_2 = 0, \tag{23}$$

where y_j is the numerical approximation to $y(x_j)$. As a rough check, for the functions $U(-1.5, x)$ and $V(-1.5, x)$ tabulated in Table 19.1, of the Handbook[1], the relative error of the difference formula, defined as

$$\frac{|c_0y_0 + c_1y_1 + c_2y_2|}{|c_0y_0| + |c_1y_1| + |c_2y_2|},$$

evaluates to 0.00748 and 0.00227 respectively. For the royal road method, the approximate formula obtained is

$$1.12500y_0 - 0.95833y_1 + 1.04167y_2 = 0; \tag{24}$$

in this case the relative error evaluates to 0.02007 and 0.00514 respectively.

However, the advantage claimed for using an exact difference method, and then to approximate the integral, is not the modest factor of about 3 shown in this example, but the possibility of approximating the integral much more accurately by using non-grid values of y, obtained by a process of interpolation and defect correction. This process is exhaustively discussed by Butcher at al.[3] for the case where $Ly = y''$.

3.3. Reverse engineering

Suppose that a difference equation of the form

$$c_0 y(x_0) + c_1 y(x_1) + c_2 y(x_2) = 0 \tag{25}$$

has been derived by whatever means (for example, central differences or upwinding). Then one may try to find a second-order differential operator L with constant coefficients such that (25) is satisfied exactly when y is a solution to $Ly = 0$.

As an example we work out exactly what happens when one discretizes the operator $y''(x) + py'(x) + qy(x)$ by differences. The central difference approximation is

$$(\frac{1}{h^2} - \frac{p}{2h})y(x-h) + (q - \frac{2}{h^2})y(x) + (\frac{1}{h^2} + \frac{p}{2h})y(x+h). \tag{26}$$

The upwinding approximation for positive p is

$$\frac{1}{h^2}y(x-h) + (q - \frac{2}{h^2} - \frac{p}{h})y(x) + (\frac{1}{h^2} + \frac{p}{h})y(x+h). \tag{27}$$

We multiply (26) by $h^2/\sqrt{1-\frac{1}{4}p^2h^2}$ and (27) by $h^2/\sqrt{1+hp}$ respectively, so that the product of the first and last coefficients is 1, and compare respective coefficients with (15) to obtain:

	Central	Upwinding
μ	$\frac{1}{2h}\log\frac{1+hp/2}{1-hp/2}$	$\frac{\log(1+hp)}{2h}$,
σ	$\frac{1}{h}\operatorname{arc\,cosh}\frac{2-h^2q}{\sqrt{4-p^2h^2}}$	$\frac{1}{h}\operatorname{arc\,cosh}\frac{2+ph-qh^2}{\sqrt{4+4hp}}$.

We are now in the same position as in the previous section: we wish to solve a differential equation of the form

$$y''(x) + py'(x) + qy(x) = f(x)$$

but have available the correct coefficients for the differential operator

$$Ly := y'' + 2\mu y' + (\mu^2 - \sigma^2)y,$$

with μ and σ given in the above table. Since $2\mu \neq p$ in general, y' will appear under the integral sign in the exact difference formula.

4. References

1. M. Abramowitz and I. A. Stegun (eds). *Handbook of Mathematical Functions.* National Bureau of Standards, Washington, D.C., 1964.
2. Uri M. Ascher, Robert M. M. Mattheij, and Robert D. Russell. *Numerical Solution of Boundary Value Problems.* SIAM, Philadelphia, 1995.
3. J. C. Butcher, J. R. Cash, G. Moore, and R. D. Russell. Defect correction for two-point boundary value problems on non-equidistant meshes. *Mathematics of Computation*, 64:629–648, 1995.
4. I. Christie and A. R. Mitchell. Upwinding of high order Galerkin methods in conduction-convection problems. *Int. J. Num. Meth. Engng.*, 12:1764–1771, 1978.
5. L. Collatz. *Numerical Treatment of Differential Equations.* Springer, Berlin, 1960.
6. R. Courant and D. Hilbert. *Methods of Mathematical Physics*, volume 1. Interscience, New York, 1953.
7. Philip J. Davis. *Interpolation and Approximation.* Dover, New York, 1975.
8. Carl de Boor. *A Practical Guide to Splines.* Springer, New York, 1978.
9. A. Ghizzetti and A. Ossicini. *Quadrature Formulas.* Academic Press, New York, 1970.

AN EXPLICIT SYMPLECTIC INTEGRATOR WITH MAXIMAL STABILITY INTERVAL

M. A. LOPEZ-MARCOS, J. M. SANZ-SERNA
Departamento de Matemática Aplicada y Computación, Facultad de Ciencias, Universidad de Valladolid, Valladolid, Spain
E-mail: lopezmar@cpd.uva.es, sanzserna@cpd.uva.es

and

ROBERT D. SKEEL
Dept. of Computer Science and Beckman Inst., Univ. of Illinois at Urbana-Champaign 1304 West Springfield Avenue, Urbana, Illinois 61801-2987, USA
E-mail: skeel@cs.uiuc.edu

ABSTRACT

We derive and test a Runge-Kutta-Nyström method that possesses a stability interval that is maximal over all methods that are explicit, symplectic, effectively fourth order and use three force evaluations per step. Effective order four means that the output of the given method is going to be processed so as to enhance its accuracy and that the numerical solution after processing possesses $O(h^4)$ error bounds. By comparing the new method with a similar method of conventional order four, we show that, when designing schemes, the use of the notion of effective order leads to more efficient integrators than the notion of conventional order. The new method is less efficient than a related method, introduced by Rowlands, that uses the Hessian of the potential. This shows the interest of investigating further methods that use the Hessian of the potential.

1. Introduction

In this paper we derive and test a Runge-Kutta-Nyström (RKN) method that possesses a stability interval that is maximal over all methods that are explicit, symplectic, effectively fourth order and use three force evaluations per step. While the recent literature on symplectic integration of Hamiltonian systems is very large, see e.g. Hairer, Nørsett and Wanner[6], Sanz-Serna and Calvo[11], the stability intervals of symplectic methods have received little attention. This is surprising since symplectic integrators are primarily useful in situations where highly accurate solutions are not required and the interest lies in obtaining statistical or qualitative properties. In these situations the fastest components typically have the smallest amplitudes and need not be resolved accurately. Then stability is likely to be the factor limiting the stepsize. There is a second reason why stability intervals may be of interest: a large constant in any of the terms of the asymptotic expansion of the global error in powers of the stepsize h is likely to lead to a small stability interval. Therefore by controlling the stability interval it may be possible to obtain methods with small error coefficients at all powers of h.

Molecular dynamics simulations (Allen and Tildesley[1]) are often carried out with the simple second order, symplectic, time-reversible (selfadjoint in the terminology of Hairer, Nørsett and Wanner[6] or Sanz-Serna and Calvo[11]) Verlet method, which possesses a *scaled* stability interval of length 2. This length is known to be (Chawla and Sharma[4]) the largest possible for any symplectic, explicit method (regardless of the order of accuracy). Recall that a numerical method with stability interval of length L has a stability stepsize restriction $h \leq L$ when applied to the model problem $d^2y/dt^2 = -y$. If the method uses m force evaluations per step, then the scaled length L/m measures how many units of time per force evaluation may be advanced with the integrator within a stable simulation. Therefore, the Verlet method may well be the best choice if one is interested in explicit, symplectic integrators of order two.

There is not much point in using symplectic methods of order three: by concatenating an order three method and its adjoint one obtains a method of order four (Sanz-Serna and Calvo[11], Section 8.4.5), so that one may directly consider methods of order four. Here we construct a Runge-Kutta-Nyström (RKN) method that possesses a stability interval that is maximal over all methods (RKN or not) that are explicit, symplectic, effectively fourth order and use three force evaluations per step. The terminology effective order r is taken from Butcher[2]. It implies that the output of the given method is going to be processed so as to enhance its accuracy and that the numerical solution *after processing* possesses $O(h^r)$ global errors. López-Marcos, Sanz-Serna and Skeel[8] have showed how to process order four, symplectic, time-reversible symplectic methods at virtually no cost. Therefore when designing our method, it is meaningful to look at its effective order rather than at its conventional order.

In Section 2 we review the idea of processing. In Section 3 we study stability intervals and find the maximal stability interval for explicit, symplectic methods of effective order four that use three force evaluations per step. We present an interpretation of processing in terms of the eigenvectors and eigenvalues of the amplification matrix. In Section 4 a method is constructed that realizes the maximal stability interval. In Section 5 we study analytically the accuracy of the new method. By comparing the new method with a similar method of conventional order four, we show that, when designing schemes, the use of the notion of effective order leads to more efficient integrators than the notion of conventional order. By imposing the conditions for conventional order many degrees of freedom in the method are wasted. However it turns out that the new method is less efficient than a related method of effective order four introduced by Rowlands that uses the Hessian of the potential. It is shown in López-Marcos, Sanz-Serna and Skeel[9] that, in many problems, the cost of evaluating at a given point the potential *and* the Hessian of the potential is less than that of two evaluations of the potential. The cost per step of the Rowlands method (one evaluation of the gradient and Hessian) is significantly lower than that of the method constructed here (three evaluations of the gradient). Then, even though the new method is more accurate per step than the Rowlands method, it is less accurate per unit of cost. This may well show the interest in investigating further methods that use the Hessian of the potential. Numerical results are presented in the final Section 6.

2. Processing

Throughout the paper we consider Hamiltonian functions of the form

$$H(\mathbf{q},\mathbf{p}) = \frac{1}{2}\mathbf{p}^T M^{-1}\mathbf{p} + V(\mathbf{q}), \tag{1}$$

where the potential V is a smooth function and M is a constant, invertible, symmetric matrix. The Hamiltonian system corresponding to (1) is given by

$$\frac{d\mathbf{q}}{dt} = M^{-1}\mathbf{p}, \qquad \frac{d\mathbf{p}}{dt} = -V_{\mathbf{q}}(\mathbf{q}). \tag{2}$$

The notation $V_{\mathbf{q}}$ means the gradient of V with respect to $\mathbf{q}$. The negative of this gradient is the force.

A one-step method for the integration of (2) is given by a transformation $\psi_{h,H}$ that maps the approximation $(\mathbf{q}_n,\mathbf{p}_n)$ corresponding to a time level $t_n = nh$ into the approximation $(\mathbf{q}_{n+1},\mathbf{p}_{n+1}) = \psi_{h,H}(\mathbf{q}_n,\mathbf{p}_n)$ at the next time level t_{n+1}. Let us assume that we have been given a method $\psi_{h,H}$, that in what follows is called the basic method. When processing is used, there are two sets of variables being considered. The first set, that we denote by capital letters $(\mathbf{Q},\mathbf{P})$, corresponds to the values computed by the basic method; specifically, we compute the sequence $(\mathbf{Q}_{n+1},\mathbf{P}_{n+1}) = \psi_{h,H}(\mathbf{Q}_n,\mathbf{P}_n)$, $n = 0, 1, \ldots$, starting from $(\mathbf{Q}_0,\mathbf{P}_0)$. The second set of variables $(\mathbf{q},\mathbf{p})$ is related to the first through a transformation $(\mathbf{Q},\mathbf{P}) = \chi_{h,H}(\mathbf{q},\mathbf{p})$. It is the lower case variables that are seen as the processed numerical approximations to the solutions of (2). Thus there are three steps involved in the processed algorithm:

1. *Preprocessing:* Find, from the initial values $(\mathbf{q}(0),\mathbf{p}(0))$ the starting values for time-stepping $(\mathbf{Q}_0,\mathbf{P}_0) = \chi_{h,H}(\mathbf{q}(0),\mathbf{p}(0))$.

2. *Time-stepping:* Compute $(\mathbf{Q}_{n+1},\mathbf{P}_{n+1}) = \psi_{h,H}(\mathbf{Q}_n,\mathbf{P}_n)$, $n = 0, 1, \ldots$.

3. *Postprocessing:* If output at time $t = nh$ is desired, then find $(\mathbf{q}_n,\mathbf{p}_n) = \chi_{h,H}^{-1}(\mathbf{Q}_n,\mathbf{P}_n)$, which provides the numerical approximation to $(\mathbf{q}(nh),\mathbf{p}(nh))$.

The cost of preprocessing can be ignored, because preprocessing is performed only once in each integration. Note that

$$(\mathbf{q}_{n+1},\mathbf{p}_{n+1}) = \chi_{h,H}^{-1}(\psi_{h,H}(\chi_{h,H}(\mathbf{q}_n,\mathbf{p}_n)))$$

and therefore the processed solutions can be interpreted as unprocessed solutions computed with the method

$$\widehat{\psi}_{h,H} = \chi_{h,H}^{-1} \circ \psi_{h,H} \circ \chi_{h,H}. \qquad (3)$$

Processing is of interest if $\widehat{\psi}_{h,H}$ is a more accurate method than $\psi_{h,H}$ and the cost of postprocessing is negligible, either because output is not frequently required or because $\chi_{h,H}^{-1}$ is cheaply evaluated. Then, processing provides the accuracy of $\widehat{\psi}_{h,H}$ at the cost of the less accurate method $\psi_{h,H}$. López-Marcos, Sanz-Serna and Skeel[8] have showed how, in many cases, pre- and postprocessing may be carried out at virtually no cost. The idea of processing goes back to Butcher[2]. Further references are given by López-Marcos, Sanz-Serna and Skeel[8].

In what follows we always assume that the basic method $\psi_{h,H}$ is symplectic. Then, *formally,* the basic method provides (Sanz-Serna and Calvo[11], Section 10.1) exact solutions of a perturbed Hamiltonian system whose Hamiltonian function $\tilde{H}_h$ is a perturbation of the true Hamiltonian (1) of the system (2) being integrated. This is a characteristic feature of symplectic integrators; for a nonsymplectic method the computed points also lie on the solutions of a perturbation of the system (2), but the perturbed system is not Hamiltonian. If the preprocessor $\chi_{h,H}$ is a canonical or symplectic mapping, then the processed method (3) is also symplectic and then provides approximations that solve exactly its associated perturbed Hamiltonian system. The modified Hamiltonian of the processed method is (López-Marcos, Sanz-Serna and Skeel[8])

$$\widetilde{\widetilde{H}}_h(\mathbf{q},\mathbf{p}) = \tilde{H}_h(\mathbf{Q},\mathbf{P}) = \tilde{H}_h(\chi_{h,H}(\mathbf{q},\mathbf{p})). \qquad (4)$$

The aim of processing is then, given $\psi_{h,H}$ (i.e., given $\tilde{H}_h$), to find a symplectic transformation $\chi_{h,H}$ so that the right hand side of (4), that drives the processed solution, is as close as possible to the Hamiltonian H that drives the true solution.

Let us now focus on basic methods that are time-reversible. The corresponding modified Hamiltonian is of the form (López-Marcos, Sanz-Serna and Skeel[8])

$$\begin{aligned}\tilde{H}_h(\mathbf{Q},\mathbf{P}) \;=\;& H(\mathbf{Q},\mathbf{P}) \\ & + h^2\frac{A}{2}[\mathbf{P}^T M^{-1} V_{\mathbf{QQ}}(\mathbf{Q}) M^{-1}\mathbf{P}] \\ & + h^2\frac{B}{2}[V_{\mathbf{Q}}(\mathbf{Q})^T M^{-1} V_{\mathbf{Q}}(\mathbf{Q})] + O(h^4), \qquad (5)\end{aligned}$$

Here A and B are method-dependent constants and $V_{\mathbf{QQ}}$ is the Hessian matrix of V. The terms in square brackets are so-called elementary Hamiltonians. If $A = B = 0$, then $\tilde{H}_h$ and the true H differ in $O(h^4)$ terms and the basic method has order four; otherwise the order of accuracy is only two.

Since the preprocessor $(\mathbf{Q},\mathbf{P}) = \chi_{h,H}(\mathbf{q},\mathbf{p}) = id + O(h^2)$ has to be a symplectic transformation, it will be the exact solution flow of a Hamiltonian system. In other words, a Hamiltonian function H_χ has to exist such that $(\mathbf{Q},\mathbf{P})$ is the value at time h of the solution with initial condition $(\mathbf{q},\mathbf{p})$ of the Hamiltonian system associated with H_χ. The expansion of H_χ will be of the form

$$H_\chi = h\lambda[\mathbf{p}^T M^{-1} V_\mathbf{q}(\mathbf{q})] + O(h^3),$$

with λ an undetermined parameter; no $O(1)$ nor $O(h^2)$ contribution is included because (5) possesses no $O(h)$ nor $O(h^3)$ term. The $O(h)$ term includes the only elementary Hamiltonian of order two. We conclude that the preprocessor may be sought in the form (López-Marcos, Sanz-Serna and Skeel[8])

$$\begin{aligned} \mathbf{Q} &= \mathbf{q} + h^2\lambda[M^{-1}V_\mathbf{q}(\mathbf{q})] + O(h^4), \\ \mathbf{P} &= \mathbf{p} - h^2\lambda[V_{\mathbf{qq}}(\mathbf{q})M^{-1}\mathbf{p}] + O(h^4), \end{aligned} \tag{6}$$

and by substitution in (5) we find, in view of (4),

$$\begin{aligned} \widetilde{\widetilde{H}}_h(\mathbf{q},\mathbf{p}) &= H(\mathbf{q},\mathbf{p}) \\ &\quad + h^2\left(\frac{A}{2} - \lambda\right)[\mathbf{p}^T M^{-1} V_{\mathbf{qq}}(\mathbf{q}) M^{-1}\mathbf{p}] \\ &\quad + h^2\left(\frac{B}{2} + \lambda\right)[V_\mathbf{q}(\mathbf{q})^T M^{-1} V_\mathbf{q}(\mathbf{q})] + O(h^4). \end{aligned} \tag{7}$$

It is clear that, for the processed method to be of order four, i.e., for the basic method to be of effective order four, it is necessary and sufficient that the system $A/2 - \lambda = 0$, $B/2 - \lambda = 0$ may be solved for λ. Obviously this happens if and only if

$$A = -B, \tag{8}$$

a condition that should be compared with the condition $A = B = 0$ for the basic method to be of order four.

3. Stability Intervals

When the basic method with modified Hamiltonian (5) is applied to the integration of the harmonic oscillator $H(q,p) = (1/2)(p^2 + q^2)$, the computed points satisfy

$$\begin{bmatrix} Q_{n+1} \\ P_{n+1} \end{bmatrix} = M_\psi \begin{bmatrix} Q_n \\ P_n \end{bmatrix},$$

where M_ψ is the amplification matrix of the method, which should approximate the matrix

$$\begin{bmatrix} \cos h & \sin h \\ -\sin h & \cos h \end{bmatrix} \tag{9}$$

that advances the true solution. For stability, the eigenvalues of M_ψ must have modulus ≤ 1. Since the method is symplectic, $\det(M_\psi) = 1$, and then it is well known that stability is equivalent to M_ψ having trace of modulus ≤ 2. When this condition is satisfied, both eigenvalues have unit modulus.

Our aim is to discuss the stability of methods that are explicit, symplectic, time-reversible, use three force evaluations per step, and are effectively of order four. What does the trace of M_ψ look like for those methods? For explicit methods the entries (and hence the trace) of M_ψ are polynomials in h. For an explicit method using m force evaluations per step the trace is a polynomial $P(z)$ of degree $\leq m$ in the variable $z = h^2$: the formula for updating $\mathbf{q}$ will have nested evaluations of the force and each evaluation brings along a factor h^2. It remains to ascertain how the coefficients of $P(z)$ are constrained by the requirement of effective order four. The expansion of M_ψ in powers of h may be easily found if we recall that a step of the basic method is equivalent to advancing h units of time with the true solution of the modified Hamiltonian (5), which for the harmonic oscillator reads

$$\tilde{H}_h(Q,P) = \frac{1}{2}\left((1+Ah^2)P^2 + (1+Bh^2)Q^2\right) + O(h^4).$$

In this way we find

$$M_\psi = \begin{bmatrix} 1 - \frac{h^2}{2} + \left(\frac{1}{24} - \frac{A+B}{2}\right)h^4 + O(h^6) & h + \left(-\frac{1}{6} + A\right)h^3 + O(h^5) \\ -h + \left(\frac{1}{6} - B\right)h^3 + O(h^5) & 1 - \frac{h^2}{2} + \left(\frac{1}{24} - \frac{A+B}{2}\right)h^4 + O(h^6) \end{bmatrix}.$$

From here, we see that the condition (8) for effective order four is equivalent to

$$\operatorname{trace}(M_\psi) = 2 - h^2 + \frac{h^4}{12} + O(h^6). \tag{10}$$

Summing up, if the basic method is explicit, uses three evaluations per step and is of effective order four, the trace of M_ψ is of the form

$$P(z) = 2 - z + \frac{z^2}{12} + \alpha z^3, \quad z = h^2, \tag{11}$$

where α is a free parameter. Let us determine α so as to have the largest stability region. For $\alpha = 0$, $P(z)$ is a parabola with a minimum value $P(6) = -1$; stability is lost at $z = 12$ when $P = 2$. If $\alpha > 0$, then, for $z > 0$, the graph of P is strictly above the $\alpha = 0$ parabola. Hence $\alpha > 0$ is less stable than $\alpha = 0$. For $\alpha < 0$ and close to 0, the graph of $P(z)$ intersects the line $P = 2$ at $z_0 = 0$, $z_1 \approx 12$, and $z_2 \gg 1$ and the line $P = -2$ at $z_3 > z_2$. Stability is then restricted by the intersection at z_1. As α decreases away from 0, z_1 increases (thereby increasing stability) and z_2 and z_3 decrease. When α reaches the value $-1/576$, the points z_1 and z_2 coalesce and the equation $P = -2$ has a unique real root $z_3 \approx 32.3$, so that stability is restricted by the intersection at z_3. A further decrease in α implies a decrease in z_3 and hence in the length of the stability interval. Therefore the longest stability interval occurs at $\alpha = -1/576$ and the optimal trace is

$$P(z) = 2 - z + \frac{z^2}{12} - \frac{z^3}{576}, \quad z = h^2. \tag{12}$$

The stability interval has length $L \approx 5.69$ and scaled length $L/3 \approx 1.89$. This is within 5% of Verlet's optimal 2.

So far we have just discussed the stability of the basic method $\psi_{h,H}$. What is the stability interval of the processed method $\widehat{\psi}_{h,H}$? It is clear that it should be the same as that of the basic method, because it is really the basic method that is being used to propagate the numerical solution. From a more mathematical point of view, we note that from (3), the amplification matrices of the basic and processed methods are related through

$$M_{\widehat{\psi}} = M_{\chi}^{-1} M_{\psi} M_{\chi}, \tag{13}$$

where M_{χ} is the matrix that, for the harmonic oscillator, transforms the variables (q,p) into the variables (Q,P). Thus $M_{\widehat{\psi}}$ and M_{ψ} are related by a similarity transformation and have the same eigenvalues and the same stability properties.

We close this section with two comments on the relation (13). If we think of $\psi_{h,H}$ as given and try to find an optimal processor $\chi_{h,H}$, we see from (13) that, for the harmonic oscillator, the most we can achieve by processing is to change the eigenvectors of M_{ψ} into the exact eigenvectors, i.e., into the eigenvectors of (9), without changing the eigenvalues. The eigenvectors of the amplification matrix govern the shape of the numerical trajectories on the (q,p) plane; the eigenvalues governs the phase of the numerical solution on its trajectory. Similar considerations apply to any linear problem.

Since M_{ψ} and $M_{\widehat{\psi}}$ have the same trace, if $\widehat{\psi}_{h,H}$ is of order four, then the trace of M_{ψ} differs from the trace $2\cos h$ of the exact (9) in $O(h^5)$ terms (a fourth-order scheme introduces $O(h^5)$ errors in one step). The trace is even in h, so that actually $\text{trace}(M_{\psi}) = 2\cos h + O(h^6)$. This provides an alternative derivation of the formula (10).

4. Constructing the Method

We now show that the optimal trace polynomial (12) can be realized by an explicit, time-reversible, symplectic RKN method using three function evaluations per step. For three evaluations one may choose between two formats. In the first, the method has three stages. In the second—which we choose—the method has four stages but possesses the FSAL (first same as last) property, whereby the last force evaluation of the current step provides the first force evaluation to be used at the next step. For time-reversible methods choosing between both formats is just a matter of convenience, as we will discuss later.

With the requirements of symplecticness and time reversibility, an explicit, four-stage, FSAL RKN method is given by (Okunbor and Skeel[9], Sanz-Serna and Calvo[11], Section 8.5)

$$\mathbf{P}_n^1 = \mathbf{P}_n - h\left(\frac{1}{2} - b\right) V_{\mathbf{Q}}(\mathbf{Q}_n),$$

$$
\begin{aligned}
\mathbf{Q}_n^1 &= \mathbf{Q}_n + h\left(\frac{1}{2}-\gamma\right)M^{-1}\mathbf{P}_n^1,\\
\mathbf{P}_n^2 &= \mathbf{P}_n^1 - hbV_{\mathbf{Q}}(\mathbf{Q}_n^1),\\
\mathbf{Q}_n^2 &= \mathbf{Q}_n^1 + 2h\gamma M^{-1}\mathbf{P}_n^2,\\
\mathbf{P}_n^3 &= \mathbf{P}_n^2 - hbV_{\mathbf{Q}}(\mathbf{Q}_n^2),\\
\mathbf{Q}_{n+1} &= \mathbf{Q}_n^2 + h\left(\frac{1}{2}-\gamma\right)M^{-1}\mathbf{P}_n^3,\\
\mathbf{P}_{n+1} &= \mathbf{P}_n^3 - h\left(\frac{1}{2}-b\right)V_{\mathbf{Q}}(\mathbf{Q}_{n+1}),
\end{aligned}
\tag{14}
$$

where γ and b are free parameters. In the compact notation of Sanz-Serna and Calvo[11], Sections 8.4, 8.5, this method is described as

$$
[\frac{1}{2}-b, b, b, \frac{1}{2}+b], \qquad (\frac{1}{2}-\gamma, 2\gamma, \frac{1}{2}-\gamma, 0); \tag{15}
$$

the coefficients of the first group are used for the p variables and those of the second group for the q variables. The square brackets surrounding the p coefficients indicate that the first updating in (14) affects the p variables.

Note that if we define $\mathbf{Q}_n^* = \mathbf{Q}_n^1 + h\gamma M^{-1}\mathbf{P}_n^2$, $\mathbf{P}_n^* = \mathbf{P}_n^2$, then $(\mathbf{Q}_n^*, \mathbf{P}_n^*)$ is an approximation to the solution at $t = (n+1/2)h$, i.e., halfway through the step. Furthermore it is trivial to check that the transformation $(\mathbf{Q}_n^*, \mathbf{P}_n^*) \to (\mathbf{Q}_{n+1}^*, \mathbf{P}_{n+1}^*)$ that maps one halfway approximations into the next is in fact a RKN step with the method

$$
(b, 1-2b, b, 0), \qquad [\gamma, \frac{1}{2}-\gamma, \frac{1}{2}-\gamma, \gamma], \tag{16}
$$

that only has three stages (see formula (8.19) in Sanz-Serna and Calvo[11]). Therefore the four-stage, FSAL method (14) is related through the change of variables $(\mathbf{Q}_n, \mathbf{P}_n) \to (\mathbf{Q}_n^*, \mathbf{P}_n^*)$ to the three stage method (16). By considerations similar to those we presented towards the end of the preceding section, both methods then have the same stability properties. This proves that when looking for RKN methods that realize the optimal trace polynomial, the two formats mentioned above (three stages or four stages with FSAL) are equivalent.

After applying the method (14) to the harmonic oscillator, we find that the trace of the amplification matrix is given by

$$
2 - z + 2b(1-2\gamma)\left(\frac{1}{4}-\frac{b}{2}+\frac{\gamma}{2}\right)z^2 - \frac{b^2}{2}(1-2b)\gamma(1-2\gamma)^2 z^3,
$$

and, after comparing with (12), we have to consider the system

$$
2b(1-2\gamma)\left(\frac{1}{4}-\frac{b}{2}+\frac{\gamma}{2}\right) = \frac{1}{12}, \tag{17}
$$

$$
-\frac{b^2}{2}(1-2b)\gamma(1-2\gamma)^2 = -\frac{1}{576}. \tag{18}
$$

This has the solution

$$\gamma_{\text{opt}} = \frac{2+2^{1/3}+2^{-1/3}}{6} \approx 0.6756, \quad b_{\text{opt}} = \frac{1-2^{1/3}-2^{-1/3}}{6} \approx -0.1756. \qquad (19)$$

We recall from Section 4 that (17) is necessary and sufficient for effective order four. Hence the method (14) with parameter values (19) provides an explicit, symplectic, time-reversible RKN method with effective order four and optimal stability interval.

A final observation. From (19), $b_{\text{opt}} = 1/2 - \gamma_{\text{opt}}$ and therefore the compact notation (15) of the optimal method can be also be written as

$$[\gamma_{\text{opt}}, \frac{1}{2} - \gamma_{\text{opt}}, \frac{1}{2} - \gamma_{\text{opt}}, \gamma_{\text{opt}}], \qquad (b_{\text{opt}}, 1-2b_{\text{opt}}, b_{\text{opt}}, 0).$$

Comparison with (16) shows that the optimal method for the three-stage format can be obtained from the optimal method in the FSAL, four-stage format by swapping the roles of the $\mathbf{Q}$ and $\mathbf{P}$ variables.

5. The Accuracy of the New Method

It is of interest to compare analytically the method (14), (19) with other explicit, symplectic, time-reversible, (effectively) fourth-order methods.

To simplify the discussion we assume that the methods are applied to a linear Hamiltonian problem with potential $V(\mathbf{q}) = (1/2)\mathbf{q}^T S\mathbf{q}$, where S is a constant stiffness matrix. Then the modified Hamiltonian $\tilde{H}_h$ of the unprocessed method is of the form (López-Marcos, Sanz-Serna and Skeel[8])

$$\begin{aligned}\tilde{H}_h(\mathbf{Q},\mathbf{P}) &= H(\mathbf{Q},\mathbf{P}) \\ &\quad + h^2\frac{A}{2}[\mathbf{P}^T M^{-1}SM^{-1}\mathbf{P}] \\ &\quad + h^2\frac{B}{2}[\mathbf{Q}^T SM^{-1}S\mathbf{Q}] \\ &\quad + h^4\frac{C}{2}[\mathbf{P}^T M^{-1}SM^{-1}SM^{-1}\mathbf{P}] \\ &\quad + h^4\frac{D}{2}[\mathbf{Q}^T SM^{-1}SM^{-1}S\mathbf{Q}] + O(h^6).\end{aligned}$$

Here A, B, C and D are method-dependent constants. This differs from the earlier expansion (5) in that we have substituted for the Hessian of V its current constant value S and we have displayed the $O(h^4)$ terms in the expansion. Recall from (8) that, since we are dealing with methods of effective order four, we may set $A = -B$.

The transformation $\chi_{h,H}$ is sought in the form (López-Marcos, Sanz-Serna and Skeel[8])

$$\mathbf{Q} = \mathbf{q} + h^2\lambda[M^{-1}S\mathbf{q}] + h^4\left(\frac{\lambda^2}{2}+\mu\right)[M^{-1}SM^{-1}S\mathbf{q}] + O(h^6),$$

Table 1. Error constant and stability interval of effectively fourth-order methods

Method	Error constant E	Stability Interval L
FRCR unprocessed	0.054	1.57
FRCR processed	0.047	1.57
Optimal stability (processed)	0.00037	5.69
Rowlands (processed)	0.00098	3.46

$$\mathbf{P} = \mathbf{p} - h^2\lambda[SM^{-1}\mathbf{p}] + h^4\left(\frac{\lambda^2}{2} - \mu\right)[SM^{-1}SM^{-1}\mathbf{p}] + O(h^6).$$

Again this differs from the earlier (6) in that we have substituted for V and displayed the $O(h^4)$ terms. By substituting as required by (4), we find

$$\begin{aligned}\widetilde{\widetilde{H}}_h(\mathbf{q},\mathbf{p}) &= H(\mathbf{q},\mathbf{p}) \\ &+ h^2\left(\frac{A}{2} - \lambda\right)[\mathbf{p}^T M^{-1}SM^{-1}\mathbf{p}] \\ &- h^2\left(\frac{A}{2} - \lambda\right)[\mathbf{q}^T SM^{-1}S\mathbf{q}] \\ &+ h^4\left(\frac{C}{2} + \lambda^2 - A\lambda - \mu\right)[\mathbf{p}^T M^{-1}SM^{-1}SM^{-1}\mathbf{p}] \\ &+ h^4\left(\frac{D}{2} + \lambda^2 - A\lambda + \mu\right)[\mathbf{q}^T SM^{-1}SM^{-1}S\mathbf{q}] + O(h^6).\end{aligned}$$

Clearly we have to set $\lambda = A/2$ to achieve a processed method $\widehat{\psi}_{h,H}$ of order four. Furthermore, as in López-Marcos, Sanz-Serna and Skeel[8], we set $\mu = (C-D)/4$ so as to minimize

$$E = \left(\left(\frac{C}{2} + \lambda^2 - A\lambda - \mu\right)^2 + \left(\frac{D}{2} + \lambda^2 - A\lambda + \mu\right)^2\right)^{1/2}, \tag{20}$$

a measure of the size of the $O(h^4)$ error coefficients in the Hamiltonian $\widetilde{\widetilde{H}}_h$ of the processed method.

We now study the error constant and the stability interval of different methods. We first (Table 1) do so without taking into account the work per step and then (Table 2) report values scaled by work.

In Table 1 we have provided the size of the error constant (20) and the length L of the stability interval. The acronym FRCR refers to the three stage, fourth-order, symplectic RKN method constructed by Forest and Ruth[5] and Candy and Rozmus[3]. This is the only RKN method that with three stages achieves order four. It has been noted before (Sanz-Serna and Calvo[11], Section 9.1) that FRCR possesses large error constants. We now see in the table that its stability interval is short. The method only benefits slightly by processing.

Table 2. Scaled error constant and scaled stability interval of effectively fourth-order methods

Method	Error constant $mE^{1/4}$	Stability Interval L/m
FRCR unprocessed	1.45	0.52
FRCR processed	1.40	0.52
Optimal stability (processed)	0.42	1.90
Rowlands (processed)	0.35	1.73

A comparison of FRCR with the optimal stability method bears out the advantages of processing and of the associated concept of effective order. When looking for methods that use three force evaluations per step, the requirement of order four leads to FRCR; by relaxing this requirement to effective order four, we have been able to find a method with error constant one hundred times smaller and stability intervals almost four times larger.

'Rowlands' refers to the method of effective order four introduced by Rowlands[10] and given by (the superscripts on the square brackets refer to evaluation at $\mathbf{Q}_n$ or $\mathbf{Q}_{n+1}$)

$$\begin{aligned}
\mathbf{P}^{n+1/2} &= \mathbf{P}^n + \frac{h}{2}[-V_{\mathbf{Q}} + \frac{1}{12}h^2 V_{\mathbf{QQ}} M^{-1} V_{\mathbf{Q}}]^n, \\
\mathbf{Q}^{n+1} &= \mathbf{Q}^n + hM^{-1}\mathbf{P}^{n+1/2}, \\
\mathbf{P}^{n+1} &= \mathbf{P}^{n+1/2} + \frac{h}{2}[-V_{\mathbf{Q}} + \frac{1}{12}h^2 V_{\mathbf{QQ}} M^{-1} V_{\mathbf{Q}}]^{n+1}. \qquad (21)
\end{aligned}$$

This uses the Hessian $V_{\mathbf{QQ}}$ of the potential and has an FSAL property: the last square bracket in (21) will be reused at the next step. Therefore, per step, one needs one evaluation of $V_{\mathbf{Q}}$ and one evaluation of $V_{\mathbf{QQ}}$. For many problems the cost per step of (21) is the same as that of an RKN method using two force evaluations per step (López-Marcos, Sanz-Serna and Skeel[8]). The trace of (21) is given by (11) with $\alpha = 0$, so that $L = 2\sqrt{3}$, and the scaled stability interval has length $\sqrt{3}$. This is only 13% smaller than Verlet's 2.

It useful to point out that (21) can be seen as a limiting case of methods of the form (14). Assume that we impose the constraint (17) on the parameters γ and b in (14). This leaves a family of methods of effective order four depending on a single parameter, say γ. After expressing b as a function of γ, let γ tend to 1/2 in (14). Then a little analysis shows that the limiting scheme is given by (21). As γ decreases from the optimal value in (19) to Rowlands' 1/2, the error constant E increases and the stability interval L decreases. We see in the table that N and L are almost twice as good for the optimal method than for Rowlands. However we should take into account that while for $\gamma \neq 1/2$, (14) costs *three force evaluations* per step, the limiting scheme only costs *two*.

Table 2 contains essentially the same information that Table 1, but the error constant E and stability length L have been replaced by their normalized counterparts

$mE^{1/4}$ and L/m (m is the number of force evaluations per step) to account for the fact that different methods require different amounts of work per step. Note that $mE^{1/4}$ measures the work required to achieve a target error. From the table we conclude that Rowlands method is more efficient than the optimal stability method. This suggests the need for further investigating symplectic methods that use the Hessian of the potential.

6. Numerical Experiments

The analytical comparisons of the preceding section were based on a linear problem. We complemented our assessment of the meth ods considered by carrying out numerical experiments for a highly nonlinear problem. We integrated the liquid argon problem described in López-Marcos, Sanz-Serna and Díaz[7] (see also López-Marcos, Sanz-Serna and Skeel[8]). We implemented the optimal stability method and the Rowlands method, both with the cheap processing suggested by López-Marcos, Sanz-Serna and Skeel[8]. The FRCR method was not considered: its poor practical performance has already been discussed in Sanz-Serna and Calvo[11], Chapter 9.

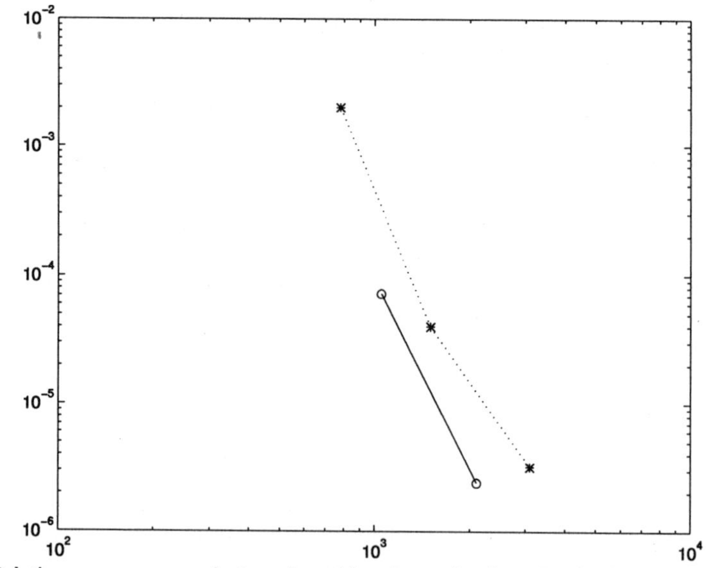

Fig. 1. Relative energy error against number of function evaluations. Rowlands method (solid line) and optimal stability method (broken line)

Figure 1 is an efficiency plot as those presented in López-Marcos, Sanz-Serna and Skeel[8]. The vertical axis gives the relative error in the Hamiltonian or total energy H. The energy error is very relevant in symplectic integration: it measures the difference between the exact and modified Hamiltonians. The horizontal axis measures work. For the RKN method with optimal stability region, the work equals the number of force evaluations (i.e., is three times as large as the number of steps). For the Rowlands method, work is measured as twice the number of steps. The solid line and circles corresponds to Rowlands method ran with $h = 0.128$ and $h = 0.064$ ($h = 0.256$ was unstable). The broken line with stars corresponds to the optimal stability interval with $h = 0.256, 0.128, 0.0064$ ($h = 0.512$ was unstable). We first note that, in this nonlinear problem, the optimal stability method is able to operate successfully with steplength $h = 0.256$, while the Rowlands method is not. This matches the fact that the linear analysis yields a shorter stability interval for Rowlands' than for the optimal stability method. The plot also shows that, again in agreement with the linear analysis, the Rowlands method is more efficient than the optimal stable scheme.

7. Acknowledgements

MALM and JMS have been partly supported by grant DGICYT PB92-254. This work was done while RDS was Iberdrola Visiting Professor at the University of Valladolid, with additional support from DOE/NSF grant DE-FG02-91ER25099/DMS-9304268, NIH grant P41RR05969 and NSF/ARPA grant ASC-9318159.

8. References

1. M. P. Allen and D. J. Tildesley, *Computer Simulation of Liquids*, Clarendon Press, Oxford, 1987.
2. J. Butcher, in *Conference on the Numerical Solution of Differential Equations,* Lecture Notes in Math., Vol. 109, Springer, Berlin, 1969.
3. J. Candy, and W. Rozmus, *J. Comput. Phys.* **92**, 230–256, 1991.
4. M. M. Chawla and S. R. Sharma, *BIT* **21**, 455–464, 1981.
5. E. Forest and R. D. Ruth, *Physica* **D43,** 105–117, 1990.
6. E. Hairer, S. P. Nørsett and G. Wanner, *Solving Ordinary Differential Equations I, Nonstiff Problems*, 2nd. Ed., Springer, Berlin, 1993.
7. M. A. López-Marcos, J. M. Sanz-Serna and J. C. Díaz, *J. Comput. Appl. Math.* (1995, to appear).
8. M. A. López-Marcos, J. M. Sanz-Serna and R. D. Skeel, in *Numerical Analysis 1995*, Proceedings of the 1995 Dundee Conference, (1996, to appear).
9. D. Okunbor and R. D. Skeel, *Math. Comput.* **59,** 439–455, 1992.
10. G. Rowlands, *J. Comput. Phys.* **97,** 235–239, 1991.
11. J. M. Sanz-Serna, and M. P. Calvo, *Numerical Hamiltonian Problems,* Chapman & Hall, London, 1994.

A TWO-STEP ITERATIVE METHOD

V.S. MANORANJAN
Department of Pure and Applied Mathematics, Washington State University, Pullman, Washington 99164-3113 U.S.A.

ABSTRACT

In multigrid methods, it is always preferred to employ smoothing techniques which are convergent. In practice, the standard Jacobi and Gauss-Seidel methods are the choices for smoothers. However, it is known that if the spectral radius conditions is violated, then the convergence is not guaranteed for these methods. In this paper we develop a simple two-step Jacobi type method which has better convergence properties and which can be employed as a convergent smoother whenever the standard iterative methods fail. We provide the applicability of the method on a variety of problems.

1. Introduction

In the recent years, there has been much interest in multigrid methods because of their accelerated convergence properties[1,8]. Multigrid algorithms are iterative solvers and are applied, in general, to large systems of linear equations obtained from the discretization of partial differential equations on either a finite difference or a finite element mesh. A multigrid algorithm is formulated in such a way it handles both the high and low frequency errors in the approximate solution. This involves the use of a "smoother" to eliminate high frequency errors and employing the same "smoother" again but, on a coarser grid to eliminate low frequency errors.

In practical applications, a simple iterative method such as the Jacobi or the Gauss-Seidel method is normally chosen as the "smoother"[6]. Although, the multigrid algorithm may work even with a divergent "smoother", provided it diverges slowly and smooths the error rapidly, practitioners always prefer employing smoothing methods which are convergent. This means that if either the Jacobi or the Gauss-Seidel method is the convergent "smoother", then the coefficient matrix of the system of linear equations should be such that the respective iteration matrix has a spectral radius less than unity[7,10]. So, if the spectral radius condition is violated in either case, one will not prefer to employ the Jacobi or the Gauss-Seidel method as a "smoother". Therefore, it is important to develop simple iterative algorithms which can be applied to systems of linear equations, where the standard iterative methods fail, and which are easily employable as convergent "smoothers" in multigrid algorithms. Further, the construction of such iterative algorithms can also be very useful in the study of convection dominated diffusion problems.

Recently, some work along these lines has appeared in the literature[2,3,5]. In[5], a Gauss-Seidel type spectrum enveloping technique was developed and in [2,3] block iterative methods for cyclically reduced non-self-adjoint linear systems were proposed.

The two-step Jacobi type iterative method, we construct in this paper is based on inner/outer iterations and has a better convergence region compared to the standard iterative method. We demonstrate the usefulness of our method on several test problems where the standard iterative method fails.

2. The Two-Step Iterative Method

Given the system of equations

$$Ax = b \tag{2.1}$$

where $A = [a_{ij}]_{i,j=1,\cdots,n}, b = (b_1, \cdots, b_n)^T, x = (x_1, \cdots, x_n)^T$, let us decompose the nonsingular matrix A as $L + D + U$, where L is a strictly lower triangular matrix, D is a diagonal matrix such that $a_{ii} \neq 0, i = 1, \cdots, n$, (i.e. D is nonsingular) and U is a strictly upper triangular matrix. Now we can multiply A by D^{-1} in order to get every element in the main diagonal to be equal to 1. *Therefore without loss of generality we consider coefficient matrices A where D is simply I, the identity matrix.*

Let $r = \rho(L + U)$, the spectral radius of $L + U$. Assuming $D = I$, we re-write equation (2.1) as

$$[(\alpha + r)I + L + U]x = b + (\alpha + r - 1)x \tag{2.2}$$

by adding the vector $(\alpha + r - 1)x$ to both sides of the equation, where α is a positive parameter.

Let $x^{(0)}$ (the initial guess) be given. Define the sequence $\{x^{(m)}\}_{m=0,1,\cdots}$ by

$$[(\alpha + r)I + L + U]x^{(m+1)} = b + (\alpha + r - 1)x^{(m)}. \tag{2.3}$$

Then the sequence $\{x^{(m)}\}_{m=0,1,\cdots}$ converges if the spectral radius $\rho(M(\alpha)) < 1$, where

$$M(\alpha) = (\alpha + r - 1)[(\alpha + r)I + L + U]^{-1}. \tag{2.4}$$

But now, since the unknown x is also on the right hand side of Equation (2.2), for each m we need to solve a system of the form

$$\tilde{A}x^{(m+1)} = \tilde{b}$$

where

$$\tilde{A} = [(\alpha + r)I + L + U]$$

and

$$\tilde{b} = b + (\alpha + r - 1)x^{(m)}.$$

This can be done by using the Jacobi iterative method.

Hence, solving the system of equations $Ax = b$ involves employing the Jacobi iterative method on systems of the form $\tilde{A}x^{(m+1)} = \tilde{b}, \forall m$ and then generating the sequence $\{x^{(m)}\}$ corresponding to (2.3).

It can be seen that if $r < 1$ and α is chosen such that $\alpha + r = 1$ then the above method reduces to the Jacobi iterative method and hence converges. Therefore, it is sufficient to study the case when $r \geq 1$.

We will call the iteration defined by equation (2.3) the outer iteration and the iteration given by the Jacobi method the inner iteration. Then we have the following results. Some of the proofs are fairly simple and they have been omitted.

Result 1. The inner iteration converges for $\alpha > 0$.

Result 2. Let $\mu_j = \beta_j \pm i\gamma_j, \gamma_j \geq 0, j = 1, \cdots, n$ be the eigenvalues of $L + U$. Then the outer iteration converges if $-1 \leq \beta_j, j = 1, \cdots, n$.

If $r = 1$ then we have $-1 \leq \beta_j, j = 1, \cdots, n$ and it follows from *Results 1 & 2* that the two-step iterative method converges. Therefore we will assume that $r > 1$.

Result 3. Let $\lambda_j = \delta_j \pm i\kappa_j, \kappa_j \geq 0, j = 1, \cdots, n$ be the eigenvalues of A. Then the outer iteration converges if $0 \leq \delta_j, j = 1, \cdots, n$.

Result 4. The outer iteration converges if A is positive definite.

From *Result 3* and the Gerschgorin Theorem [4,9] we have the following.

Result 5. If A is such that $\sum_{\substack{j=1 \\ j \neq i}}^{n} |a_{ij}| = 1, i = 1, \cdots, n$, then the outer iteration converges.

Proof. The Gerschgorin Theorem states that all the eigenvalues of A are in the set $S = \cup_{\ell=1}^{n} S_\ell$, where

$$S_\ell = \{z \in C; |z - 1| \leq \sum_{\substack{j=1 \\ j \neq \ell}}^{n} |a_{\ell j}|\}.$$

It is given that $\sum_{\substack{j=1 \\ j \neq i}}^{n} |a_{ij}| = 1$ and so, all the eigenvalues of A lie inside the circle $|z - 1| = 1$, where $z = (x + iy) \in C$ and hence they have nonnegative real parts.

If A is skew-Hermitian than all the eigenvalues are pure imaginary and we have another result.

Result 6. The outer iteration converges if A is skew-Hermitian.

In the case of the tridiagonal matrix, tridiag $(a, 1, b)$, we can show the following.

Result 7. Let

$$A = \begin{bmatrix} 1 & b & & & & \\ a & 1 & b & & & \\ & a & 1 & b & & \\ & & \ddots & \ddots & \ddots & \\ & & & a & 1 & b \\ & & & & a & 1 \end{bmatrix} \in \Re^{n \times n},$$

then the outer iteration converges if $ab \leq 1/4$.

Proof. The eigenvalues of A are given by

$$\lambda_k = 1 + 2(ab)^{1/2} \cos\left(\frac{k\pi}{n+1}\right), k = 1, \cdots, n.$$

If $ab > 0$ we have that λ_k is real for every k. Since $\cos\left(\frac{k\pi}{n+1}\right) > -1, k = 1, \cdots, n$ we have that $\lambda_k > 1 - 2(ab)^{1/2}$. Therefore if $1 - 2(ab)^{1/2} \geq 0$ then $\lambda_k > 0$ and consequently the outer iteration converges if $ab \leq 1/4$.

If $ab < 0$ then $Re\lambda_k = 1 (> 0)$ for all k and therefore the outer iteration will converge. □

The conditions for convergence obtained so far depend only on the real parts of the eigenvalues of $L + U$. However, it is possible to obtain conditions for convergence which involve the imaginary parts of the eigenvalues (of $L + U$) too. The following result gives such a condition.

Result 8. If all the eigenvalues of $L + U$ are outside the circle $|z + r| = r - 1$ then there exists a $\alpha_0 > 0$ such that the two-step iterative method converges for all $\alpha \in (0, \alpha_0)$.

Proof. One can easily show that $\rho(M(\alpha)) < 1$ if

$$\frac{(\alpha + r - 1)^2}{(\alpha + r + \beta_j)^2 + \gamma_j^2} < 1.$$

If $\beta_j \geq -1$ then the eigenvalues are outside of the circle $|z + r| = r - 1$ and we know that the above inequality holds. So, we study only those eigenvalues for which $\beta_j < -1$. Consider

$$\begin{aligned} (\alpha + r + \beta_j)^2 + \gamma_j^2 &= (\alpha + r - 1 + \beta_j + 1)^2 + \gamma_j^2 \\ &= (\alpha + r - 1)^2 + 2(\alpha + r - 1)(\beta_j + 1) + (\beta_j + 1)^2 + \gamma_j^2. \end{aligned}$$

We will have $\rho(M(\alpha)) < 1$ if

$$2(\alpha + r - 1)(\beta_j + 1) + (\beta_j + 1)^2 + \gamma_j^2 > 0$$

or

$$\alpha < \frac{-2(r-1)(\beta_j+1) - (\beta_j+1)^2 - \gamma_j^2}{2(\beta_j+1)} = \alpha_0, \text{ say .}$$

In order to choose a $\alpha(> 0)$ according to the above inequality, we need to have $\alpha_0 > 0$. i.e., we should have the condition

$$(\beta_j + 1)^2 + 2(r-1)(\beta_j + 1) + \gamma_j^2 > 0$$

or equivalently

$$(\beta_j + r)^2 + \gamma_j^2 > (r-1)^2.$$

This means that the algorithm will converge if the eigenvalues of $L + U$ are outside the circle $|z + r| = r - 1$ and α is chosen such that $0 < \alpha < \alpha_0$. □

In the following section, we present some results obtained by carrying out numerical experiments using the proposed two-step method.

3. Numerical Results

We performed the numerical experiments on four different linear systems of equations $A_i x = b_i (i = 1, \cdots, 4)$. The initial vectors for the outer and the inner iterations were $x^{(0)} = (0, 0 \cdots, 0)^T$ and $x_{(0)} = x^{(m)}$ respectively. The respective stopping criteria of the iterations were $\|x^{(m+1)} - x^{(m)}\|_2 < \varepsilon_1$ and $\|x_{(k+1)} - x_{(k)}\|_2 < \varepsilon_2$. In all the computations, a Fortran implementation of the method was used.

The matrices A_i of the systems $A_i x = b_i$ are given by

$$A_1 = \begin{bmatrix} 1 & 1 & 0 \\ -2 & 1 & 1 \\ 0 & -2 & 1 \end{bmatrix},$$

$$A_2 = \begin{bmatrix} 1 & 5/23 & 7/23 & 14/23 & 16/23 \\ 17/24 & 1 & 1/24 & 1/3 & 5/8 \\ 11/25 & 18/25 & 1 & 2/25 & 9/25 \\ 10/21 & 4/7 & 19/21 & 1 & 1/7 \\ 2/11 & 3/11 & 13/22 & 10/11 & 1 \end{bmatrix},$$

$$A_3 = \begin{bmatrix} 1 & 5 & 4 & 4 & 3 \\ 1/4 & 1 & -1/2 & 1 & -1/4 \\ 0 & 2/3 & 1 & 2/3 & 1/3 \\ 3/10 & 8/10 & 3/10 & 1 & 9/10 \\ 1 & 4 & 1 & 2 & 1 \end{bmatrix},$$

and

$$A_4 = \begin{bmatrix} 1 & 1 & 1 & 3 \\ -1 & 1 & -1 & 1 \\ -1 & -1 & 1 & 1 \\ 1 & -1 & 1 & 1 \end{bmatrix}.$$

The corresponding b_i's are

$$b_1 = \begin{bmatrix} 2 \\ 0 \\ -1 \end{bmatrix}, b_2 = \begin{bmatrix} 65/23 \\ 65/24 \\ 13/5 \\ 65/21 \\ 65/22 \end{bmatrix}, b_3 = \begin{bmatrix} 17 \\ 3/2 \\ 8/3 \\ 33/10 \\ 9 \end{bmatrix}, \text{ and } b_4 = \begin{bmatrix} 6 \\ 0 \\ 0 \\ 2 \end{bmatrix}.$$

In the tables that follow we present the eigenvalues of the matrices $(L + U)$ and $(I + U)^{-1}L$ with respect to the test problems and the associated spectral radii.

Table 1

	Eigenvalues of $L+U$	Spectral radius $\rho(L+U)$	Convergence Jacobi
A_1	0	2	no
	$\pm\, 2i$		
A_2	1.8383	1.8383	no
	$-0.4649 \pm 0.7578i$		
	$-0.4542 \pm 0.1663i$		
A_3	3.5607	3.5607	no
	$-0.8572 \pm 0.1112i$		
	$-0.9230 \pm 0.8147i$		
A_4	1	1.7709	no
	1.5943		
	$-1.2972 \pm 1.2056i$		

Table 2

	Eigenvalues of $(I+U)^{-1}L$	Spectral radius $\rho((I+U)^{-1}L)$	Convergence Gauss-Seidel
A_1	0	4	no
	0		
	4		
A_2	0	0.1611	yes
	$-0.0241 \pm 0.0910i$		
	-0.0132		
	-0.1611		
A_3	0	1.5144	no
	0.7011		
	-0.9328		
	$-1.209 \pm 0.9119i$		
A_4	0	3.3830	no
	0.0874		
	$-2.5437 \pm 2.2303i$		

Figures 1 and 2 show the relative positions of the eigenvalues of the matrix $(L+U)$ with respect to the circles $|z| = r, |z+r| = r-1$ and the line $Re z = -1$. The dark circles indicate these positions.

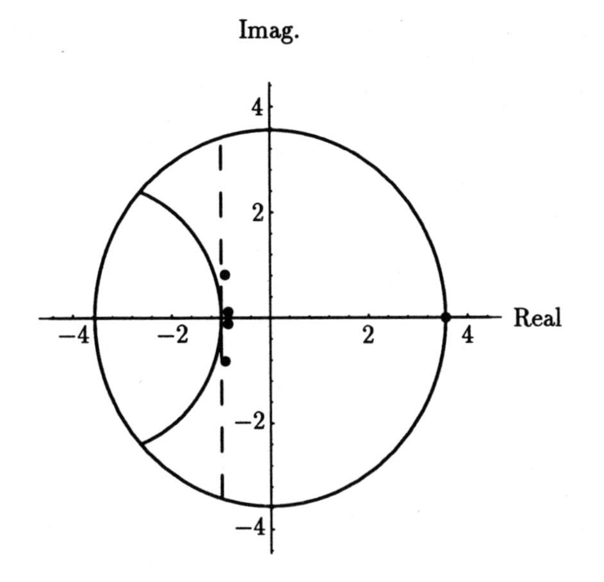

Figure 1: Eigenspectrum of $L_3 + U_3$

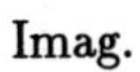

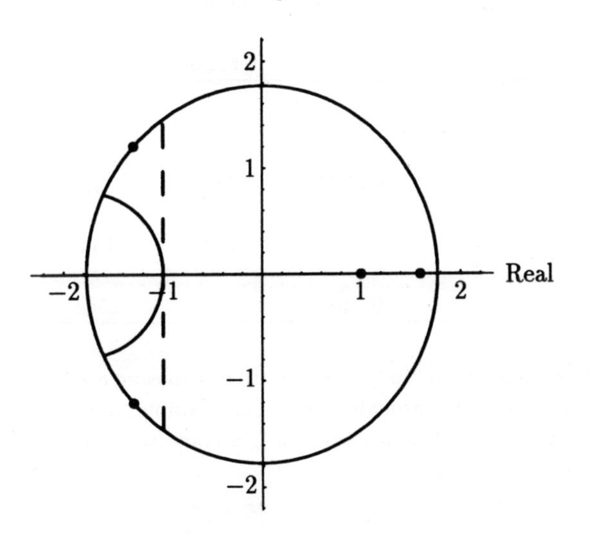

Figure 2: Eigenspectrum of $L_4 + U_4$

Although our numerical results are consistent with and illustrate the theoretical points we made in section 2, the question is, how well can we choose α? Our theoretical results indicate that there is an optimal value of α which makes the convergence of the two-step method faster. We determined (numerically) this optimal value of α for the test problems, and in Table 3 we present the numbers of outer/inner iterations for convergence.

Table 3

Matrix	Optimal α	outer iteration	inner iteration
A_1	2.2	40	289
A_2	1.9	43	259
A_3	1.1	305	1930
A_4	0.38	59	2552

In Table 4, we give results for α chosen such that $\rho(J(\alpha)) = \rho(M(\alpha))$.

Table 4

Matrix	Optimal α	outer iteration	inner iteration
A_1	1.0	28	333
A_2	0.6723	26	322
A_3	0.1877	180	3102
A_4	0.4408	54	2565

Moreover, our numerical study indicates that, choosing a small value for α always leads to convergence at a reasonable rate. For example, $\alpha = \frac{1}{r}$, where r is the estimated value of the spectral radius $\rho(L+U)$, is found to be a good choice.

4. Conclusions

In this paper we have developed a simple two-step Jacobi type iterative method for solving systems of linear equations. We obtained the conditions for convergence for various cases and demonstrated the applicability of the method on problems where the other known simple methods fail. In regard to multigrid techniques, whenever the standard Jacobi and Gauss-Seidel methods fail as convergent "smoothers", this two-step Jacobi type method can be easily implemented as the new convergent "smoother". Further, this method is also amenable to parallel adaptation.

5. References

1. Briggs, W.L., *A Multigrid Tutorial* (SIAM, Philadelphia, 1987).
2. Elman, H.C. & Golub, G.H., Iterative methods for cyclically reduced non-self-adjoint linear systems, *Math. Comp.* **54**, 671-700, (1990).
3. Elman, H.C. & Golub, G.H., Iterative methods for cyclically reduced non-self-adjoint linear systems II, *Math. Comp.* **56**, 215-242, (1991).
4. Golub, G.H. & Van Loan, C.F., *Matrix Computations* (The John Hopkins University Press, Baltimore, 1989).
5. Manoranjan, V.S. & Drake, R., A spectrum enveloping technique for convection-diffusion computations, *IMA J. Numer. Anal.*, **13**, 431-443, (1993).
6. McCormick, S., *Multigrid Tutorial*, CBMS Regional Research Conference on Multigrid Adaptive Methods for Partial Differential Equations, Washington, D.C., (1991).
7. Varga, R. *Matrix Iterative Analysis* (Prentice-Hall, Englewood Cliffs, N.J., 1962).
8. Wesseling, P., *An Introduction to Multigrid Methods* (John Wiley & Sons, New York, 1992).
9. Wilkinson, J.H., *The Algebraic Eigenvalue Problem* (Clarendon Press, Oxford, 1965).
10. Young, D., *Iterative Solution of Large Linear Systems* (Academic Press, New York, 1971).

COERCIVITY FOR ONE-DIMENSIONAL CELL VERTEX APPROXIMATIONS

K.W. MORTON
Oxford University Computing Laboratory
Wolfson Building, Parks Road, Oxford, OX1 3JD
E-mail: Bill.Morton@comlab.ox.ac.uk

ABSTRACT

Previous error analysis for the cell vertex scheme has been limited to situations where the cell residuals can be set to zero. However, in practical use for compressible flow computations it is necessary to extend the method by the use of distribution matrices and the careful addition of artificial viscosity terms. In this paper we make a start on the error analysis that is required for this more general method. The chosen example is a one-dimensional convection-diffusion problem with an expansion critical or turning point.

1. Introduction

Consider approximating the vector of unknowns $\mathbf{w}(x,y)$ which satisfy the system of conservation laws

$$\frac{\partial \mathbf{f}}{\partial x} + \frac{\partial \mathbf{g}}{\partial y} = \mathbf{S}, \tag{1}$$

together with appropriate boundary conditions. When $\mathbf{f} \equiv \mathbf{f}(\mathbf{w})$ and $\mathbf{g} \equiv \mathbf{g}(\mathbf{w})$, a first order system of equations results which may be of hyperbolic type, of elliptic type or of mixed type; examples are given by the Euler equations for steady, inviscid, compressible gas flow or the unsteady St. Venant equations describing one-dimensional river flow. In both examples one of the most useful and practical schemes of approximation consists of associating the unknowns with the vertices of a mesh (quadrilateral or rectangular in the two cases), applying Gauss' theorem to each cell of the mesh, and using the trapezoidal rule along each edge to approximate the resulting line integrals. For the unsteady, one-dimensional hyperbolic system this is called the box difference scheme and is associated with the names of Wendroff [14], Preissmann [12] and Thomée [13]; for the steady Euler equations it is called the cell vertex finite volume method and has been advocated and developed by Ni [11], Hall [3], Paisley [8] and many others subsequently.

A second order equation system is obtained if $\mathbf{f} \equiv \mathbf{f}(\mathbf{w}, \nabla\mathbf{w})$ and $\mathbf{g} \equiv \mathbf{g}(\mathbf{w}, \nabla\mathbf{w})$, as would be the case for the Navier–Stokes equations of viscous fluid flow or for any system of convection-diffusion equations. Then two extensions are possible: one can introduce a subsidiary equation system $\mathbf{Z} = \nabla\mathbf{w}$ and approximate this by the same scheme, as was done by Keller and Cebeci [4,5] for the boundary layer equations; or one can recover an approximation to $\nabla\mathbf{w}$ at each vertex for direct substitution into the flux functions $\mathbf{f}$ and $\mathbf{g}$ and the same cell vertex equations as in the first order case, as has been developed in Mackenzie [6], Crumpton *et al.* [1] and subsequent papers. The latter is the more generally applicable scheme as it can deal with the complete range of problems, for example, corresponding to all values of the dimensionless Péclet number which parametrises the relative importance of convective and diffusive phenomena. Thus if Ω_α is a typical mesh cell and $\mathbf{R}_\alpha(\cdot)$ denotes the discrete operator on that cell that is derived from (1) in the manner described above, the basic cell vertex scheme gives an approximation $\mathbf{W}$ by setting

$$\mathbf{R}_\alpha(\mathbf{W}) = \mathbf{0} \quad \forall\alpha, \tag{2}$$

which are called the *cell residual* equations. If we write $\mathbf{F}_i$ for the approximation to $\mathbf{f}(\mathbf{w}, \nabla\mathbf{w})$ obtained at the vertex $\mathbf{x}_i = (x_i, y_i)$ from $\mathbf{W}$ and the recovered gradient, then with the vertices of Ω_α numbered 1,2,3,4 in anticlockwise order we have

$$\begin{aligned}\mathbf{R}_\alpha(\mathbf{W}) := \frac{1}{2V_\alpha}\{&(\mathbf{F}_1 - \mathbf{F}_3)\delta y_{24} + (\mathbf{F}_2 - \mathbf{F}_4)\delta y_{31} \\ &-(\mathbf{G}_1 - \mathbf{G}_3)\delta x_{24} - (\mathbf{G}_2 - \mathbf{G}_4)\delta x_{31}\} - \mathbf{S}_\alpha,\end{aligned} \tag{3}$$

where $\delta y_{24} := y_2 - y_4$ etc., V_α is the measure of the cell Ω_α and $\mathbf{S}_\alpha$ is the average of the source function $\mathbf{S}$ over the cell.

There is a predominant difficulty with this cell vertex scheme, however, which occurs with both first order and second order equations; namely, the natural association of the unknowns with the vertices and the equations with the cells means that it is often not possible to ensure that there are equal numbers of each after boundary conditions have been applied. Hence, in order to compute approximations to the Euler or Navier–Stokes equations, it is normal practice to form *nodal residuals* at each vertex corresponding to an unknown vector $\mathbf{W}_i$ by combining the cell residuals from neighbouring cells, and then possibly applying some *artificial viscosity* or *artificial dissipation*; the equations that are actually solved then take the form

$$\mathbf{N}_i(\mathbf{W}) := \frac{\sum_{(\alpha)} V_\alpha[D_{\alpha,i}\mathbf{R}_\alpha + \mathbf{A}_{\alpha,i}]}{\sum_{(\alpha)} V_\alpha} = \mathbf{0} \quad \forall i, \tag{4}$$

where the $D_{\alpha,i}$ are called *distribution matrices* (or distribution factors in the scalar case) and the $\mathbf{A}_{\alpha,i}$ term represents the artificial viscosity.

Unfortunately, all of the error analysis that has been carried out for the cell vertex scheme applies only to situations where the simple form (2) can be used. The purpose of this paper is to initiate the analysis that has to be used when the more general form (4) is needed.

Distribution matrices are required when the number of cell equations (2) together with the boundary conditions exceed the number of unknowns, so that some of the cell residuals have to be combined. Artificial viscosity terms are needed when the opposite situation occurs and the number of equations is too few. Furthermore, the averaging along a cell edge that leads to the cell residual (3) means that the cell equations (2) suffer from the spurious chequer-board mode; and the averaging involved in the use of distribution matrices can also introduce further spurious modes. Thus the artificial viscosity is also required to control these modes.

Some of these phenomena can be studied even with one-dimensional problems, and this is the subject of this paper. By adopting the general form (4) even for the simple cases when the distribution matrices and artificial viscosity terms are not needed, we avoid the problem of having too few equations; then the well-posedness of the system reduces to establishing a coercivity condition, typically of the form

$$(\mathbf{N}(\mathbf{U}) - \mathbf{N}(\mathbf{V}), \mathbf{U} - \mathbf{V}) \geq \sigma^2 \|\mathbf{U} - \mathbf{V}\|_h^2 \tag{5}$$

where $(\cdot,\cdot)$ is the l_2 inner product and $\|\cdot\|_h$ is some suitable discrete norm.

We begin with the consideration of a pure convection problem with an expansion critical point or turning point. This can be dealt with in the form (2) by splitting the cell residual for the turning point cell, as in Morton *et al.* [10]; but in section 2 we use the form (4) and establish coercivity by appropriate choices of distribution factors and artificial dissipation terms. Then in section 3 we add diffusion to this problem. The coercivity analysis of Morton and Stynes [9] was valid in this case only for a mesh that became more refined in the flow direction, and this is an awkward restriction. However, in García-Archilla and Mackenzie [2] it was shown that the method has second order accuracy on any mesh, so in this section we show how a small amount of artificial dissipation could be used to establish coercivity on any mesh. Finally, in section 4 we consider the effect on the accuracy of the scheme of adding these terms to the simple formulation given by (2) and (3).

2. Coercivity for pure convection at an expansion critical point

In conservation law form the convection-diffusion problem in one dimension can be written

$$-\epsilon(au')' + (bu)' = S \quad \text{on } (0,1). \tag{6}$$

We can suppose that $a(x)$ is strictly positive and interest is focussed on problems which are uniformly well posed as $\epsilon \to 0$. In the absence of turning points, e.g. for $b(x) > 0$, such well-posedness is easily established for both the continuous and the discrete problem. We suppose instead that we have a single turning point or critical point, at $\xi \in (0,1)$ where $b(\xi) = 0$; then in the case of an expanding convective flow field well-posedness of (6) is again simple to show, so in studying the discrete problem we assume that

$$b'(x) \geq \gamma > 0 \quad \text{on } [0,1]. \tag{7}$$

Indeed, in this case the solution of the reduced problem for (6), i.e. when $\epsilon = 0$, can be given explicitly as

$$u(x) = [1/b(x)] \int_\xi^x S(t)\mathrm{d}t, \quad x \neq \xi, \tag{8}$$

$$u(\xi) = S(\xi)/b'(\xi). \tag{9}$$

For $\epsilon \neq 0$, a Dirichlet boundary condition at $x = 0$ or $x = 1$ will lead to a boundary layer there.

The cell vertex method generally gives an accurate monotone approximation in a boundary layer; and the accuracy of the scheme in the absence of turning points has been thoroughly analysed in Mackenzie and Morton [7] and Morton and Stynes [9]. Here we extend the latter analysis to the turning point problem described above when $\epsilon = 0$; the case of $\epsilon > 0$ will be dealt with in the next section.

On a nonuniform mesh $0 = x_0 < x_1 < \ldots < x_{J-1} < x_J = 1$, with interval lengths $h_j = x_j - x_{j-1}$, we suppose that the turning point is in the kth interval, namely

$$x_{k-1} \leq \xi < x_k, \quad \text{with } k > 1. \tag{10}$$

The cell residuals are given by

$$\begin{aligned} R_{j-\frac{1}{2}}(U) := h_j^{-1}[(b_j U_j - \epsilon a_j U_j') - (b_{j-1}U_{j-1} - \epsilon a_{j-1} U_{j-1}')] \\ - \tfrac{1}{2}(S_{j-1} + S_j), \quad j = 1, 2, \ldots, J. \end{aligned} \tag{11}$$

Since in this section we consider only the limiting case $\epsilon \to 0$, the form of recovery used to obtain U_j' is irrelevant. In going to the more general formulation of (4) we will use upwind distribution factors for all but the kth cell; that is, we set

$$D_{j-\frac{1}{2},j-1} = 1 - s_{j-\frac{1}{2}}, \quad D_{j-\frac{1}{2},j} = 1 + s_{j-\frac{1}{2}}, \tag{12}$$

with

$$s_{j-\frac{1}{2}} = \operatorname{sign} b_{j-\frac{1}{2}} \quad \text{for } j \neq k, \tag{13}$$

where $b_{j-\frac{1}{2}}$ is an average value of b in the cell; but we leave $s_{k-\frac{1}{2}}$ as a free parameter. Then, as in (4) the nodal residuals are defined and the solution is determined by setting for $i = 0, 1, \ldots, J$

$$\begin{aligned} N_i(U) := (h_i + h_{i+1})^{-1}[h_i(D_{i-\frac{1}{2},i}R_{i-\frac{1}{2}} + A_{i-\frac{1}{2},i}) \\ + h_{i+1}(D_{i+\frac{1}{2},i}R_{i+\frac{1}{2}} + A_{i+\frac{1}{2},i})] = 0; \end{aligned} \tag{14}$$

for the boundary points we need to set $h_0 = h_1, h_{J+1} = h_J$ and $R_{-\frac{1}{2}} = R_{J+\frac{1}{2}} = 0$ here. We will leave until later the specification of the artificial viscosity terms, and we will denote by $N_i^0(U)$ the nodal residuals with these terms set to zero, as well as $\epsilon = 0$.

In order to demonstrate coercivity, we introduce the inner product

$$(f,g) := \sum_{j=0}^{J} \tfrac{1}{2}(h_j + h_{j+1}) f_j g_j; \tag{15}$$

and we use this to define the bilinear form

$$B^h(V,W) := (N(U+V) - N(U), W), \tag{16}$$

with $B^0(\cdot,\cdot)$ resulting when N^0 is used. In this latter case it is clear that

$$\begin{aligned} B^0(V,V) = {} & \sum_0^{k-2} V_j \Delta_+(bV)_j + \sum_{k+1}^{J} V_j \Delta_-(bV)_j \\ & + \tfrac{1}{2}[(1 - s_{k-\frac{1}{2}})V_{k-1} + (1 + s_{k-\frac{1}{2}})V_k]\Delta_-(bV)_k, \end{aligned} \tag{17}$$

the coercivity of which can be best studied with the form given in the following lemma. Here Δ_- and Δ_+ are the usual backward and forward undivided difference operators.

Lemma 2.1 *For the bilinear form without artificial viscosity, we have*

$$\begin{aligned} 2B^0(V,V) = {} & -\, b_0 V_0^2 - \sum_1^{k-1} b_j (\Delta_- V_j)^2 + \sum_{k+1}^{J} b_{j-1}(\Delta_- V_j)^2 + b_J V_J^2 \\ & + \sum_1^{k-1} (\Delta_- b_j) V_{j-1}^2 + (\Delta_- b_k) V_k V_{k-1} + \sum_{k+1}^{J} (\Delta_- b_j) V_j^2 \\ & + s_{k-\frac{1}{2}} (\Delta_- V_k) \Delta_-(bV)_k. \end{aligned} \tag{18}$$

Proof The terms under the summation signs in (17) can be broken up to give, for example,

$$\begin{aligned} V_j \Delta_+(bV)_j &= (\Delta_+ b_j) V_j^2 + b_{j+1} V_j (V_{j+1} - V_j) \\ &= (\Delta_+ b_j) V_j^2 - \tfrac{1}{2} b_{j+1} [(V_{j+1} - V_j)^2 - (V_{j+1}^2 - V_j^2)] \\ &= \tfrac{1}{2}(\Delta_+ b_j) V_j^2 - \tfrac{1}{2} b_{j+1} (\Delta_- V_{j+1})^2 + \tfrac{1}{2} \Delta_+ (b_j V_j^2). \end{aligned}$$

Hence we have

$$\sum_0^{k-2} V_j \Delta_+(bV)_j = \tfrac{1}{2} \sum_0^{k-2} (\Delta_+ b_j) V_j^2 - \tfrac{1}{2} \sum_1^{k-1} b_j (\Delta_- V_j)^2 + \tfrac{1}{2}(b_{k-1} V_{k-1}^2 - b_0 V_0^2),$$

with a similar sum to the right of the turning point. With $s_{k-\frac{1}{2}} = 0$, the terms in $2B^0(V,V)$ associated with the kth cell are then just

$$\begin{aligned} & b_{k-1} V_{k-1}^2 - b_k V_k^2 + (V_{k-1} + V_k)(b_k V_k - b_{k-1} V_{k-1}) \\ & \quad = (b_k - b_{k-1}) V_k V_{k-1}, \end{aligned}$$

which gives all the terms in (18), except those dependent on $s_{k-\frac{1}{2}}$ which are obtained from (17). □

All the terms in (18) are positive definite except the product V_kV_{k-1} and the terms depending on $s_{k-\frac{1}{2}}$. In general, the freedom to choose $s_{k-\frac{1}{2}}$ is inadequate to force the whole expression for $B^0(V,V)$ to be positive definite and artificial dissipation terms are needed. However, some special cases are worth noting before we consider the general case. If $b_{k-1} = 0$, so that $b_k > 0$, the two sets of terms give

$$b_k[V_kV_{k-1} + s_{k-\frac{1}{2}}(V_k^2 - V_kV_{k-1})],$$

so that taking $s_{k-\frac{1}{2}} = 1$ gives the positive definite term $b_kV_k^2$; but then $B^0(V,V) = 0$ if $V_i = 0$ for $i \neq k-1$ while $V_{k-1} \neq 0$, which reflects the fact that the system of residual equations contains no equation to determine V_{k-1}. Similarly, if $b_k = 0$ and $b_{k-1} < 0$ then taking $s_{k-\frac{1}{2}} = -1$ makes $B^0(V,V) \geq 0$, but V_k is not determined by the residual equations. Between these two extremes, the case $b_k + b_{k-1} = 0$ gives

$$b_k[2V_kV_{k-1} + s_{k-\frac{1}{2}}(V_k^2 - V_{k-1}^2)],$$

which cannot be made positive definite by any choice of $s_{k-\frac{1}{2}}$.

Thus we consider the addition of artificial viscosity, both second order and fourth order; note that in the latter case it is more commonly referred to as artificial dissipation. As used for modelling the Navier–Stokes equations in Crumpton *et al.* [1], the terms for substituting in (14) have the form

$$\begin{aligned} A_{i-\frac{1}{2},i} &= \tau^{(2)}_{i-\frac{1}{2}}(U_i - U_{i-1}) - \tau^{(4)}_{i-\frac{1}{2}}(\delta^2U_i - \delta^2U_{i-1}) \\ A_{i-\frac{1}{2},i-1} &= \tau^{(2)}_{i-\frac{1}{2}}(U_{i-1} - U_i) - \tau^{(4)}_{i-\frac{1}{2}}(\delta^2U_{i-1} - \delta^2U_i), \end{aligned} \tag{19}$$

where $\delta^2 = \Delta_+\Delta_- = \Delta_-\Delta_+$. Note that the sum of these two terms associated with a single cell is zero, so that conservation is unaffected by their inclusion; and the net addition to $N_i(U)$ at the node x_i is

$$-(h_i + h_{i+1})^{-1}\Delta_+[h_i\tau^{(2)}_{i-\frac{1}{2}}\Delta_-U_i - h_i\tau^{(4)}_{i-\frac{1}{2}}\Delta_-\delta^2U_i]. \tag{20}$$

For simplicity, we suppose that $\tau^{(2)}$ and $\tau^{(4)}$ are both zero in the intervals near the boundaries; then summing by parts over the contributing interior cells, we have

$$\begin{aligned} 2[B^c(V,V) - B^0(V,V)] &= -\sum V_j\Delta_+[h_j(\tau^{(2)}_{j-\frac{1}{2}}\Delta_-V_j - \tau^{(4)}_{j-\frac{1}{2}}\Delta_-\delta^2V_j)] \\ &= \sum h_j[\tau^{(2)}_{j-\frac{1}{2}}(\Delta_-V_j)^2 - \tau^{(4)}_{j-\frac{1}{2}}(\Delta_-V_j)(\Delta_-\delta^2V_j)], \end{aligned} \tag{21}$$

where the notation $B^c(\cdot,\cdot)$ denotes that only the convection terms are included. In the following lemma we first consider the effect of the second order terms, using a discrete norm which is motivated by the right-hand side of (18), namely

$$\|V\|_h^2 := |b_0|\, V_0^2 + |b_J|\, V_J^2 + \sum_0^J (\Delta b)_j V_j^2$$
$$+ \sum_1^J \left|b_{j-\frac{1}{2}}\right| (\Delta_- V_j)^2, \tag{22}$$

where

$$(\Delta b)_j := \begin{cases} \Delta_+ b_j & \text{for } j \le k-2 \\ \frac{1}{2}\Delta_- b_k \equiv \frac{1}{2}\Delta_+ b_{k-1} & \text{for } j = k-1, k \\ \Delta_- b_j & \text{for } j \ge k+1 \end{cases} \tag{23}$$

and $\left|b_{j-\frac{1}{2}}\right| := \min(|b_{j-1}|\,, |b_j|)$. We also suppose in this lemma that $s_{k-\frac{1}{2}} = 0$.

Lemma 2.2 *Addition of second order artificial viscosity of the form (19) with*

$$\tau^{(2)}_{k-\frac{1}{2}} \ge D_- b_k \equiv (b_k - b_{k-1})/h_k, \tag{24}$$

and $\tau^{(2)}_{j-\frac{1}{2}} \ge 0$ *for* $j \ne k$*, is sufficient to ensure that with* $s_{k-\frac{1}{2}} = 0$ *we have*

$$B^c(V,V) \ge \tfrac{1}{2} \|V\|_h^2 . \tag{25}$$

Proof The only nonpositive definite term in (18), arising from the kth interval, can be rewritten by using

$$V_k V_{k-1} = \tfrac{1}{2}[V_{k-1}^2 + V_k^2 - (\Delta_- V_k)^2]$$

to give, with $s_{k-\frac{1}{2}} = 0$,

$$B^0(V,V) = \tfrac{1}{2}\|V_h\|^2 - \tfrac{1}{2}\left[\left|b_{k-\frac{1}{2}}\right| + \tfrac{1}{2}\Delta_- b_k\right](\Delta_- V_k)^2. \tag{26}$$

Since $\left|b_{k-\frac{1}{2}}\right| \le \frac{1}{2}\Delta_- b_k$, the coefficient of $(\Delta_- V_k)^2$ here can be dominated by the coefficient $\frac{1}{2} h_k \tau^{(2)}_{k-\frac{1}{2}}$ in the corresponding term provided by (21) if (24) is satisfied. □

Now let us consider whether the amount of artificial viscosity can be reduced by choosing $s_{k-\frac{1}{2}}$ more carefully and also modifying the norm (22). Apart from the terms $(\Delta_- V_{k-1})^2$ and $(\Delta_+ V_k)^2$, the only terms involving V_{k-1} and V_k in (18) can be written, with $\beta := -b_{k-1}/b_k$ and $s := s_{k-\frac{1}{2}}$, as

$$b_k[(1+\beta)V_k V_{k-1} + s(V_k^2 - \beta V_{k-1}^2 - (1-\beta)V_k V_{k-1})]; \tag{27}$$

and with $\tau := h_k \tau^{(2)}_{k-\frac{1}{2}}$ the artificial viscosity term that is added is $\tau(V_k - V_{k-1})^2$. Choosing s to minimise the magnitude of τ needed to make the result positive definite yields the following result.

Lemma 2.3 *The minimum second order artificial viscosity needed to render $B^c(V,V)$ positive definite is given by*

$$\tau^{(2)}_{k-\frac{1}{2}} > h_k^{-1}\frac{|b_k b_{k-1}|}{|b_k|+|b_{k-1}|}, \tag{28}$$

with $\tau^{(2)}_{j-\frac{1}{2}} = 0$ for $j \neq k$, which occurs when the distribution factors in the turning point cell are given by

$$s_{k-\frac{1}{2}} = \frac{b_k + b_{k-1}}{|b_k|+|b_{k-1}|}. \tag{29}$$

Proof The coefficient of either V_k^2 or V_{k-1}^2 in (27) is negative, according to the sign of $s \equiv s_{k-\frac{1}{2}}$. If $\beta < 1$, we choose $s \geq 0$; then with the artificial viscosity term added the coefficients of both V_k^2 and V_{k-1}^2 are positive if $\tau > \beta b_k s = |b_{k-1}|\, s$. The whole expression is positive if this condition is strengthened to

$$\tau > \tfrac{1}{4}b_k[(1+\beta)(s^2+1) - 2(1-\beta)s], \tag{30}$$

and the right-hand side here has its minimum of $b_k\beta/(1+\beta)$ at $s = (1-\beta)/(1+\beta)$. This gives the result of (28) and (29), which is only changed when $\beta > 1$ by the change in sign of s. □

We have introduced absolute value signs in the denominators of (28) and (29) to emphasise the fact that the form for s in (29) can be used for all $s_{j-\frac{1}{2}}$ to generalise (13) to the turning point cell. Note that, with this choice, any value of $\tau^{(2)}_{k-\frac{1}{2}} > 0$ will restore coercivity when $b_{k-1} = 0$; and in the worst case, when $b_k + b_{k-1} = 0$, the amount needed is a quarter of that in Lemma 2.2.

However, fourth order dissipation is to be preferred because of its smaller effect on the smooth solution that is generally expected for the present problem. We show in the following lemma how it also can restore coercivity to $B^c(V,V)$.

Lemma 2.4 *Suppose the distribution factors are given by (12) with*

$$s_{j-\frac{1}{2}} = \frac{b_j + b_{j-1}}{|b_j|+|b_{j-1}|}. \tag{31}$$

Then $B^c(V,V)$ is made positive definite by adding fourth order artificial dissipation solely in the turning point cell with

$$(4-2\sqrt{3})\frac{|b_k b_{k-1}|}{|b_k|+|b_{k-1}|} < h_k \tau^{(4)}_{k-\frac{1}{2}} < (4+2\sqrt{3})\frac{|b_k b_{k-1}|}{|b_k|+|b_{k-1}|}. \tag{32}$$

Proof Using the same notation as in (27) except for $\tau := h_k \tau^{(4)}_{k-\frac{1}{2}}$, and adding in the $(\Delta_- V_{k-1})^2$ and $(\Delta_+ V_k)^2$ terms, we have from (18) and (21)

$$\begin{aligned} 2B^c(V,V) \geq\; & b_k\{(\Delta_+V_k)^2 + \beta(\Delta_-V_{k-1})^2 + s(V_k^2 - \beta V_{k-1}^2) \\ & + [(1+\beta) - s(1-\beta)]V_k V_{k-1}\} \\ & + \tau(\Delta_-V_k)(2\Delta_-V_k - \Delta_-V_{k-1} - \Delta_+V_k). \end{aligned} \tag{33}$$

From a Cauchy–Schwarz inequality we have

$$|(\Delta_- V_k)(\Delta_+ V_k)| \leq \tfrac{1}{2}\left[\left(\frac{\tau}{2b_k}\right)(\Delta_- V_k)^2 + \left(\frac{2b_k}{\tau}\right)(\Delta_+ V_k)^2\right], \tag{34}$$

with a similar bound for $|(\Delta_- V_k)(\Delta_- V_{k-1})|$. Hence, by introducing

$$R := \frac{1}{b_k} + \frac{1}{|b_{k-1}|} \equiv \frac{1}{b_k}(1 + \frac{1}{\beta}),$$

we obtain, from balancing the first and last pairs of terms in (33),

$$\begin{aligned} 2B^c(V,V) \geq{}& b_k\{s(V_k^2 - \beta V_{k-1}^2) + [(1+\beta) - s(1-\beta)]V_k V_{k-1}\} \\ &+ \tau(2 - \tfrac{1}{4}R\tau)(V_k - V_{k-1})^2. \end{aligned} \tag{35}$$

Now this is in exactly the same form as was obtained using second order artificial dissipation, and leading to the inequality (28). Noting that the right-hand side of (28) is just $(Rh_k)^{-1}$, we obtain the condition

$$R\tau(2 - \tfrac{1}{4}R\tau) > 1 \tag{36}$$

which is satisfied by the inequalities in (32). □

The upper bound on $\tau^{(4)}_{k-\frac{1}{2}}$ that is required by (32) may seem a little restrictive, because of its dependence on b near the turning point, but it is not so in practice. It comes about through the Cauchy–Schwarz inequality (34) when b_k is small, or the corresponding case when b_{k-1} is small. However, the fourth order dissipation would normally be applied over a patch of cells and this would overcome this problem. To illustrate the situation, let us write $\tau_j := h_j \tau^{(4)}_{j-\frac{1}{2}}$ and set $\tau_j = 0$ except for $j = k-1, k, k+1$. Then with simple Cauchy–Schwarz bounds we have from (21)

$$\begin{aligned} 2[B^c(V,V) - B^0(V,V)] &= \sum_{k-1}^{k+1} \tau_j(\Delta_- V_j)(2\Delta_- V_j - \Delta_- V_{j-1} - \Delta_- V_{j+1}) \\ &\geq \sum_{k-1}^{k+1} \tau_j[(\Delta_- V_j)^2 - \tfrac{1}{2}(\Delta_- V_{j-1})^2 - \tfrac{1}{2}(\Delta_- V_{j+1})^2]. \end{aligned}$$

By choosing

$$\tau_{k-1} = \tau_{k+1} = \tfrac{1}{2}\tau_k, \tag{37}$$

this collapses to

$$2[B^c(V,V) - B^0(V,V)] \geq \tau_k[\tfrac{1}{2}(\Delta_- V_k)^2 - \tfrac{1}{4}(\Delta_- V_{k-2})^2 - \tfrac{1}{4}(\Delta_- V_{k+2})^2]. \tag{38}$$

Now from the assumption (7), it is clear that

$$|b_{k-2}| \geq h_{k-1}\gamma, \quad b_{k+1} \geq h_{k+1}\gamma;$$

and hence the negative terms in (38) can be dominated by the corresponding terms in $B^0(V,V)$ given by (18) if

$$\max\left(\tau^{(4)}_{k-\frac{3}{2}}, \tau^{(4)}_{k+\frac{1}{2}}\right) \le 2\gamma, \quad \text{i.e.} \quad \tau^{(4)}_{k-\frac{1}{2}} \le 4\gamma \min\left(\frac{h_{k-1}}{h_k}, \frac{h_{k+1}}{h_k}\right), \tag{39}$$

which is a very unrestrictive bound. Comparing (38) with (35), we have in fact constructed a positive definite $B^c(V,V)$ so long as

$$\frac{|b_k b_{k-1}|}{|b_k| + |b_{k-1}|} < h_k \tau^{(4)}_{k-\frac{1}{2}} \le 4\min(|b_{k-2}|, b_{k+1}), \tag{40}$$

rather than the conditions (32).

The particular choice of distribution factors given by (31) leads to a minor modification to the discrete norm (22) if a lower bound to $B^c(V,V)$ is sought, as in (25). If $\tau^{(2)}_{k-\frac{1}{2}}$ is taken to be double the value on the right of (28), then the quadratic form in (27) is modified to $b_k V_k^2 + |b_{k-1}| V_{k-1}^2$; that is, (27) with s given by (29) can be written as

$$b_k V_k^2 + |b_{k-1}| V_{k-1}^2 - 2\frac{|b_k b_{k-1}|}{|b_k| + |b_{k-1}|}(\Delta_- V_k)^2. \tag{41}$$

Hence we introduce $\|V\|^2_{h*}$ as follows,

$$\|V\|^2_{h*} \equiv \|V\|^2_h, \quad \text{except that} \quad (\Delta b)_{k-1} = |b_{k-1}|, (\Delta b)_k = b_k. \tag{42}$$

Also, since $\left|b_{k-\frac{1}{2}}\right|$ is not less than the coefficient of $2(\Delta_- V_k)^2$ in (41), it is easily checked that

$$B^c(V,V) \ge \tfrac{1}{2}\|V\|^2_{h*} \quad \text{if} \quad \tau^{(2)}_{k-\frac{1}{2}} \ge 3\left|b_{k-\frac{1}{2}}\right|/h_k. \tag{43}$$

Similar results can be obtained using fourth order dissipation, although the coefficients of $(\Delta_- V_{k-1})^2$ and $(\Delta_- V_{k+1})^2$ (or those of terms further from the turning point) are reduced by the arguments of (33) and (34). For example, it is straightforward to show that the choice of coefficients given by (37) gives the following result,

$$B^c(V,V) \ge \tfrac{1}{4}\|V\|^2_{h*} \tag{44}$$

$$\text{if} \quad 5\min(|b_{k-1}|, b_k) \le \tau_k \le 2\max(|b_{k-2}|, b_{k+1}); \tag{45}$$

note that the dissipation can be spread more widely if these conditions cannot be met. We shall use these lower bounds in the error analysis of section 4.

If artificial dissipation is to be applied more generally, as is often needed for more complex problems, its form and the variation of its coefficients needs to be considered more carefully, for example whether it should be based on divided rather than undivided differences. This will not be a major consideration in the present paper. In the next two sections, however, where the effect of artificial dissipation on the coercivity of dissipation terms and on the truncation error are considered, we will need to take account of the wider application of such terms.

3. Coercivity of diffusion terms on a general mesh

In Mackenzie and Morton [7] an energy analysis of the error obtained with a first order approximation to the diffusive fluxes $\epsilon a_j U_j'$ in (11) was carried out on a general mesh. And in Morton and Stynes [9] a similar analysis was performed with a second order approximation on a mesh that was restricted to be graded in the flow direction, i.e. $b_j > 0 \; \forall j \Rightarrow h_j \geq h_{j+1}$. However, it has been shown by García-Archilla and Mackenzie [2] that this latter scheme is second order convergent even on a random mesh. In this section we will therefore show how the addition of artificial viscosity can be used to render this scheme coercive on any mesh.

For simplicity we will apply homogeneous Neumann boundary conditions at both $x = 0$ and $x = 1$, so the nodal residual equations (14) are used to determine all the unknown nodal values as in the pure convection case; only the more general definition of the cell residuals given by (11) need to be substituted in (14), and here we shall assume that $a_j = 1 \;\; \forall j$. The definitions of (15) and (16) remain valid as well as the expansion for $B^0(V, V)$ given by (17), but the diffusion terms now need to be added to give the full bilinear form

$$B^h(U, V) = B^c(U, V) + \epsilon B^d(U, V). \tag{46}$$

Combining (11),(12),(14) and (16) with the boundary conditions $U_0' = U_J' = 0$, extended by $U_{-1}' = U_{J+1}' = 0$, gives

$$B^d(U, V) = -\sum_0^J V_j \left[\tfrac{1}{2}(1 + s_{j-\frac{1}{2}})\Delta_- U_j' + \tfrac{1}{2}(1 - s_{j+\frac{1}{2}})\Delta_+ U_j'\right]. \tag{47}$$

Summing this by parts and using the identity

$$\Delta_+ \left[(1 + s_{j-\frac{1}{2}})V_j\right] + \Delta_-[(1 - s_{j+\frac{1}{2}})V_j] \equiv (1 + s_{j+\frac{1}{2}})\Delta_+ V_j + (1 - s_{j-\frac{1}{2}})\Delta_- V_j,$$

we then obtain

$$B^d(U, V) = \sum_1^J (\Delta_- V_j) \left[\tfrac{1}{2}(1 + s_{j-\frac{1}{2}})U_{j-1}' + \tfrac{1}{2}(1 - s_{j-\frac{1}{2}})U_j'\right]. \tag{48}$$

Now the general form used to give the gradients U_j' in terms of the nodal values is

$$U_j' = \alpha_j D_+ U_j + (1 - \alpha_j) D_- U_j, \tag{49}$$

where D_+ and D_- denote the divided forward and backward difference operators, respectively. Thus we finally obtain the following quadratic form when $U = V$,

$$B^d(V,V) = \sum_1^J c_{j-\frac{1}{2}}(D_-V_j)^2 + \sum_1^{J-1} d_j(D_-V_j)(D_+V_j), \tag{50}$$

where

$$c_{j-\frac{1}{2}} = h_j[\tfrac{1}{2}(1-s_{j-\frac{1}{2}})(1-\alpha_j) + \tfrac{1}{2}(1+s_{j-\frac{1}{2}})\alpha_{j-1}] \tag{51}$$
$$d_j = h_j[\tfrac{1}{2}(1-s_{j-\frac{1}{2}})\alpha_j] + h_{j+1}[\tfrac{1}{2}(1+s_{j+\frac{1}{2}})(1-\alpha_j)], \tag{52}$$

and we set $\alpha_0 = 0, \alpha_J = 1$.

The positive definiteness of $B^d(V,V)$ is most easily considered when the cell vertex scheme reduces to the three-point fully upwind difference scheme away from the turning point, for then only d_{k-1} and d_k are nonzero in the above expansions. This scheme occurs when the distribution factors given by (13), implying that $s_{j-\frac{1}{2}} = -1$ for $j \leq k-1$ and $s_{j-\frac{1}{2}} = 1$ for $j \geq k+1$, are coupled with the gradient choice given by

$$\alpha_j = 0 \quad \text{for} \quad j < k-1, \quad \alpha_j = 1 \quad \text{for} \quad j > k. \tag{53}$$

Then, writing $s_{k-\frac{1}{2}} = s$, we have

$$d_{k-1} = \alpha_{k-1}h_{k-1} + \tfrac{1}{2}(1+s)(1-\alpha_{k-1})h_k, \quad d_k = \tfrac{1}{2}(1-s)\alpha_k h_k + (1-\alpha_k)h_{k+1}, \tag{54}$$

with all other d_j equal to zero; no choice of α_{k-1}, α_k and s can make both of these zero. Moreover, from (51) we see that the corresponding two cross product terms would have to be dominated by terms given by

$$c_{k-\frac{3}{2}} = (1-\alpha_{k-1})h_{k-1}, \quad c_{k-\frac{1}{2}} = [\tfrac{1}{2}(1-s)(1-\alpha_k) + \tfrac{1}{2}(1+s)\alpha_{k-1}]h_k,$$
$$c_{k+\frac{1}{2}} = \alpha_k h_{k+1}. \tag{55}$$

Assuming each c_j is positive, the necessary and sufficient condition for this domination to hold is that

$$4c_{k-\frac{3}{2}}c_{k-\frac{1}{2}}c_{k+\frac{1}{2}} \geq c_{k+\frac{1}{2}}d_{k-1}^2 + c_{k-\frac{3}{2}}d_k^2. \tag{56}$$

This condition is a direct generalisation of the familiar $4ac \geq b^2$ condition for a quadratic in two variables, and is readily derived from the conditions for the tridiagonal matrix associated with the quadratic form (50) to be positive definite; for future reference it is worth noting what these are, namely that the determinants Δ_j of the principal minors be positive, where these quantities are given by the recursion $\Delta_0 = 1, \Delta_1 = c_{\frac{1}{2}}$ and

$$\Delta_j = c_{j-\frac{1}{2}}\Delta_{j-1} - \tfrac{1}{4}d_{j-1}^2\Delta_{j-2}. \tag{57}$$

Suppose we take $s = 1$ and $\alpha_k = 1$, thus extending the upwinding on the right down to the turning point cell and making $d_k = 0$. Then (56) reduces to

$$4\alpha_{k-1}(1-\alpha_{k-1})h_{k-1}h_k \geq [\alpha_{k-1}h_{k-1} + (1-\alpha_{k-1})h_k]^2,$$

which can only be satisfied if

$$\alpha_{k-1}h_{k-1} = (1-\alpha_{k-1})h_k, \quad \text{i.e.} \quad \alpha_{k-1} = h_k/(h_{k-1}+h_k). \tag{58}$$

This corresponds to the scheme for calculating the gradient at x_{k-1} called Method B by Mackenzie and Morton [7].

However, the main practical interest is in the coercivity of the second order accurate scheme that is obtained by using this formula everywhere; that is, by interpolating a quadratic to U_{j-1}, U_j, U_{j+1} at each triplet of mesh points in order to obtain U_j', which gives

$$\alpha_j = h_j/(h_j + h_{j+1}), \; 1-\alpha_j = h_{j+1}/(h_j + h_{j+1}). \tag{59}$$

This in turn gives for the coefficients (51) and (52) in the quadratic form (50),

$$c_{j-\frac{1}{2}} = \tfrac{1}{2}(1 - s_{j-\frac{1}{2}})\frac{h_j h_{j+1}}{h_j + h_{j+1}} + \tfrac{1}{2}(1 + s_{j-\frac{1}{2}})\frac{h_{j-1}h_j}{h_{j-1} + h_j} \tag{60}$$

$$d_j = \tfrac{1}{2}(1 - s_{j-\frac{1}{2}})\frac{h_j^2}{h_j + h_{j+1}} + \tfrac{1}{2}(1 + s_{j+\frac{1}{2}})\frac{h_{j+1}^2}{h_j + h_{j+1}}. \tag{61}$$

To avoid too much complication, we will use the simple compact conditions

$$d_j^2 \leq c_{j-\frac{1}{2}}c_{j+\frac{1}{2}} \quad \text{for} \quad 2 \leq j \leq J-2, \tag{62}$$

$$d_1^2 \leq 2c_{\frac{1}{2}}c_{\frac{3}{2}}, \quad d_{J-1}^2 \leq 2c_{J-\frac{3}{2}}c_{J-\frac{1}{2}} \tag{63}$$

which are sufficient to ensure that all the determinants in (57) are non-negative and hence that $B^d(\cdot,\cdot)$ is positive semi-definite.

Now the addition of fourth order artificial dissipation supplements $\epsilon B^d(V,V)$, as in (21), by

$$\tfrac{1}{2}\sum h_j \tau_{j-\frac{1}{2}}^{(4)}(\Delta_- V_j)[2\Delta_- V_j - \Delta_- V_{j+1} - \Delta_- V_{j-1}] \tag{64}$$

if, as we have so far, we use undivided differences for its definition. The result is that the coefficients in (50) are modified to give

$$c_{j-\frac{1}{2}}^* = c_{j-\frac{1}{2}} + \epsilon^{-1}h_j^3\tau_{j-\frac{1}{2}}^{(4)}, \quad d_j^* = d_j - \tfrac{1}{2}\epsilon^{-1}h_j h_{j+1}(h_j\tau_{j-\frac{1}{2}}^{(4)} + h_{j+1}\tau_{j+\frac{1}{2}}^{(4)}). \tag{65}$$

This strongly suggests replacing (50) by the equivalent quadratic form in the undivided differences, and hence introducing the notation

$$\tau_{j-\frac{1}{2}} := \epsilon^{-1} h_j \tau^{(4)}_{j-\frac{1}{2}}, \quad \tilde{c}_{j-\frac{1}{2}} = h_j^{-2} c_{j-\frac{1}{2}}, \quad \tilde{d}_j = (h_j h_{j+1})^{-1} d_j; \tag{66}$$

then the condition(62) for positive definiteness becomes

$$[\tilde{d}_j - \tfrac{1}{2}(\tau_{j-\frac{1}{2}} + \tau_{j+\frac{1}{2}})]^2 \leq (\tilde{c}_{j-\frac{1}{2}} + \tau_{j-\frac{1}{2}})(\tilde{c}_{j+\frac{1}{2}} + \tau_{j+\frac{1}{2}} \quad \forall j. \tag{67}$$

It is clear that this can always be satisfied with sufficiently large values of $\tau^{(4)}$; what we now need to do is estimate how large these values have to be.

Taking a constant value

$$h_j \tau^{(4)}_{j-\frac{1}{2}} \equiv \epsilon \tau_{j-\frac{1}{2}} = \epsilon \tau \quad \forall j, \tag{68}$$

ensures, from summing (64) by parts, that increasing the artificial viscosity always increases the positive definiteness of $B^d(\cdot,\cdot)$; also the quadratic terms in (67) then cancel and the conditions become

$$\tau \geq \frac{\tilde{d}_j^2 - \tilde{c}_{j-\frac{1}{2}} \tilde{c}_{j+\frac{1}{2}}}{2\tilde{d}_j + \tilde{c}_{j-\frac{1}{2}} + \tilde{c}_{j+\frac{1}{2}}} \quad \forall j. \tag{69}$$

To bound the right-hand side, we introduce a notation for the mesh ratios and their upper and lower bounds,

$$\gamma \leq \gamma_j := h_{j+1}/h_j \leq \Gamma \quad \forall j. \tag{70}$$

Then for $s_{j-\frac{1}{2}} = s_{j+\frac{1}{2}} = 1$, we obtain

$$\tilde{c}_{j-\frac{1}{2}} = \frac{1}{h_j(1+\gamma_{j-1})}, \quad \tilde{d}_j = \frac{\gamma_j}{h_j(1+\gamma_j)},$$

which leads to the condition, after a little algebra, that we need

$$\tau \geq \frac{1}{h_j(1+\gamma_j)} \frac{\gamma_j^3 + \gamma_j^3 \gamma_{j-1} - 1 - \gamma_j}{2\gamma_j^2 + 2\gamma_j^2 \gamma_{j-1} + \gamma_j + \gamma_j^2 + 1 + \gamma_{j-1}} \quad \text{for } j \geq k+1.$$

This implies that no artificial dissipation is needed here if the mesh is decreasing towards the boundary, as was shown by Morton and Stynes [9]; a similar result holds for $j \leq k-2$. However, for the following lemma we will use the obviously sufficient condition $\tau \geq \frac{1}{2}\tilde{d}_j$.

Note that so far we have been imprecise about the boundary conditions for the artificial dissipation, and have previously assumed that it was set to zero near the boundaries. However, for the present purposes it is clearly convenient to apply it to every cell by satisfying (68). It is easily checked that this can be achieved by setting $\delta^2 V_0 = 0 = \delta^2 V_J$ in its definition, and that then all the summing by parts that has been applied is valid. Thus in (68), (69) and below it is understood that $j = 1, 2, \ldots, J$.

Lemma 3.5 *The addition of fourth order dissipation with coefficients given by (68) such that*

$$\tau h_j \geq \tfrac{1}{2} \quad \forall j, \tag{71}$$

is sufficient to make $B^d(\cdot,\cdot)$ positive definite.

Proof In the worst case, which is when $s_{j-\frac{1}{2}} = -1$ and $s_{j+\frac{1}{2}} = 1$ and might occur for $j = k$, we have from (61) and (65) that

$$\tilde{d}_j = \frac{1}{h_j + h_{j+1}} \left(\frac{h_{j+1}}{h_j} + \frac{h_j}{h_{j+1}} \right) \leq \min(h_j^{-1}, h_{j+1}^{-1}).$$

Then the condition $\tau \geq \frac{1}{2}\tilde{d}_j$ gives (71). □

The condition (71) is clearly far from sharp; in particular, the fact that from (68) we then have

$$\tau^{(4)}_{j-\frac{1}{2}} \geq \tfrac{1}{2}\epsilon h_j^{-2}$$

is inconvenient where the mesh is very fine. This strongly suggests that divided differences should be used in defining the artificial dissipation, instead of (19).

4. Coercivity-based error bounds

We denote the linear interpolant of the exact solution u of (6) by u^I, the difference $U - u^I$ by E and the gradient recovery error $u'(x_j) - (u^I)'_j$ by η_j. Then since the cell vertex approximation is given by $N(U) = 0$, we obtain from (16)

$$\begin{aligned} B^h(E,E) &= (N(u^I + E) - N(u^I), E) \\ &= -(N(u^I), E). \end{aligned} \tag{72}$$

The term $N(u^I)$ has the form of a truncation error, in part arising from the cell residual and in part from the artificial dissipation terms. For the former, we assume that we replace the trapezoidal rule in (11) by exact integration of the source term and take $a(x) \equiv 1$; then we have

$$\begin{aligned} h_j R_{j-\frac{1}{2}}(u^I) &= \Delta_-(b_j u_j^I - \epsilon(u^I)'_j) - \int_{x_{j-1}}^{x_j} S(x)\mathrm{d}x \\ &= \Delta_-(b_j u_j^I - \epsilon(u^I)'_j) - \Delta_-(b_j u(x_j) - \epsilon u'(x_j)) \\ &= \epsilon\Delta_-(u'(x_j) - (u^I)'_j) =: \epsilon\Delta_-\eta_j. \end{aligned} \tag{73}$$

If we also use only fourth order artificial dissipation we define, as in (19),

$$A^I_{i-\frac{1}{2},i} := -\tau^{(4)}_{i-\frac{1}{2}}\Delta_-\delta^2 u^I_i, \quad \text{etc.} \tag{74}$$

in order to obtain the second contribution to the truncation error.

From (14), and using the distribution factors given by (31) we therefore obtain

$$\begin{aligned}(N(u^I),E) &= \tfrac{1}{2}\sum_0^J E_j[h_j(D_{j-\frac{1}{2},j}R_{j-\frac{1}{2}}(u^I)+A^I_{j-\frac{1}{2},j}) \\ &\qquad + h_{j+1}(D_{j+\frac{1}{2},j}R_{j+\frac{1}{2}}(u^I)+A^I_{j+\frac{1}{2},j})] \\ &=: S_R + S_A, \end{aligned} \tag{75}$$

where, as in (17),

$$\begin{aligned} S_R &= \epsilon\sum_0^J E_j[\tfrac{1}{2}(1+s_{j-\frac{1}{2}})\Delta_-\eta_j + \tfrac{1}{2}(1-s_{j+\frac{1}{2}})\Delta_-\eta_{j+1}] \\ &= \epsilon\Big\{\sum_0^{k-2} E_j\Delta_+\eta_j + \sum_{k+1}^J E_j\Delta_-\eta_j \\ &\qquad + \left(\frac{|b_{k-1}|\,E_{k-1}+b_kE_k}{|b_{k-1}|+b_k}\right)\Delta_-\eta_k\Big\}, \end{aligned} \tag{76}$$

and, if artificial dissipation is applied in all cells by setting $\delta^2 u^I_0$ and $\delta^2 u^I_J$ to zero,

$$S_A = \tfrac{1}{2}\sum_0^J E_j\Delta_+(h_j\tau^{(4)}_{j-\frac{1}{2}}\Delta_-\delta^2u^I_j), \tag{77}$$

with $\tau^{(4)}_{-\frac{1}{2}} = \tau^{(4)}_{J+\frac{1}{2}} = 0$. In order to bound the ratio of this inner product with $\|E\|_{h*}$, we could sum each of these by parts and make the maximum use of the terms in (22) and (42). For simplicity, however, we generally will use only the forms (76) and (77) to obtain

$$\left|(N(u^I),E)\right| \le \|E\|_{h*}(\epsilon\|\eta\|_\Delta + \tfrac{1}{2}\left\|u^I\right\|_\tau), \tag{78}$$

where, from (23) and (42),

$$\|\eta\|^2_\Delta = \sum_0^{k-2}\frac{(\Delta_+\eta_j)^2}{\Delta_+b_j} + \frac{(\Delta_-\eta_k)^2}{|b_{k-1}|+b_k} + \sum_{k+1}^J\frac{(\Delta_-\eta_j)^2}{\Delta_-b_j}, \tag{79}$$

and

$$\left\|u^I\right\|^2_\tau = \sum_0^J(\Delta b)^{-1}_j\left|\Delta_+(h_j\tau^{(4)}_{j-\frac{1}{2}}\Delta_-\delta^2u^I_j)\right|^2, \tag{80}$$

with the notational convention at the boundaries as described above. We conclude by combining these expressions with the bounds derived in sections 3 and 4 to give the following result.

Theorem 4.6 *Suppose the cell vertex method defined by (14), (12) and (31) is applied to the expansion critical point problem given by (6) and (7). Then the addition of fourth order artificial dissipation to satisfy (45) and (71) enforces coercivity on any mesh and yields the following error bound*

$$\|E\|_{h*} \le \epsilon[4\|\eta\|_\Delta + \tau_d \left\|u^I\right\|_{(4)}] + \tau_c \left\|u^I\right\|_k, \tag{81}$$

where $\|\eta\|_\Delta$ *is given by (79),* $\tau_d = h_{\min}^{-1}$, $\tau_c = 10\left|b_{k-\frac{1}{2}}\right|$,

$$\left\|u^I\right\|_4^2 = (\Delta b)_0^{-1}(\delta^2 u_1^I)^2 + (\Delta b)_J^{-1}(\delta^2 u_{J-1}^I)^2 + \sum_1^{J-1}(\Delta b)_j^{-1}(\delta^4 u_j^I)^2, \tag{82}$$

with $\delta^2 u_0^I = 0 = \delta^2 u_J^I$, *and*

$$\left\|u^I\right\|_k^2 = \tfrac{1}{4}\left|b_{k-\frac{3}{2}}\right|^{-1}(\Delta_-\delta^2 u_{k-1}^I)^2 + \left|b_{k-\frac{1}{2}}\right|^{-1}(\Delta_-\delta^2 u_k^I)^2 + \tfrac{1}{4}\left|b_{k+\frac{1}{2}}\right|^{-1}(\Delta_-\delta^2 u_{k+1}^I)^2. \tag{83}$$

Proof Artificial dissipation satisfying (45) ensures that $B^c(E,E) \ge \frac{1}{4}\|E\|_{h*}^2$, and a further amount satisfying the conditions of Lemma 3.5 ensures that $B^d(E,E) \ge 0$. Hence we have $B^h(E,E) \ge \frac{1}{4}\|E\|_{h*}^2$. To obtain (81) we use (72) and (75)–(77), substituting the two sets of artificial dissipation terms into (77) before applying the Cauchy–Schwarz inequality separately. For the diffusion inner product, the bounds are obtained as in (78)–(80) so as to give (82), with τ_d obtained from (71). However, for the convection inner product it is more convenient to sum by parts to give, from (37),

$$-\tfrac{1}{2}\tau_k[(\Delta_-E_{k-1})\tfrac{1}{2}\Delta_-\delta^2 u_{k-1}^I + (\Delta_-E_k)\Delta_-\delta^2 u_k^I + (\Delta_-E_{k+1})\tfrac{1}{2}\Delta_-\delta^2 u_{k-1}^I].$$

Then applying a Cauchy–Schwarz inequality to this, with τ_k given by (45) and $\tau_c = 2\tau_k$, yields (83). □

Let us consider briefly the order of accuracy implied by this bound, remembering that we can assume that the solution u is quite smooth. The coefficients $(\Delta b)_j^{-1}$ in the norms will normally be $O(h^{-1})$; since we will have $\Delta\eta = O(h^3)$ on a smooth mesh, we can expect $\|\eta\|_\Delta$ to be $O(h^2)$. Similarly, with $\delta^4 u^I = O(h^4)$ and $\tau_d = O(h^{-1})$, we can expect $\tau_d\left\|u^I\right\|_{(4)}$ to be $O(h^2)$, although more care may be needed in the treatment of the $\delta^2 u_1^I$ and $\delta^2 u_{J-1}^I$ terms in (82), as well as the terms involving $(\Delta b)_j^{-1}$ for $j = k-1$ and k. However, both of these terms are multiplied by the factor ϵ. Finally, the locally applied artificial viscosity near the turning point gives a term of the order $\left|b_{k-\frac{1}{2}}\right|^{\frac{1}{2}}\Delta_-\delta^2 u^I$ which is $O(h^{3.5})$. Thus in both cases the addition of fourth order artificial viscosity has greatly widened the range of conditions under which the scheme is coercive, without affecting its accuracy.

5. References

1. Crumpton, P.I., Mackenzie, J.A. and Morton, K.W. (1993), Cell vertex algorithms for the compressible Navier-Stokes equations, *J. Comput. Phys.* **109**(1), 1–15, 1993.
2. García-Archilla, B. and Mackenzie, J.A. (1995), Analysis of a supraconvergent cell vertex finite-volume method for one-dimensional convection-diffusion problems, *IMA J. Numer. Anal.* **15**, 101–115, 1995.
3. Hall, M.G. (1985), Cell-vertex multigrid schemes for solution of the Euler equations, *in* K.W. Morton and M.J. Baines, eds, Proceedings of the Conference on Numerical Methods for Fluid Dynamics, University of Reading, Clarendon Press, Oxford, pp. 303–345, 1985.
4. Keller, H.B. and Cebeci, T. (1971), Accurate numerical methods for boundary layer flow I: two-dimensional laminar flows, *in* Lecture Notes in Physics, Proceedings of Second International Conference on Numerical Methods in Fluid Dynamics, Springer-Verlag, Berlin, pp. 92–100.
5. Keller, H.B. and Cebeci, T. (1972), Accurate numerical methods for boundary layer flows II: two-dimensional turbulent flows, *AIAA J.* **10**(9), 1193–1199, 1972.
6. Mackenzie, J.A. (1991), Cell vertex finite volume methods for the solution of the compressible Navier-Stokes equations, PhD thesis, Oxford University.
7. Mackenzie, J.A. and Morton, K.W. (1992), Finite volume solutions of convection-diffusion test problems, *Math. Comp.* **60**(201), 189–220, 1992.
8. Morton, K.W. and Paisley, M.F. (1989), A finite volume scheme with shock fitting for the steady Euler equations, *J. Comput. Phys.* **80**, 168–203, 1989.
9. Morton, K.W. and Stynes, M. (1994), An analysis of the cell vertex method, M^2AN **28**(6), 699–724, 1994.
10. Morton, K.W., Rudgyard, M.A. and Shaw, G.J. (1994), Upwind iteration methods for the cell vertex scheme in one dimension, *J. Comput. Phys.* **114**(2), 209–226, 1994.
11. Ni, R.-H. (1981), A multiple grid scheme for solving the Euler equations, *AIAA Journal* **20**(11), 1565–1571, 1981.
12. Preissmann, A. (1961), Propagation des intumescences dans les canaux et rivières, *in* 1st Congrès de l'Assoc. Française de Calc., AFCAL, Grenoble, pp. 433–442, 1961.
13. Thomée, V. (1962), A stable difference scheme for the mixed boundary problem for a hyperbolic first order system in two dimensions, *J. Soc. Indust. Appl. Math.* **10**, 229–245,1962.
14. Wendroff, B. (1960), On centered difference equations for hyperbolic systems, *J. Soc. Indust. Appl. Math.* **8**, 549–555,1960.

A METHOD OF LINES PACKAGE, BASED ON MONOMIAL SPLINE COLLOCATION, FOR SYSTEMS OF ONE DIMENSIONAL PARABOLIC DIFFERENTIAL EQUATIONS

THEODORE B. NOKONECHNY, PATRICK KEAST
Department of Mathematics, Statistics and Computing Science, Dalhousie University
Halifax, Nova Scotia, B3H 4J5, Canada
E-mail: nokon@cs.dal.ca, keast@cs.dal.ca

and

PAUL H. MUIR
Department of Mathematics and Computing Science, Saint Mary's University
Halifax, Nova Scotia, Canada
E-mail: muir@cs.dal.ca

ABSTRACT

We consider systems of parabolic partial differential equations in one space variable, for which we describe a method of lines algorithm based on monomial spline collocation for the discretization of the spatial domain. While the usual application of this technique transforms the system of partial differential equations into a system of time dependent ordinary differential equations which can be integrated by standard initial value solvers, our approach leads to coupled systems of differential-algebraic equations. These equations are solved using a well known package, DASSL, which we have modified to take advantage of the special structure of the Jacobians which arise.

1. The Problem Class

In this paper we consider systems of $NPDE$ parabolic partial differential equations (PDEs). The systems considered are of the form

$$\frac{\partial u}{\partial t}(t,x) = f(t,x,u(t,x),u_x(t,x),u_{xx}(t,x)), \tag{1}$$

for $x_a < x < x_b$, $t_0 < t \leq t_{out}$, with initial conditions at $t = t_0$ given by

$$u(t_0,x) = u_0(x), \quad x_a \leq x \leq x_b, \tag{2}$$

subject to the separated boundary conditions at $x = x_a$ and $x = x_b$ given by

$$b_{x_a}(t,u(t,x_a),u_x(t,x_a)) = 0, \quad t_0 \leq t \leq t_{out}, \tag{3}$$

$$b_{x_b}(t,u(t,x_b),u_x(t,x_b)) = 0, \quad t_0 \leq t \leq t_{out}, \tag{4}$$

where

$$u(t,x) = [u_1(t,x),u_2(t,x), \ \ldots \ ,u_{NPDE}(t,x)]^T, \tag{5}$$

$$u_x(t,x) = \left[\frac{\partial u_1}{\partial x}(t,x), \frac{\partial u_2}{\partial x}(t,x), \ \dots\ , \frac{\partial u_{NPDE}}{\partial x}(t,x)\right]^T, \quad (6)$$

$$u_{xx}(t,x) = \left[\frac{\partial^2 u_1}{\partial x^2}(t,x), \frac{\partial^2 u_2}{\partial x^2}(t,x), \ \dots\ , \frac{\partial^2 u_{NPDE}}{\partial x^2}(t,x)\right]^T, \quad (7)$$

$$f(t,x,u,u_x,u_{xx}) = [f_1(t,x,u,u_x,u_{xx}), \ \dots\ , f_{NPDE}(t,x,u,u_x,u_{xx})]^T, \quad (8)$$

$$u_0(x) = [u_{0,1}(x), u_{0,2}(x), \ \dots\ , u_{0,NPDE}(x)]^T, \quad (9)$$

$$b_{x_a}(t,u,u_x) = [b_{x_a,1}(t,u,u_x), \ \dots\ , b_{x_a,NPDE}(t,u,u_x)]^T, \quad (10)$$

$$b_{x_b}(t,u,u_x) = [b_{x_b,1}(t,u,u_x), \ \dots\ , b_{x_b,NPDE}(t,u,u_x)]^T. \quad (11)$$

In (8), (10), and (11), for brevity, we suppress the dependence on t, x of $u = u(t,x)$, $u_x = u_x(t,x)$, and $u_{xx}(t,x)$. The domain and range of the individual elements of (5)-(11) are

$$u_i(t,x),\ \frac{\partial u_i}{\partial x}(t,x),\ \frac{\partial^2 u_i}{\partial x^2}(t,x) : [t_0,\infty)\times[x_a,x_b] \rightarrow \mathbf{R}, \quad (12)$$

$$f_i(t,x,u,u_x,u_{xx}) : [t_0,\infty)\times[x_a,x_b]\times\mathbf{R}^{NPDE}\times\mathbf{R}^{NPDE}\times\mathbf{R}^{NPDE} \rightarrow \mathbf{R}, \quad (13)$$

$$u_{0,i}(x) : [x_a,x_b] \rightarrow \mathbf{R}, \quad (14)$$

$$b_{x_a,i}(t,u,u_x),\ b_{x_b,i}(t,u,u_x) : [t_0,\infty)\times\mathbf{R}^{NPDE}\times\mathbf{R}^{NPDE} \rightarrow \mathbf{R}, \quad (15)$$

for $i = 1,\dots,NPDE$. Each $u_i(t,x)$ is assumed to have sufficient differentiability appropriate for (1). All components of f in (8) and of b_{x_a} and b_{x_b} in (10)-(11) are assumed to be differentiable with respect to their arguments. The problem class includes all systems of parabolic PDEs in one space variable which have (fixed) separated boundary conditions. For a treatment of PDE systems which include a moving boundary and/or coupled ODEs, the reader is referred to Berzins and Dew[5]. It is assumed that the system (1)-(4) is such that the existence of a unique solution is assured.

Our code is not intended to solve problems which have rapidly changing boundary layers, since it uses a currently fixed spatial mesh which is not suited to such problems. However, if this method is combined with a mesh refinement scheme, then boundary layers may be dealt with. Much recent work has been done on adaptive MOL algorithms which attempt to adjust the spatial discretization. Based on some estimate of the spatial and/or temporal errors, these methods attempt to adapt to the problem by changes in one or more of the time steps, the spatial mesh, and the order of approximation. Verwer et al[20,21] discuss a process called '*static-regridding*', in which the grid is adapted at discrete times. Lawson et al[12] and Berzins et al[6] describe a combined spatial and temporal approach. In Huang and Russell[10] a moving mesh procedure is used to control spatial errors, with the *number* of mesh points being kept fixed, while their *location* is adapted to the solution behaviour. Flaherty and Moore[9,16] use a combined order and spatial adaptivity. The above references, and the references therein, provide a good background to the literature of adaptive methods for parabolic differential equations.

We discuss a Method of Lines (MOL) approach to (1), in which the spatial variable is discretized, resulting in a system of Initial Value Ordinary Differential Algebraic Equations (ODE/DAEs). The spatial discretization technique we will employ will be collocation at Gaussian points in each subinterval of the given mesh. There are several MOL codes available, some based on finite differences for the spatial discretization, and others based on C^0 collocation, or collocation using B-Splines. Commercial packages which provide MOL codes include for example, the NAG library and IMSL. Among public domain codes available from netlib are, for example, PDECOL[14] and EPDCOL[11] (both using B-Splines), PDEONE[13] (which uses finite differences), and PDECHEB[5] (C^0 collocation), all of which are available through netlib from the Transactions on Mathematical Software collection. The paper by Carroll[7] includes a comparison of several one dimensional MOL codes. In this paper our code will be compared to EPDCOL, a modified form of PDECOL.

There are three principal components in the MOL approach to the solution of a system of parabolic equations, namely:

(a) The spatial discretization.

(b) The solver used for the resulting ODE/DAE system.

(c) The techniques used to solve the linear systems which arise.

For (a) and (c) we will use techniques not used in any other MOL code which is generally available. In Section 2 we describe the discretization which we use, and show that a system of DAEs is produced. In Section 3, we discuss the DAE system, and the Jacobian which arises in the solution of the nonlinear systems occurring in the time step. In Section 4 we look at the linear algebraic problems which arise. These have a special form and our package takes full advantage of this, using the package ABDPACK[15]. For (b) we use the program DASSL[19], in a modified form to allow interaction with ABDPACK. In Section 5, the user interface for the software is explained, and we show the results of applying the new software to a problem taken from the paper in which PDECOL[14] appeared.

2. The Spatial Discretization

The spatial discretization will be carried out by collocating at the Gauss-Legendre points. This is, in itself, nothing new; Gauss-Legendre collocation is used in both PDECOL and EPDCOL. However, both of these codes use B-Splines as the basis functions, whereas we will use a monomial basis. Our choice of monomial splines rather than B-splines[8] is motivated by earlier developments in software for boundary value ordinary differential equations. The widely used code COLSYS[1], developed about 20 years ago, makes use of B-splines. However, later investigation by Ascher et al[4] indicated that monomial splines were a preferred choice, and this led to the development of a modified version of COLSYS called COLNEW. This code uses monomials and is reported to give improved performance both in execution time and storage requirements.

2.1. The Basis Functions

First we describe the monomial spline basis used for the spatial discretization. The notation used is essentially that used in Ascher, Mattheij and Russell[2]. The spatial mesh $X = \{x_i\}_{i=1}^{NINT+1}$ is a partition of the interval $[x_a, x_b]$, defined by

$$x_a = x_1 < x_2 < \ldots < x_{NINT} < x_{NINT+1} = x_b. \tag{16}$$

We will use $\mathcal{M}_X^{\kappa,m}$ to denote the set of $C^{m-1}[x_a, x_b]$ monomial splines of order κ (degree $(\kappa - 1)$) defined on the mesh X. Thus

$$\mathcal{M}_X^{\kappa,m} = \{v \in C^{m-1}[x_a, x_b] : v|_{[x_i, x_{i+1}]} \in \mathcal{P}_\kappa,\ i = 1, \ldots, NINT\}, \tag{17}$$

where $\mathcal{P}_\kappa$ is the set of all polynomials of order κ or less, and $\kappa > m \geq 0$. The dimension of the space $\mathcal{M}_X^{\kappa,m}$ is $\kappa \cdot NINT - m(NINT - 1) = (\kappa - m)NINT + m$. In our code we assume that $m = 2$, but for generality we keep m as a variable here.

In order to define the collocation points we require two additional pieces of notation. First, define the mesh step size sequence $H = \{h_i\}_{i=1}^{NINT}$ by

$$h_i = x_{i+1} - x_i. \tag{18}$$

Second, let $\{\rho_r\}_{r=1}^{k}$ be the k Gauss-Legendre points (the zeros of the degree k Legendre polynomial) on the interval $[0, 1]$ where

$$0 < \rho_1 < \rho_2 < \ldots < \rho_k < 1, \tag{19}$$

where $k = \kappa - m$. We now define the collocation point sequence $\Xi = \{\xi_j\}_{j=1}^{NCPTS}$ by

$$\begin{aligned}
\xi_1 &= x_1, && (20)\\
\xi_{l+r} &= x_i + h_i\rho_r, \quad \text{where } l = (i-1)k + 1, &&\\
&\qquad \text{for } i = 1, \ldots, NINT, \quad r = 1, \ldots, k, && (21)\\
\xi_{NCPTS} &= x_{NINT+1}. && (22)
\end{aligned}$$

Note that Ξ is an increasing partition of $[x_a, x_b]$ and the inclusion of x_1 and x_{NINT+1} with the collocation points is done for notational convenience as in Madsen et al[14]. The number of collocation points is $NCPTS = k \cdot NINT + m$, which is the dimension of the space $\mathcal{M}_X^{\kappa,m}$.

The function $u(t,x)$ is approximated by $U(t,x)$, which is a linear combination of functions from the space $\mathcal{M}_X^{k+m,m}$ with coefficients which are functions of t. The basis functions we use for $\mathcal{M}_X^{k+m,m}$ are monomials locally, each with support over only one subinterval. If we consider the restriction of $U(t,x)$ to the subinterval $[x_i, x_{i+1}]$, for $i = 1, \ldots, NINT$, we have $U_i(t,x) = U(t,x)|_{[x_i,x_{i+1}]}$ where

$$U_i(t,x) = \sum_{j=1}^{m} \phi_j(x - x_i) y_{i,j}(t) + h_i^m \sum_{j=1}^{k} \psi_j \left(\frac{x - x_i}{h_i} \right) z_{i,j}(t). \tag{23}$$

Further, $U_i(t,x) : [t_0, \infty) \times [x_i, x_{i+1}] \to \mathbf{R}^{NPDE}$, and $y_{i,j}(t), z_{i,j}(t) : [t_0, \infty) \to \mathbf{R}^{NPDE}$, are unknown functions of time. The set of m functions,

$$\{\phi_j(x)\}_{j=1}^{m} = \left\{ \frac{x^{j-1}}{(j-1)!} \right\}_{j=1}^{m}, \tag{24}$$

gives the local representation of the first m (canonical) monomial basis functions. The k functions, $\{\psi_j(x)\}_{j=1}^{k}$ are chosen to satisfy the following conditions:

1. $\psi_j(x) : [0,1] \to \mathbf{R}$ is of order $k+m$, for $j = 1, \ldots, k$.
2. $\frac{d^{l-1}\psi_j}{dx^{l-1}}(0) = 0$ for $j = 1, \ldots, k$, and $l = 1, \ldots, m$.

These requirements do not completely define $\{\psi_j(x)\}_{j=1}^{k}$, and consequently this set is not unique. In order to specify the set uniquely, one choice is to have

$$\frac{d^{l-1}\psi_j}{dx^{l-1}}(1) = \delta_{j-k+m,l}, \quad j, l = 1, \ldots, k, \tag{25}$$

where $\delta_{i,j}$ is the Kronecker delta. This would imply that $\{\psi_j(x)\}_{j=1}^{k} = \{\frac{x^{m+j-1}}{(m+j-1)!}\}_{j=1}^{k}$, so that the basis functions ψ_j are defined in the same way as the functions ϕ_j. A second choice is given by

$$\frac{d^m \psi_j}{dx^m}(\rho_r) = \delta_{j,r}, \quad j, r = 1, \ldots, k. \tag{26}$$

The choice of (26), known as a Runge-Kutta representation[4] is used in the monomial spline package employed by our PDE code, and has been shown to have benefits over (25), see Ascher et al[2].

2.2. The Collocation Equations

We require the approximate function $U(t,x)$ to satisfy the PDE at the $NCPTS$ collocation points $\Xi = \{\xi_j\}_{j=1}^{NCPTS}$. Special treatment is required for the boundary conditions, and therefore the collocation points ξ_1 and ξ_{NCPTS} will be considered in detail later. For the collocation points on the i-th subinterval we require $U(t,x)$ to satisfy the PDE (1). For $i = 1,\dots,NINT$, this gives

$$\frac{\partial U_i}{\partial t}(t,\xi_{l+r}) = f\left(t, U_i(t,\xi_{l+r}), \frac{\partial U_i}{\partial x}(t,\xi_{l+r}), \frac{\partial^2 U_i}{\partial x^2}(t,\xi_{l+r})\right), \tag{27}$$

where $l = (i-1)k+1$ and $r = 1,\dots,k$. The left hand side of (27) expands to

$$\sum_{j=1}^{m} \phi_j(\xi_{l+r} - x_i)\frac{dy_{i,j}}{dt}(t) + h_i^m \sum_{j=1}^{k} \psi_j\left(\frac{\xi_{l+r} - x_i}{h_i}\right)\frac{dz_{i,j}}{dt}(t). \tag{28}$$

Thus, the collocation equations of (27) simplify to

$$\sum_{j=1}^{m} \phi_j(h_i\rho_r)\frac{dy_{i,j}}{dt}(t) + h_i^m \sum_{j=1}^{k} \psi_j(\rho_r)\frac{dz_{i,j}}{dt}(t) = f_{i,r}(t), \tag{29}$$

for $r = 1,\dots,k$ and $i = 1,\dots,NINT$, where

$$f_{i,r}(t) = f\left(t, U_i(t,\xi_{l+r}), \frac{\partial U_i}{\partial x}(t,\xi_{l+r}), \frac{\partial^2 U_i}{\partial x^2}(t,\xi_{l+r})\right), \tag{30}$$

and $l = (i-1)k+1$. These lead to the initial value ODE system

$$V_i\frac{d\vec{y}_i}{dt}(t) + W_i\frac{d\vec{z}_i}{dt}(t) = \vec{f}_i(t), \quad i = 1,\dots,NINT, \tag{31}$$

where, using the symbol $\otimes$ for Kronecker Product,

$$V_i = [\phi_j(h_i\rho_r); j = 1,\dots,m; r = 1,\dots,n] \otimes I_{NPDE} \in \mathbf{R}^{k\times m} \otimes \mathbf{R}^{NPDE\times NPDE}, \tag{32}$$

$$W_i = [h_i^m\psi_j(\rho_r); j,r = 1,\dots,k] \otimes I_{NPDE} \in \mathbf{R}^{k\times k} \otimes \mathbf{R}^{NPDE\times NPDE}, \tag{33}$$

$$\vec{y}_i(t) = \left[y_{i,1}^T(t),\dots,y_{i,m}^T(t)\right]^T \in \mathbf{R}^m \otimes \mathbf{R}^{NPDE}, \tag{34}$$

$$\vec{z}_i(t) = \left[z_{i,1}^T(t),\dots,z_{i,k}^T(t)\right]^T \in \mathbf{R}^k \otimes \mathbf{R}^{NPDE}, \tag{35}$$

$$\vec{f}_i(t) = \left[f_{i,1}^T(t),\dots,f_{i,k}^T(t)\right]^T \in \mathbf{R}^k \otimes \mathbf{R}^{NPDE}, \tag{36}$$

and I_{NPDE} is the identity matrix in $\mathbf{R}^{NPDE\times NPDE}$.

The initial value ODEs which result from the collocation process can then be summarized by the system:

$$A_c\frac{dY}{dt}(t) = F_c(t, Y(t)), \tag{37}$$

where , with $N = \kappa \cdot NINT + m = (k+m)NINT + m$,

$$Y(t) = \left[\vec{y_1}^T, \vec{z_1}^T, \vec{y_2}^T, \vec{z_2}^T, \ldots, \vec{y}_{NINT}^T, \vec{z}_{NINT}^T, \vec{y}_{NINT+1}^T\right]^T \in \mathbf{R}^N \otimes \mathbf{R}^{NPDE}, \quad (38)$$

$$F_c(t, Y(t)) = \left[\vec{0}_1^T, \vec{f_1}^T, \vec{0_2}^T, \vec{f_2}^T, \vec{0_2}^T, \ldots, \vec{f}_{NINT}^T, \vec{0_2}^T, \vec{0_1}^T\right]^T \in \mathbf{R}^N \otimes \mathbf{R}^{NPDE}. \quad (39)$$

Further, $\vec{0_1} \in \mathbf{R}^{NPDE}, \vec{0_2} \in R^{NPDE \cdot m}$, are vectors of zeros, and $A_c \in \mathbf{R}^{N \times N} \otimes \mathbf{R}^{NPDE \times NPDE}$. In the components of the two vectors $Y(t)$ and $F_c(t, Y(t))$ we have suppressed the dependence on t. The blocks of zeros in the vector F_c are matched by sets of rows of zeros in A_c. The first and last blocks of zeros correspond to the (as yet not imposed) boundary conditions; the other blocks of zeros correspond to the continuity conditions as described in the next section. Consequently, the structure of A_c can be inferred from (31) and is described later, in conjunction with the Jacobian matrix defined in Sections 3 and 4.

2.3. The Continuity Equations

The continuity conditions satisfied by $U(t, x)$ must be explicitly imposed, since they are not built into the basis functions, as is the case with the B-splines. In our case, $(m = 2)$, this means applying continuity and first derivative conditions at each of the mesh points x_i, $i = 2, \ldots, NINT + 1$, but we will, as before, express these in terms of a variable m. These conditions result in algebraic constraints on the initial value ODE system (37). One way to deal with these constraints is to differentiate them with respect to time and combine them with the initial value ODEs arising from the collocation equations (37) as is done with the boundary equations in PDECOL. However, we have chosen to leave the continuity conditions as algebraic constraints resulting in a DAE system.

The m conditions to be satisfied by $U(t, x)$ at each internal mesh point and at the right hand end point are:

$$U_i(t, x_{i+1}) = U_{i+1}(t, x_{i+1}), \quad (40)$$

$$\frac{\partial U_i}{\partial x}(t, x_{i+1}) = \frac{\partial U_{i+1}}{\partial x}(t, x_{i+1}), \quad (41)$$

$$\vdots$$

$$\frac{\partial^{m-1} U_i}{\partial x^{m-1}}(t, x_{i+1}) = \frac{\partial^{m-1} U_{i+1}}{\partial x^{m-1}}(t, x_{i+1}), \quad (42)$$

for $i = 1, \ldots, NINT$. Recalling the definition of $U_i(t, x)$ in (23) and the properties of the local monomial spline basis elements, we can write this system as:

$$-C_i \vec{y_i}(t) - D_i \vec{z_i}(t) + I \vec{y}_{i+1} = \vec{0}_{NPDE}, \quad i = 1, \ldots, NINT, \quad (43)$$

where

$$C_i = \left[\frac{d^{r-1}\phi_j}{dx^{r-1}}(h_i)\right] \otimes I_{NPDE} \in \mathbf{R}^{m\times m} \otimes \mathbf{R}^{NPDE\times NPDE}, \tag{44}$$

$$D_i = \left[h_i^{m+1-r}\frac{d^{r-1}\psi_j}{dx^{r-1}}(1)\right] \otimes I_{NPDE} \in \mathbf{R}^{m\times k} \otimes \mathbf{R}^{NPDE\times NPDE}, \tag{45}$$

$$I = I_m \otimes I_{NPDE} \in \mathbf{R}^{m\times m} \otimes \mathbf{R}^{NPDE\times NPDE}. \tag{46}$$

The algebraic constraints resulting from the continuity equations can be summarized by the system

$$B_c Y(t) = \vec{0}_N \otimes \vec{0}_{NPDE}, \tag{47}$$

where $\vec{0}_N$ is the zero vector of $\mathbf{R}^N$, and $B_c \in \mathbf{R}^{N\times N} \otimes \mathbf{R}^{NPDE\times NPDE}$. Once again, the structure of B_c can be inferred from (43) and is described later in Sections 3 and 4.

2.4. The Boundary Conditions

For the collocation method, the final requirement is that $U(t,x)$ satisfies the boundary conditions. The boundary conditions give:

$$b_{x_a}(t, U(t,\xi_1), U_x(t,\xi_1)) = \vec{0}_{NPDE}, \tag{48}$$

$$b_{x_b}(t, U(t,\xi_{NCPTS}), U_x(t,\xi_{NCPTS})) = \vec{0}_{NPDE}. \tag{49}$$

These equations represent a set of non-linear algebraic constraints on the vector $Y(t)$ which we denote as

$$F_B(t, Y(t)) = \vec{0}_N \otimes \vec{0}_{NPDE}. \tag{50}$$

Once again, there is the option of differentiating the boundary conditions, (48)-(49), with respect to t, or of applying them directly. In our code we choose to differentiate them, and this gives:

$$\frac{\partial b_{x_a}}{\partial U}(t,U,U_x)\frac{\partial U}{\partial t} + \frac{\partial b_{x_a}}{\partial U_x}(t,U,U_x)\frac{\partial U_x}{\partial t} + \frac{\partial b_{x_a}}{\partial t}(t,U,U_x) = \vec{0}_{NPDE}, \tag{51}$$

$$\frac{\partial b_{x_b}}{\partial U}(t,U,U_x)\frac{\partial U}{\partial t} + \frac{\partial b_{x_b}}{\partial U_x}(t,U,U_x)\frac{\partial U_x}{\partial t} + \frac{\partial b_{x_b}}{\partial t}(t,U,U_x) = \vec{0}_{NPDE}. \tag{52}$$

In (51) U and U_x are evaluated at (t,ξ_1), and in (52) they are evaluated at (t,ξ_{NCPTS}). Simplification of (51)-(52), using the definition of $U_i(t,x)$ in (23), yields the following linear initial value ODEs

$$\left[\ \frac{\partial b_{x_a}}{\partial U}(t,U,U_x)\quad \frac{\partial b_{x_a}}{\partial U_x}(t,U,U_x)\ \right]\frac{d\vec{y}_1}{dt}(t) = -\frac{\partial b_{x_a}}{\partial t}(t,U,U_x) \tag{53}$$

$$\left[\ \frac{\partial b_{x_b}}{\partial U}(t,U,U_x)\quad \frac{\partial b_{x_b}}{\partial U_x}(t,U,U_x)\ \right]\frac{d\vec{y}_{NINT+1}}{dt}(t) = -\frac{\partial b_{x_b}}{\partial t}(t,U,U_x) \tag{54}$$

where $\vec{y}_1(t)$ and $\vec{y}_{NINT+1}(t)$ are defined by (34) with $m = 2$. For future convenience, we define

$$TOP = \left[\ \frac{\partial b_{x_a}}{\partial U}(t,U,U_x) \ \ \frac{\partial b_{x_a}}{\partial U_x}(t,U,U_x) \ \right], \tag{55}$$

$$BOT = \left[\ \frac{\partial b_{x_b}}{\partial U}(t,U,U_x) \ \ \frac{\partial b_{x_b}}{\partial U_x}(t,U,U_x) \ \right]. \tag{56}$$

Recalling the definition of $Y(t)$ in (38), we can define a matrix $A_{dBdt} \in \mathbf{R}^{N \times N} \otimes \mathbf{R}^{NPDE \times NPDE}$ so that (53) and (54) are described by the system

$$A_{dBdt}\frac{dY}{dt}(t) = F_{dBdt}(t, Y(t)), \tag{57}$$

where

$$F_{dBdt}(t,Y(t)) = \left[-\frac{\partial b_{x_a}^T}{\partial t}(t,U,U_x), 0, \ldots, 0, -\frac{\partial b_{x_b}^T}{\partial t}(t,U,U_x)\right]^T \in \mathbf{R}^N \otimes \mathbf{R}^{NPDE}. \tag{58}$$

The matrix A_{dBdt} will be very sparse with only the top and bottom blocks having non-zero entries as shown later in the following sections.

3. The DAE System

There are four options for constructing the initial value ODE or DAE system depending on how we treat the boundary and continuity conditions. We may decide to include the two sets of conditions as algebraic conditions, or we may choose to differentiate with respect to time one or the other, or both, of the sets, and form an ODE system. It seems more natural to include the continuity conditions as algebraic constraints, and we have done so for $t > t_0$. Similarly it seems reasonable to impose the boundary conditions as algebraic constraints, thus producing a system of DAE initial value problems. The code DASSL[19] is designed to handle such systems. But a problem arises when the code starts at t_0.

Consider the generic initial value DAE system

$$AY'(t) + BY(t) - F(t, Y(t)) = \vec{0}_N \otimes \vec{0}_{NPDE}, \tag{59}$$

where $Y(t_0)$ and $Y'(t_0)$ are known, and $Y(t_{out})$ is desired. In order to start DASSL it is necessary to give accurate values for $Y(t_0)$ and $Y'(t_0)$. The values of $Y(t_0)$ can be obtained by interpolating $u_0(x)$ on the set Ξ. In the case of an explicit ODE, $Y'(t_0)$ can be obtained directly from the ODE by solving a linear system (implicitly inverting A). But if the boundary and continuity conditions are imposed as algebraic constraints then $A = A_c$ and $B = B_c$, so that A is singular, having large blocks of zeros corresponding to non-zeros in B_c, and having TOP and BOT all zeros. If, however, we differentiate with respect to t both the continuity and boundary conditions and apply (59) at $t = t_0$, then $A = A_c + A_{dBdt} + B_c$ is non-singular, and $Y'(t_0)$ can be obtained.

For $t > t_0$ we keep the continuity conditions as algebraic constraints, and differentiate the boundary conditions, adding them to the system of ODEs in (59), therefore, giving

$$A = A_c + A_{dBdt}, \quad B = B_c, \quad F(t, Y(t)) = F_c(t, Y(t)) + F_{dBdt}(t, Y(t)). \tag{60}$$

DASSL also requires the Jacobian or iteration matrix $\mathcal{J}$ given by

$$\mathcal{J} = \alpha \frac{\partial G}{\partial Y'} + \frac{\partial G}{\partial Y} \in \mathbf{R}^{N \times N} \otimes \mathbf{R}^{NPDE \times NPDE}, \tag{61}$$

where α is chosen by DASSL to accelerate convergence, and where G is the residual given by

$$G(t, Y(t), Y'(t)) \equiv AY'(t) + BY(t) - F(t, Y(t)) = \vec{0}_N \otimes \vec{0}_{NPDE}, \tag{62}$$

It is clear from the definition of G in (62) that

$$\mathcal{J} = \alpha A + B - \frac{\partial F}{\partial Y}. \tag{63}$$

The structure of $\mathcal{J}$ is determined by the support of the basis functions. To aid in describing the iteration matrix $\mathcal{J}$ we will make the following definitions:

$$\widehat{TOP} = \alpha\, TOP - \left[\begin{array}{cc} \frac{\partial F_1}{\partial Y_1} & \frac{\partial F_1}{\partial Y_2} \end{array} \right], \tag{64}$$

$$\widehat{BOT} = \alpha\, BOT - \left[\begin{array}{cc} \frac{\partial F_N}{\partial Y_{N-1}} & \frac{\partial F_N}{\partial Y_N} \end{array} \right], \tag{65}$$

$$\hat{V}_i = \alpha V_i - \left[\frac{\partial F_{I+r}}{\partial Y_{J+j}} \right], \tag{66}$$

$$\text{for } r = 1, \ldots, k, \; j = 1, \ldots, m, \tag{67}$$

$$\hat{W}_i = \alpha W_i - \left[\frac{\partial F_{I+r}}{\partial Y_{J+j}} \right], \tag{68}$$

$$\text{for } r = 1, \ldots, k, \; j = m, \ldots, k+m, \tag{69}$$

where $I = (i-1)(k+m) + 1$ and $J = (i-1)(k+m)$.

A special note regarding $\widehat{TOP}$ and $\widehat{BOT}$ is appropriate. Since it is unrealistic to expect the user to provide $\frac{\partial \frac{\partial b_{x_a}}{\partial t}}{\partial U}$, $\frac{\partial \frac{\partial b_{x_a}}{\partial t}}{\partial U_x}$, $\frac{\partial \frac{\partial b_{x_b}}{\partial t}}{\partial U}$, and $\frac{\partial \frac{\partial b_{x_b}}{\partial t}}{\partial U_x}$, we assume that these functions are identically zero. This results in an iteration matrix $\mathcal{J}$ which is not exact if this assumption is false. However, the dependence of $\mathcal{J}$ on the boundary information is assumed to be weak, so that the only affect of our assumption is possibly to slow convergence of the Newton iterations. This assumption has also been used in PDECOL [14] and EPDCOL[11].

4. The Linear Algebra component

The iteration matrix $\mathcal{J}$ can be shown to have the structure, (called almost block diagonal), illustrated in Figure 1, for the case when $NINT = 4$, $m = 2$ and $k = 3$, i.e. using splines of degree 4, or order 5. Generally, each element in this matrix is a full

Figure 1: The iteration matrix $\mathcal{J}$.

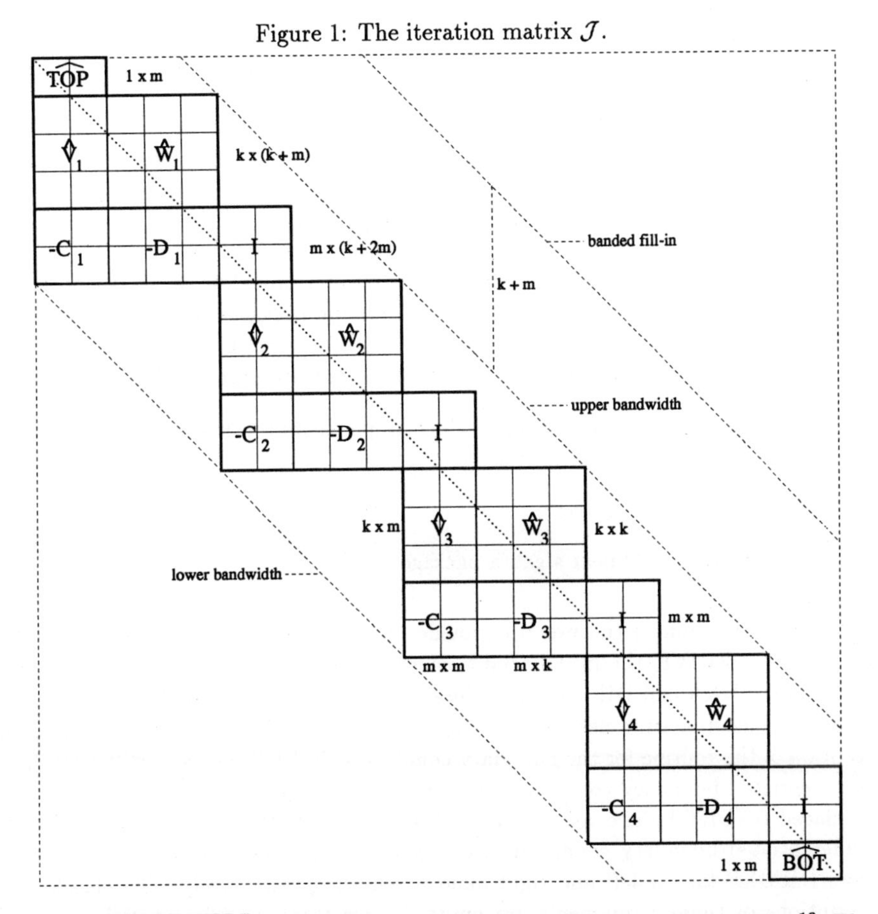

matrix in $\mathbf{R}^{NPDE \times NPDE}$. A detailed description of this is given in Nokonechny[18]. The package ABDPACK[15] is designed to perform a decomposition of this type, without additional storage being required. The code DASSL, has two options for the linear algebra component, namely full matrices and band matrices. If the band option is used on the matrix $\mathcal{J}$ considerable fill-in occurs (see Majaess et al[15]), as shown in Figure 1. It is therefore important to treat such systems using linear algebra software adapted to the particular structure shown in the figure. We have modified DASSL to include a third linear algebra option allowing us to use ABDPACK. It should be noted that the changes required to DASSL are greatly simplified as a result of careful structuring of the original code.

5. The Software

In this section we describe a package called MSCPDE (**M**onomial **S**pline **C**ollocation for **P**arabolic Partial **D**ifferential **E**quations) implementing the algorithm described earlier.

5.1. The Components of Package

There are four principal components of the package:

(i) The control program, MSCPDE. This subroutine provides the interface between the user's calling program and the underlying packages required to solve the PDE. Additionally, the subroutine MSCPDE manages the allocation of storage to the various arrays required, from a block of workspace provided by the user.

(ii) The monomial spline package, which implements the low level details of the basis functions (see Muir et al[17]).

(iii) DASSL, the DAE solver.

(iv) ABDPACK, the linear algebra package.

MSCPDE requires the user to provide information about the system of PDEs (1), F and DERIVF, the initial conditions (2), UINIT, and the boundary conditions (3)-(4), DIFBXA and DIFBXB, in differentiated form, as indicated in (51)-(52). We have retained the same user interface as PDECOL[14] and EPDCOL[11], with the exception of the routine for the boundary conditions. In PDECOL and EPDCOL both differentiated boundary conditions are defined in the one routine BNDRY. We have replaced BNDRY by two routines DIFBXA and DIFBXB so that the left and right differentiated boundary conditions are specified by two separate routines DIFBXA and DIFBXB for the left and right hand boundaries, respectively. The detailed descriptions of these components are omitted. For these we refer to Nokonechny[18]. The code itself, with sample driver programs, is available through anonymous ftp at ftp.cs.dal.ca.

As an example we consider the system of partial differential equations given by:

$$\frac{\partial u}{\partial t} = v^2\frac{\partial^2 u}{\partial x^2} + 2u\frac{\partial u}{\partial x}\frac{\partial v}{\partial x} - uv - u^2 + 10, \tag{70}$$
$$\frac{\partial v}{\partial t} = u^2\frac{\partial^2 v}{\partial x^2} + 2v\frac{\partial u}{\partial x}\frac{\partial v}{\partial x} + uv - v^2, \tag{71}$$

with initial conditions given by

$$u(0,x) = 0.5(x+1), \tag{72}$$
$$v(0,x) = \pi, \tag{73}$$

and subject to the boundary conditions

$$u(t,0) = 0.5 \tag{74}$$
$$v(t,0) = \pi \tag{75}$$
$$\frac{\partial u}{\partial x}(t,1) + \sin(u(t,1)v(t,1)) = 0.5 \tag{76}$$
$$\frac{\partial v}{\partial x}(t,1) - \cos(u(t,1)v(t,1)) = 1 \tag{77}$$

This problem is example 1 in Madsen and Sincovec[14]. The results in Table 1, below, essentially reproduce the results of that paper. The number of collocation points per subinterval, K, is 2, and the number of subintervals used, $NINT$, is 20. The requested absolute and relative tolerances, $ATOL$ and $RTOL$ are 0.01 and 0.0001 respectively.

T = 0.001 Steps = 26

PDE Component 1

0.50000E+00	0.51924E+00	0.53776E+00	0.55569E+00	0.57318E+00
0.59035E+00	0.60729E+00	0.62409E+00	0.64080E+00	0.65746E+00
0.67408E+00	0.69069E+00	0.70729E+00	0.72389E+00	0.74048E+00
0.75707E+00	0.77366E+00	0.79025E+00	0.80683E+00	0.82342E+00
0.84001E+00	0.85660E+00	0.87319E+00	0.88979E+00	0.90640E+00
0.92302E+00	0.93968E+00	0.95639E+00	0.97318E+00	0.99009E+00
0.10072E+01				

PDE Component 2

0.31416E+01	0.31333E+01	0.31331E+01	0.31332E+01	0.31334E+01
0.31335E+01	0.31336E+01	0.31337E+01	0.31337E+01	0.31338E+01
0.31338E+01	0.31339E+01	0.31340E+01	0.31340E+01	0.31341E+01
0.31341E+01	0.31342E+01	0.31342E+01	0.31343E+01	0.31343E+01
0.31344E+01	0.31344E+01	0.31345E+01	0.31345E+01	0.31346E+01
0.31346E+01	0.31347E+01	0.31348E+01	0.31348E+01	0.31349E+01
0.31350E+01				

Table 1: K = 2; NINT = 30; ATOL = 0.01; RTOL =0.0001

T = 0.01 Steps = 44

PDE Component 1

0.50000E+00	0.52477E+00	0.54917E+00	0.57295E+00	0.59603E+00
0.61840E+00	0.64011E+00	0.66122E+00	0.68178E+00	0.70185E+00
0.72152E+00	0.74082E+00	0.75983E+00	0.77860E+00	0.79720E+00
0.81569E+00	0.83413E+00	0.85258E+00	0.87112E+00	0.88982E+00
0.90876E+00	0.92801E+00	0.94767E+00	0.96784E+00	0.98863E+00
0.10102E+01	0.10325E+01	0.10559E+01	0.10805E+01	0.11065E+01
0.11340E+01				

PDE Component 2

0.31416E+01	0.30943E+01	0.30722E+01	0.30633E+01	0.30606E+01
0.30603E+01	0.30610E+01	0.30620E+01	0.30630E+01	0.30640E+01
0.30649E+01	0.30658E+01	0.30667E+01	0.30675E+01	0.30683E+01
0.30691E+01	0.30699E+01	0.30707E+01	0.30716E+01	0.30724E+01
0.30733E+01	0.30742E+01	0.30752E+01	0.30762E+01	0.30773E+01
0.30784E+01	0.30796E+01	0.30810E+01	0.30824E+01	0.30839E+01
0.30858E+01				

T = 0.1 Steps = 77

PDE Component 1

0.50000E+00	0.54716E+00	0.59704E+00	0.64874E+00	0.70148E+00
0.75459E+00	0.80753E+00	0.85987E+00	0.91129E+00	0.96155E+00
0.10105E+01	0.10579E+01	0.11038E+01	0.11481E+01	0.11908E+01
0.12319E+01	0.12714E+01	0.13093E+01	0.13457E+01	0.13805E+01
0.14140E+01	0.14460E+01	0.14766E+01	0.15060E+01	0.15341E+01
0.15610E+01	0.15868E+01	0.16114E+01	0.16351E+01	0.16577E+01
0.16794E+01				

PDE Component 2

0.31416E+01	0.30064E+01	0.29093E+01	0.28408E+01	0.27940E+01
0.27641E+01	0.27471E+01	0.27406E+01	0.27424E+01	0.27510E+01
0.27653E+01	0.27844E+01	0.28076E+01	0.28344E+01	0.28645E+01
0.28975E+01	0.29332E+01	0.29713E+01	0.30118E+01	0.30546E+01
0.30994E+01	0.31464E+01	0.31954E+01	0.32464E+01	0.32995E+01
0.33545E+01	0.34115E+01	0.34705E+01	0.35315E+01	0.35947E+01
0.36598E+01				

Table 1(continued): K = 2; NINT = 30; ATOL = 0.01; RTOL = 0.0001

T = 1.0 Steps = 95

PDE Component 1

0.50000E+00 0.54703E+00 0.59686E+00 0.64856E+00 0.70130E+00
0.75436E+00 0.80713E+00 0.85915E+00 0.91007E+00 0.95963E+00
0.10077E+01 0.10540E+01 0.10987E+01 0.11417E+01 0.11829E+01
0.12224E+01 0.12602E+01 0.12964E+01 0.13311E+01 0.13643E+01
0.13960E+01 0.14264E+01 0.14554E+01 0.14833E+01 0.15099E+01
0.15354E+01 0.15599E+01 0.15834E+01 0.16059E+01 0.16275E+01
0.16482E+01

PDE Component 2

0.31416E+01 0.30031E+01 0.29038E+01 0.28345E+01 0.27884E+01
0.27604E+01 0.27467E+01 0.27443E+01 0.27509E+01 0.27649E+01
0.27848E+01 0.28096E+01 0.28387E+01 0.28712E+01 0.29068E+01
0.29451E+01 0.29858E+01 0.30286E+01 0.30735E+01 0.31202E+01
0.31687E+01 0.32189E+01 0.32708E+01 0.33243E+01 0.33795E+01
0.34363E+01 0.34947E+01 0.35548E+01 0.36167E+01 0.36803E+01
0.37457E+01

T = 10.0 Steps = 100

PDE Component 1

0.50000E+00 0.54703E+00 0.59686E+00 0.64856E+00 0.70130E+00
0.75435E+00 0.80712E+00 0.85914E+00 0.91006E+00 0.95962E+00
0.10076E+01 0.10540E+01 0.10987E+01 0.11416E+01 0.11828E+01
0.12224E+01 0.12602E+01 0.12964E+01 0.13311E+01 0.13643E+01
0.13960E+01 0.14264E+01 0.14554E+01 0.14832E+01 0.15099E+01
0.15354E+01 0.15599E+01 0.15833E+01 0.16058E+01 0.16274E+01
0.16482E+01

PDE Component 2

0.31416E+01 0.30032E+01 0.29038E+01 0.28345E+01 0.27884E+01
0.27605E+01 0.27468E+01 0.27443E+01 0.27510E+01 0.27649E+01
0.27848E+01 0.28097E+01 0.28387E+01 0.28713E+01 0.29069E+01
0.29452E+01 0.29858E+01 0.30287E+01 0.30735E+01 0.31203E+01
0.31688E+01 0.32190E+01 0.32709E+01 0.33244E+01 0.33796E+01
0.34363E+01 0.34948E+01 0.35549E+01 0.36168E+01 0.36804E+01
0.37458E+01

Table 1(continued): K = 2; NINT = 30; ATOL = 0.01; RTOL = 0.0001

It may be observed that the results in Table 1 are similar to those in Sincovec and Madsen[14] with the number of steps taken by MSCPDE being comparable to or fewer than the number taken by PDECOL. However, the choice of tolerances is somewhat different since PDECOL only requires the user to specify a relative tolerance, whereas DASSL (and hence MSCPDE) also allows an absolute tolerance to be set. Some care must be taken in the choice of RTOL for MSCPDE, since too small a value results in a larger number of time steps being taken, without any appreciable difference in accuracy. Since the two codes are being asked to satisfy different requests, it is up to the user of MSCPDE to select ATOL and RTOL wisely.

6. References

1. U. M. Ascher, J. Christiansen and R. D. Russell *Algorithm 569: COLSYS: Collocation Software for boundary value ODEs*, ACM TOMS (1981) p. 223.
2. U. M. Ascher, R.M.M. Mattheij and R. D. Russell in *Numerical Solution of Boundary Value Problems for Ordinary Differential Equations*, SIAM Classics in Applied Mathematics, Philadelphia, 1995.
3. G. Bader and U. Ascher, *A New Basis Implementation for a Mixed Order Boundary Value ODE Solver*, SIAM J. Sci. Stat. Anal. (1987) p. 483.
4. U. Ascher, S. Pruess and R. D. Russell, *On Spline Basis Selection for Solving Differential Equations*, SIAM J. Numer. Anal. (1983) p. 121.
5. M. Berzins and P. M. Dew, *Algorithm 690: Chebyshev Polynomial Software for Elliptic-Parabolic Systems of PDEs*, ACM TOMS (1991) p. 178.
6. M. Berzins, J. Lawson and J. Ware, *Spatial and Temporal Error Control in the Adaptive Solution of Systems of Conservation Laws*, in Advances in Computer Methods for Partial Differential Equations, VII, edited by R. Vichnevetsky, D. Knight and G. Richter, IMACS (1992) p. 60.
7. J. Carroll, *A composite integration scheme for the numerical solution of systems of parabolic PDEs in one space dimension*, J. Comp. and Appl. Math. (1993) p. 327.
8. C. de Boor, *Package for Calculating with B-Splines*, SIAM J. Numer. Anal. (1977) p. 441.
9. J. E. Flaherty and P. K. Moore, *Integrated Space-time Adaptive hp-Refinement Methods for Parabolic Systems*, App. Num. Math. (1995) p. 317.
10. W. Huang, Y. Ren and R. D. Russell, *Moving Mesh Methods Based on Moving Mesh Partial Differential Equations*, J. Comp. Phys. (1994) p. 279.
11. P. Keast and P. H. Muir, *Algorithm 688: EPDCOL: A More Efficient PDECOL Code*, ACM TOMS (1991) p. 153.
12. J. Lawson, M. Berzins and P.M. Dew *Balancing Space and Time Errors in the Method of Lines*, SIAM J. Sci. Statist. Comput. (1991) p. 573.
13. N. K. Madsen and R. F. Sincovec, *Algorithm 494: PDEONE, Solutions of*

Systems of Partial Differential Equations, ACM TOMS (1975) p. 261.

14. N. K. Madsen and R. F. Sincovec, *Algorithm 540 PDECOL, General Collocation Software for Partial Differential Equations*, ACM TOMS (1979) p. 326.
15. F. Majaess, P. Keast, Graeme Fairweather, and Karin R. Bennett, *Algorithm 704: ABDPACK and ABBPACK-FORTRAN Programs for the Solution of Almost Block Diagonal Linear Systems Arising in Spline Collocation at Gaussian Points with Monomial Basis Functions*, ACM TOMS (1992) p. 205.
16. P. K. Moore and J. E. Flaherty, *Adaptive Local Overlapping Grid Methods for Parabolic Systems in Two Space Dimension*, J. Comp. Phys. (1992) p. 54.
17. P.H. Muir and K. Remington, *Software for Manipulating Monomial Spline Bases*, unpublished software.
18. T. B. Nokonechny, *The Method of Lines Using Monomial Spline Collocation for Parabolic Partial Differential Equations*, MSC thesis, Dalhousie University, Halifax, Canada, 1995.
19. L. Petzold, DASSL, electronically available through netlib, at netlib.att.com.
20. R. A. Trompert and J. G. Verwer, *Analysis of the Implicit Euler Local Uniform Refinement*, SIAM J. Sci. Comput. (1993) p. 259.
21. J. G. Verwer and R. A. Trompert, *Methods of Lines and Local Uniform Grid Refinement*, in Advances in Computer Methods for Partial Differential Equations, VII, edited by R. Vichnevetsky, D. Knight and G. Richter, IMACS (1992) p. 752.

SOLVING LEAST SQUARES PROBLEMS ON PARALLEL VECTOR PROCESSORS

M.R. OSBORNE
Centre for Mathematics and its Applications,
School of Mathematical Sciences, Australian National University
Canberra, A.C.T 0200, Australia
E-mail: mike@thrain.anu.edu.au

ABSTRACT

The solution of the standard least squares problems of data analysis type on parallel vector processors is considered. The key features are identified as a large numbers of observations, a relatively small numbers of explanatory variables, and a dense design matrix. It is shown that some care is needed in computing with long vectors, but if this is exercised then computational error is dominated asymptotically by an *eps*cond(A) term provided the algorithm has optimal error structure. A tableau based implementation of the modified Gram–Schmidt algorithm is shown to be effective on both parallel and vector computers. An implementation of the Gauss–Newton method is studied for nonlinear least squares problems. This is organised to call the modified Gram–Schmidt algorithm in each iteration.

1. Introduction

The linear least squares problem is:

$$\min_{\mathbf{x}} \mathbf{r}^T\mathbf{r} : \quad \mathbf{r} = A\mathbf{x} - \mathbf{b}, \tag{1}$$

where $A: R^n \to R^p$ is assumed to have maximum rank $p \ll n$, and $\mathbf{b}: R^n \to R$. Here the standard setting of data analysis is assumed. In this $\mathbf{b}$ is assumed to be a vector of measurements having the form of a system response $\mathbf{y}$ plus a measurement error $\mathbf{e}^*$. The responses are assumed to have the form

$$y_{ij} = \sum_{s=1}^{p} x_s \phi_{js}(t_i),\ 1 \le j \le k_i,\ 1 \le i \le N,\ n = \sum_{i=1}^{N} k_i, \tag{2}$$

where the ϕ_{js} are assumed to be smooth (enough) functions of the variable t (for example time) which indexes the measurements and will be assumed scaled so that $0 \le t \le 1$. For simplicity in presentation it will be assumed that $k_i = 1$ so that only a single response is observed at each time point and $n = N$, and that the errors are independent with distribution

$$\mathbf{e}^* \sim N(0, \sigma^2 I_n). \tag{3}$$

Provided the error covariance structure is known this can always be achieved by a preliminary scaling.

2. The Linear Least Squares Problem

2.1. Basic solution properties

There are two main approaches to the numerical solution of the linear least squares problem:

(1) Solution by means of the normal equations. The condition that the sum of squares of residuals is minimized is

$$0 = \nabla_x \mathbf{r}^T \mathbf{r} = \mathbf{r}^T A. \tag{4}$$

Substituting for $\mathbf{r}$ gives the normal equations

$$A^T A \mathbf{x} = A^T \mathbf{b}. \tag{5}$$

This gives the solution to Eq. (1)

$$\hat{\mathbf{x}} = (A^T A)^{-1} A^T \mathbf{b}, \tag{6}$$

$$\hat{\mathbf{r}} = \{I - P\}\mathbf{b}. \tag{7}$$

where $P = A(A^T A)^{-1} A^T$ is the orthogonal projection onto the range of A.

(2) Solution by means of orthogonal factorization. This factorization gives

$$A = \begin{bmatrix} Q_1 & Q_2 \end{bmatrix} \begin{bmatrix} U \\ 0 \end{bmatrix} \tag{8}$$

where $Q = \begin{bmatrix} Q_1 & Q_2 \end{bmatrix}$ is orthogonal, and $U : R^p \to R^p$ is upper triangular. U is nonsingular provided A has full rank p. Then

$$\|\mathbf{r}\|_2^2 = \| \begin{bmatrix} U \\ 0 \end{bmatrix} \mathbf{x} - \begin{bmatrix} Q_1^T \mathbf{b} \\ Q_2^T \mathbf{b} \end{bmatrix} \|_2^2$$

is minimized when

$$\hat{\mathbf{x}} = U^{-1} Q_1^T \mathbf{b}], \tag{9}$$

$$\|\hat{\mathbf{r}}\|_2^2 = \|Q_2^T \mathbf{b}\|_2^2. \tag{10}$$

The equivalence of these results when A has full rank is immediate.

2.2. Recall the statistics

Statistical properties of the solution of Eq. (1) can provide a guide to how large a value of n is desirable and how the magnitudes of important quantities can grow.

Using Eq. (3) and computing means and variances gives

$$\mathcal{E}\{\mathbf{r}\} = \mathcal{E}\{(I-P)\mathbf{e}^*\} = 0, \tag{11}$$

$$\mathcal{E}\{\mathbf{r}^T\mathbf{r}\} = \mathcal{E}\{\mathrm{trace}((I-P)\mathbf{e}^*(\mathbf{e}^*)^T(I-P))\} = (n-p)\sigma^2. \tag{12}$$

Also, under natural conditions on the response function,

$$\frac{1}{n}(A^TA)_{ij} = \frac{1}{n}\sum_{k=1}^{n}\phi_i(t_k)\phi_j(t_k) \to \int_0^1 \phi_i(t)\phi_j(t)dw(t) \tag{13}$$

where $w(t)$ is a weight function summarising the selection of the observation points t_i. It follows that $\frac{1}{n}A^TA$ tends to a bounded limiting matrix which can be assumed to be positive definite with almost no loss of generality under the assumption that A has full rank p for n finite. *An immediate consequence is that* cond(A) *is bounded for all n in the spectral and Frobenius norms.*

The error in the computed solution satisfies

$$\begin{aligned}\hat{\mathbf{x}} - \mathbf{x}^* &= (A^TA)^{-1}A^T\mathbf{e}^*, \\ &= (\frac{1}{n}A^TA)^{-1}\frac{1}{n}A^T\mathbf{e}^*, \\ &= (\frac{1}{\sqrt{n}}U)^{-1}\frac{1}{\sqrt{n}}Q_1^T\mathbf{e}^*. \end{aligned} \tag{14}$$

It follows from Eq. (13) that the first term in the error expression is bounded, while the second satisfies

$$\frac{1}{\sqrt{n}}Q_1^T\mathbf{e}^* \sim N(0, \frac{\sigma^2}{n}I_p).$$

Thus the error standard deviation tends to zero like $n^{-1/2}$ reflecting the characteristic stochastic rate of convergence. It follows either that large n or small σ^2 is required if confidence in significant figures in the answer is to be possible.

2.3. The perturbation analysis

The dependence of the stochastic error on $n^{-1/2}$ makes clear that large values of n are desirable. This leaves open the possibility that the solution process could be dominated by numerical error. To explore this possibility consider the perturbation analysis for the solution of Eq. (1) first given by Golub and Wilkinson [4]. This starts by embedding Eq. (1) in the family of least squares problems

$$\min_{\mathbf{x}} \mathbf{r}(s)^T\mathbf{r}(s);\ \mathbf{r}(s) = (A+sE)\mathbf{x} - \mathbf{b} - s\mathbf{z},$$

where E, $\mathbf{z}$ are arbitrary but fixed quantities defining the general perturbation of Eq. (1) and s is a parameter determining their scale. It follows that for s small enough the solutions satisfy

$$(A+sE)(\mathbf{x}-\hat{\mathbf{x}}) = s\{\frac{d\mathbf{r}}{ds} - (E\hat{\mathbf{x}} - \mathbf{z})\} + O(s^2).$$

Thus

$$\mathbf{x}-\hat{\mathbf{x}}=s(A^TA)^{-1}\{A^T\frac{d\mathbf{r}}{ds}-A^T(E\hat{\mathbf{x}}-\mathbf{z})\}+O(s^2). \tag{15}$$

The least squares condition can be used to evaluate $A^T\frac{d\mathbf{r}}{ds}$. This gives

$$0=(A+sE)^T\{\hat{\mathbf{r}}+s\frac{d\mathbf{r}}{ds}+O(s^2)\}$$

so that

$$A^T\frac{d\mathbf{r}}{ds}=-E^T\hat{\mathbf{r}}. \tag{16}$$

Using this result gives

$$\mathbf{x}-\hat{\mathbf{x}}=-s\{U^{-1}Q_1^T(E\hat{\mathbf{x}}-\mathbf{z})+(A^TA)^{-1}E^T\hat{\mathbf{r}}\}+O(s^2). \tag{17}$$

The first term gives the anticipated 'epscond(A)' dependence where eps is of the same order as s provided $\|\frac{1}{\sqrt{n}}(E\hat{\mathbf{x}}-\mathbf{z})\|$ is uniformly bounded in n. The second term is estimated by '$eps\|\mathbf{r}\|$cond$(A)^2$' under similar conditions. It is small if Eq. (1) is close to consistent, but this is not a realistic condition in the data analysis context. Eq. (14) shows that in exact arithmetic the solution error gets small with high probability for bounded σ provided n is large enough. It is this situation which corresponds to realistic practice and should properly be addressed. Thus we seek to determine the behaviour of the solution of the least squares problem under perturbation as n gets large.

Remark 1 *The form of error expression Eq. (17) cannot be improved. For this reason, algorithms for solving the linear least squares problem are said to have* optimal error structure *if the solution error is dominated by terms having the same dependence on condition number and residual vector as that in Eq. (17) .*

The interesting term in Eq. (17) is

$$(A^TA)^{-1}E^T\hat{\mathbf{r}}=(\frac{1}{n}A^TA)^{-1}\frac{1}{n}\tilde{E}^T\mathbf{e}^*,\ \tilde{E}=(I-P)E. \tag{18}$$

To study this for large n an appropriate tool is the law of large numbers which states that almost surely

$$\frac{1}{n}\sum_{i=1}^{n}X_ie_i^*\to 0 \tag{19}$$

under our assumptions provided the X_i are bounded, Typically the result is a random variable with the characteristic standard deviation of $O(n^{-1/2})$. Here E is assumed fixed. Essentially this means that problems with different replications of the e_i^* encounter the same fixed sequence of E components as $n\to\infty$. This means that it is an extension to apply the result to perturbations generated by rounding errors – in real situations the E_{ij} would not be independent of the e_k^*. Also, as the E_{ij} result from accumulated roundoff error their size must reflect the scale of the computation which implies a dependence on n. Thus there must be some rescaling to take account of this before the law of large numbers can be applied. Assuming that Eq. (19) suffices to indicate order of magnitude, which is equivalent to assuming that the level of interaction is relatively small (weak mixing), then it provides some indication of allowable limits to the growth of the elements of E. If a scale factor exceeding $O(n^{1/2})$ can be factored out leaving an $O(1)$ result then Eq. (17) will be dominated asymptotically by the $\text{cond}(A)^2$ term. However, if this scale is significantly smaller than $O(n^{1/2})$ then the $\text{cond}(A)$ term will dominate. This has interesting implications for the performance of the actual computations. For example, in computing $\hat{\mathbf{x}}$ from Eq. (9) the Wilkinson model for the standard sequential computation would give

$$fl(\sum_j Q_{ij}b_j) = Q_{i1}b_1 \prod_{k=1}^{n}(1+\nu_k^1) + Q_{i2}b_2 \prod_{k=2}^{n}(1+\nu_k^2) + \cdots .$$

This suggests a worst case bound of $O(n)$ for the scale factor in E. However, this can be reduced to $\log n$ by a device such as recursive doubling, and this would reverse the order of importance of the error perturbation terms.

Remark 2 *This result provides an interesting addendum to the results of Golub and Wilkinson* [4]. *Provided, for an algorithm possessing optimal error structure,*

(1) the limiting normal matrix has bounded condition number,

(2) the law of large numbers can be used to estimate Eq. (18), and

(3) appropriate measures are taken to minimise worst case rounding error growth,

then in probability the interesting term Eq. (18) does not dominate the error analysis for large enough n. This result puts the least squares error analysis more nearly on an equal footing with that for consistent systems of linear equations. It must be stressed that this result does not require small residuals.

Remark 3 *Typically recursive doubling is used in implementing reduction operations on multiprocessor computers. Thus the implementation requirements of this result are likely to be met on massively parallel computer systems.*

3. Solving the linear least squares problem

3.1. The standard methods

Methods based on the explicit solution of the normal equations typically use the Choleski factorization of the normal matrix

$$A^T A \to LL^T,\ L^T\mathbf{x} = L^{-1}A^T\mathbf{b}$$

to solve Eq. (1). The cost measured in units of floating point multiplies plus adds – here this unit is referred to as a fmad to distinguish it from the current usage which equates a flop with one floating point operation – is made up as follows: the cost of forming the upper triangle of A^TA ($np^2/2$ fmads), the cost of the Choleski factorization ($p^3/6$ fmads), and the cost of the forward and back substitutions (p^2 fmads). It proves the cheapest of the methods, but it does not have optimal error behaviour. Basically this comes about because the operations of forming the normal matrix and computing the Choleski factorization are performed independently of the computation of the right hand side. Thus the solution is not the solution of a strict least squares problem. This has an immediate consequence in an unfavourable error dependence on $eps\text{cond}(A)^2$.

The method based on orthogonal factorization is usually implemented using Householder transformations [3]. The resulting algorithm does have optimum error structure [5]. It costs more ($np^2 - p^3/3$ fmads) and does not give either Q_1 or $\mathbf{r}$ explicitly. It does have the advantage that if Q is computed from the individual transformations then this can be done in such a way that the resulting matrix is very close to orthogonal.

3.1.1. The modified Gram Schmidt method

It follows from Eq. (8) that

$$Q_1 = AU^{-1}.$$

Thus, if the columns of A are successively orthogonalised and scaled then the result must be equivalent to the above calculation. Modified Gram Schmidt (MGS) orthogonalises the current pivotal column ($\kappa_i(A)$ say) to $\kappa_{i+1}(A), \cdots, \kappa_p(A)$ at each step $i = 1, 2, \cdots, p$. Augmenting A to form a tableau M, and then applying the sequence of orthogonalisation and scaling steps columnwise gives

$$M = \begin{bmatrix} A & -\mathbf{b} \\ I & 0 \end{bmatrix} \rightarrow \begin{bmatrix} Q_1 & \mathbf{r} \\ U^{-1} & \mathbf{x} \end{bmatrix}. \tag{20}$$

This follows from the above observation for the first p columns of M, and it is clear from the construction that $\mathbf{r}$ results in column $p+1$. But this result is obtained also by postmultiplying the initial tableau by $\begin{bmatrix} \mathbf{x}^T & 1 \end{bmatrix}^T$. This explains the final form. The cost is $np^2 + p^3/6$ fmads if the zero structure in U^{-1} is observed, and $np^2 + p^3/2$ if it is not (but note that the more expensive calculation can work with fixed array lengths). The MGS procedure is amenable to very succinct sequential code as the Matlab fragment in Figure 1 shows. This form of the MGS algorithm perhaps appeared first in an unpublished Ph.D thesis by Burrus. It is used extensively in Longley [6]. It is a forward sweep operation which vectorizes on long vectors. At each stage the solution of the least squares problem based on the subset of columns $1, 2, \cdots, i$ is obtained. Also, as each partially processed column is treated identically,

```
M=[A -b; eye(p) zeros(1,p)'];
FOR i=1:p
    di=1./M(1:n,i)'*M(1:n,i);
    FOR j=i+1:p+1
        rij=M(1:n,i)'*M(1:n,j)*di;
        M(1:n+i,j)=M(1:n+i,j)-rij*M(1:n+i,i);
        END
    di=sqrt(di);
    M(1:n+i,i)=di*M(i:n+i,i);
END
```

Fig. 1. Matlab code for MGS

column pivoting strategies are easily incorporated and do not affect vectorization.

3.2. Optimal error structure

In its usual form MGS when properly implemented has optimal error structure [1]. In particular, this requires that the back substitution be carried out as

$$\hat{\mathbf{x}} = U^{-1}\mathbf{v}, \quad v_i = \mathbf{q}_i^T \mathbf{r}^{(i)}, \quad i = 1, 2, \cdots, p, \tag{21}$$

where $\mathbf{q}_i$ is the current pivotal row of the tableau, and $\mathbf{r}^{(i)}$ is the corresponding residual vector in column $p+1$. Our algorithm differs from that considered in Björck [1] only in performing a bordering calculation to go from $\hat{\mathbf{x}}_{i-1}$ to $\hat{\mathbf{x}}_i$ in the current step. Recently Björck has analysed MGS again [2]. He starts from the interesting observation that the Householder method applied to $\begin{bmatrix} 0 & A^T \end{bmatrix}^T$, $\begin{bmatrix} 0 & \mathbf{b}^T \end{bmatrix}^T$ gives exactly the same computed quantities as MGS.

Here experimental evidence is presented to show that the tableau form of MGS has optimal error structure. The example class considered is constructed by starting with Q_1 having p orthonormal columns, U upper triangular with free parameter(s) which control conditioning, and $\mathbf{r}$ constructed to be orthogonal to the columns of Q_1 and having an adjustable scale. Then if $\hat{\mathbf{x}}$ is imposed, $\mathbf{b}$ can be computed and the problem and its solution defined. It proves convenient to define Q_1 using the first p terms of the finite Fourier series on n points:

$$(Q_1)_{ij} = \sqrt{\frac{2}{n+1}} \sin i\pi \frac{j}{n+1}, \; i = 1, 2, \cdots, p, \; j = 1, 2, \cdots, n. \tag{22}$$

The form of U is chosen so that it becomes very ill conditioned as the parameter s is increased:

$$U_{ij} \quad = 0 \quad \text{if } i > j,$$

Table 1. MGS has optimal error structure

figures	e	s	cond
0	.01	1.7	7.24e9
0	.001		
1	.0001		
2	.00001		
0	.01	1.6	1.51e9
1	.001		
2	.0001		
3	.00001		
1	.01	1.5	3.13e8
2	.001		
3	.0001		
4	.00001		

$$
\begin{aligned}
&= \left(\tfrac{1}{i}\right)^s \quad \text{if } i = j,\ s \text{ prescribed,} \\
&= \tfrac{j}{2j-i} \quad \text{for } i = 1, 2, \cdots, j-1.
\end{aligned} \tag{23}
$$

The residual vector is chosen as a multiple e of an ignored term in the finite Fourier series (so that $\|\mathbf{r}\| = e$):

$$
r_i = e\sqrt{\frac{2}{n+1}} \sin(2p-1)\pi\frac{i}{n+1},\ i = 1, 2, \cdots, n. \tag{24}
$$

Finally, the solution vector is given by

$$
\hat{x}_i = 1,\ i = 1, 2, \cdots, p. \tag{25}
$$

Numerical results for the case $n = 1000$, $p = 10$ are given in Table 1. What is reported is the number of correct figures in the computed answer together with the values of e, and s, and the computed Frobenius condition number. Values are chosen so that an $eps\mathrm{cond}(A)^2$ term would dominate and give no correct figures if the $\|\mathbf{r}\|$ term were not present. The effect of reducing e is to increase the number of correct figures obtained in exactly the manner expected of optimal error structure.

3.3. Estimating condition number

One of the consequences of the result that each sweep gives the solution of a subset least squares problem is that a recurrence can be given for the Frobenius condition number

$$\text{cond}_F(A) = \left\{\text{trace}(A^T A)\right\}^{1/2} \left\{\text{trace}(A^T A)^{-1}\right\}^{1/2}. \tag{26}$$

If the columns of A are normalised to have length 1 initially then

$$\text{cond}_F(A) = p^{1/2} \|U^{-T}\|_F. \tag{27}$$

After i steps of the MGS procedure only columns $i+1, \cdots, p+1$ change in the susequent computation. Setting

$$\begin{aligned} c_0 &= 0, \\ c_i &= c_{i-1} + \sum_{j=n+1}^{n+i} M(j,i)^2, \; i = 1, 2, \cdots, n, \end{aligned}$$

then

$$\text{cond}_F(A) = (pc_p)^{1/2}. \tag{28}$$

Also, after i st¡eps,

$$\text{cond}_F(A_i) = (ic_i)^{1/2}$$

is available to test for linear independence of the columns in the current subproblem.
Remark 4 *The additional cost of computing* $\text{cond}_F(A)$ *is small in sequential calculations, but the vector lengths in the recurrence are typically short in the model problem considered here. Thus it is less attractive but still small on a vector computer. On parallel systems it would depend on the manner of spreading the data how effectively this computation is implemented. Also, it is possible to batch this computation with that of the scalar products one step ahead. This would involve a dummy scalar product calculation on the last condition number step, but avoids the additional communications associated with the separate computation of each of the sequence of calculations required to monitor the conditioning of the subset least squares problems.*

3.3.1. Parallel implementation

On a distributed memory parallel vector processor such as the Fujitsu VPP500 the main constraint on the implementation is imposed by the needs to balance computing with communication requirements. Communication is required:

(1) to compute the scalar products $\sum_{j=1}^{n} M(j,i)M(j,k)$ the contribution to the sum computed on each processor must be summed. It may also be necessary to collect partial condition number estimates over processors and

(2) in each case the final result is required on all processors.

Typically this requires either

(1) efficient communication and broadcast of very small amounts of data , or

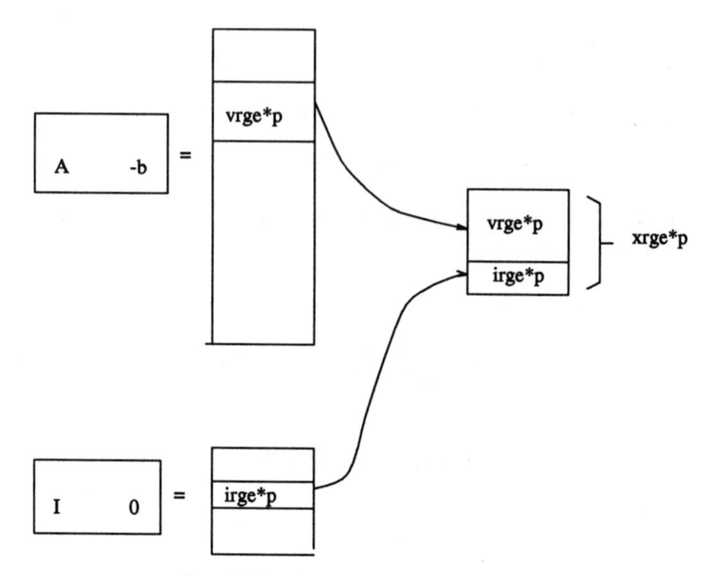

Fig. 2. Distribution of data to processors

(2) overlap of communication and computing. This can be difficult to implement if the summing of the partial sums cannot be performed asynchronously on associated communications processors.

The form of data distribution that has been used is illustrated in Figure 2. In order to treat each of the processors in basically the same manner the banded sections have dimensions as close as possible the same. The dimensions are specified in Figure 3. The annotated code on each processor to perform the basic MGS procedure is shown in Figure 4. The annotations are in VPP Fortran. Note that this fragment batches

```
np=number of processors
vr=mod(n,np)
pr=mod(p,np)
then (on processor i, 1<=i<=np)
vrge=int(n/np)+(if i<=vr then 1 else 0)
irge=int(p/np)+(
     (if vr+pr<=np and vr<i<=vr+pr) or
     (vr+pr>np and (i>vr or i<=vr+pr-np))
      then 1 else 0)
```

Fig. 3. Bounds for data distribution

```
!XOCL SPREAD DO /(PP)
      do 1000 L=1,nproc
         do 22 j=i,pp1
            ri(j)=0.
            do 21 k=1,vrge
21             ri(j)=ri(j)+A(k,i)*A(k,j)
22          continue
1000     continue
!XOCL END SPREAD SUM(ri)
      di=1./ri(i)
      do 23 j=i+1,pp1
         ri(j)=ri(j)*di
         do 24 k=1,xrge
24          A(k,j)=A(k,j)-ri(j)*A(k,i)
23       continue
      do 25 k=1,xrge
25       A(k,i)=di*A(k,i)
```

Fig. 4. Parallel code for distributed data

the scalar product calculation at each stage taking into account that the SUM function can accept a vector argument. Here the code could easily be rearranged to overlap computing and communication but the annotated Fortran does not appear to permit this. Also the code does not take into account zeros in the appended matrix so that it corresponds to the larger of the two complexity estimates for MGS. The percentage change is proportional to $p/(n * nproc)$ which is typically very small.

An alternative approach would have been to spread the data by columns rather than by rows. Then at each orthogonalisation sweep the pivotal column would have to be sent to each processor. This provides some possibility of the overlapping of communication with computation that appears difficult to achieve in the layout by rows without modification of the SUM function so that the reduction operation is performed on the communications processors. If this were done then the small volume of communication in the row oriented algorithm would be a definite advantage, The disadvantages of the column form are that it seems that the norm of the pivotal column would have to be computed on each processor, and some care is needed to avoid communications overhead in computing the condition number estimate.

3.4. Timing results

Numerical results are reported for an easy parameter case of the example considered in section 3.2. The computations have been carried out on 4 processors of

Table 2. Execution speeds for MGS

n	1000	10000	100000
rate in mflops	170	1200	2400

Table 3. Timing breakdowns for MGS algorithm

n	1000		100000	
i	scalar product	saxpy	scalar product	saxpy
10	.106e3	.97e1	.751e3	.512e3
9	.99e2	.103e2	.678e3	.487e3
8	.98e2	.92e1	.629e3	.417e3
7	.96e2	.92e1	.553e3	.390e3
6	.96e2	.74e1	.512e3	.325e3
5	.93e2	.81e1	.436e3	.289e3
4	.93e2	.65e1	.394e3	.227e3
3	.90e2	.73e1	.321e3	.194e3
2	.90e2	.56e1	.278e3	.130e3
1	.88e2	.58e1	.204e3	.100e3
0	.84e2		.84e2	

a VPP500 for the array dimensions $p = 10,\ n = 10^3, 10^4, 10^5$. The speeds achieved are recorded in Table 2. A more detailed breakdown is given in Table 3 where the timings for each processor are given for $n = 10^3$, $n = 10^5$ for the scalar products plus communication (that is, for the code between the XOCL's), and for the remaining code in the program fragment in Figure 4 which is dominated by the saxpys. The communications also includes the batching in of the condition number estimation so the final scalar product ($i = 0$ in Table 3) gives an indication of the overhead, in this case about 80 microseconds. The timings in Table 3 imply a reasonably complicated structure. In particular, the even numbered saxpys appear to be distinctly faster than the odd suggesting some loop unrolling over at least two steps of the j loop. This can be seen also in Figure 5 which plots speed estimates for the individual saxpy loops. This shows that $i = 4k + 1$ is particularly unfortunate.

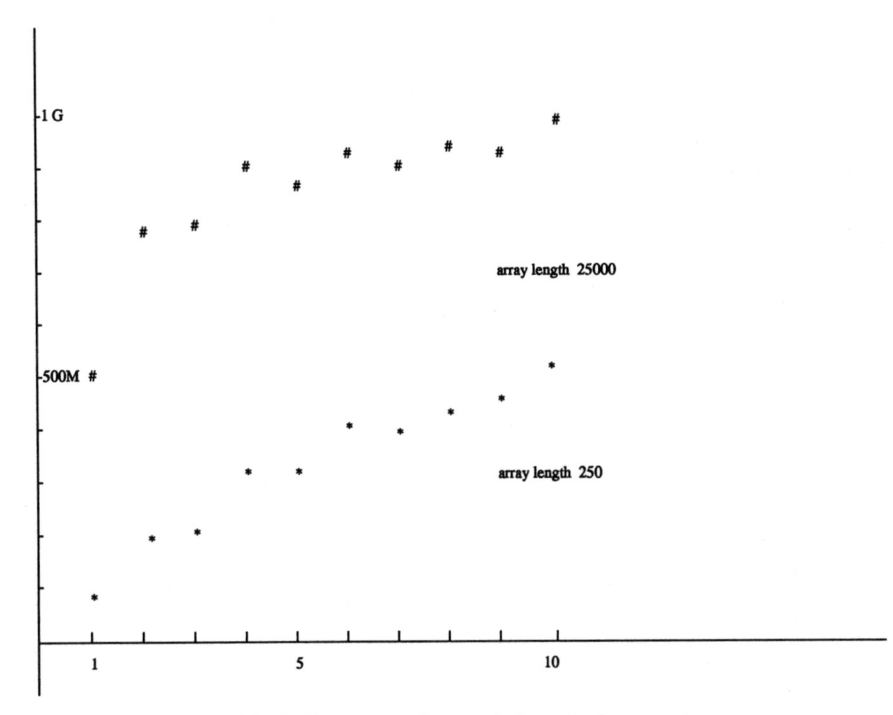

Fig. 5. Saxpy execution speeds by outer loop count

4. Nonlinear problems

4.1. Basic properties

The form of nonlinear least squares problem considered is the obvious generalisation of the linear data analysis problem. That is, given a model $y(t) = g(t, \mathbf{x})$ for a system response where g is a smooth enough function on $S \subset R \times R^p \to R$, and observations $b_i = y_i + e_i^*$, $i = 1, 2, \cdots, n$ estimate $\mathbf{x}$ by minimizing the sum of squares of residuals $f_i(\mathbf{x}) = b_i - g(t_i, \mathbf{x})$. This is written

$$\min_{\mathbf{x} \in S} F(\mathbf{x}); \; F(\mathbf{x}) = \sum_{i=1}^{n} f_i(\mathbf{x})^2 = \|\mathbf{f}\|_2^2. \tag{29}$$

Here it is assumed that the constraint $\mathbf{x} \in \mathbf{S}$ does not affect the computation. Apart from this, the major simplification in restricting attention to Eq. (29) occurs in the glossing over of possible scaling difficulties. In the data analysis context it is likely that problem scaling would be reflected in the structure of the covariance matrix for $\mathbf{e}^*$. If the covariance matrix

$$V = \mathcal{E}\{\mathbf{e}^*(\mathbf{e}^*)^T\}$$

is known a priori then

$$\mathbf{f} \leftarrow V^{-1/2}\mathbf{f} \tag{30}$$

should be acceptably scaled. The analogue of the full rank condition on the design matrix A in Eq. (1) is that $A = \nabla_x \mathbf{f}(\mathbf{x})$ has full rank in S.

The condition for a stationary point of Eq. (29) is

$$0 = \nabla_x F = 2\mathbf{f}^T A. \tag{31}$$

Also important is the associated linear sub problem

$$\min_{\mathbf{x} \in S} \mathbf{r}^T\mathbf{r}; \quad \mathbf{r} = \mathbf{f}(\mathbf{x}) + A\mathbf{h}. \tag{32}$$

Solution of this problem is linked to the solution of Eq. (31) by the equivalences

$$\nabla_x F = 0, \ \mathbf{f} = \mathbf{r}, \ \|\mathbf{r}\|_2 = \|\mathbf{f}\|_2. \tag{33}$$

These follow directly. First note that by the least squares condition applied to Eq. (32) that

$$\|\mathbf{r}\|_2^2 = \mathbf{r}^T\mathbf{f} \ \Rightarrow \ \|\mathbf{r}\|_2 \leq \|\mathbf{f}\|_2.$$

Also

$$\begin{aligned} \mathbf{f}^T\mathbf{r} &= \|\mathbf{f}\|_2^2 + \mathbf{f}^T A\mathbf{h}, \\ &= \|\mathbf{f}\|_2^2 + (\mathbf{f} - \mathbf{r})^T A\mathbf{h}, \\ &= \|\mathbf{f}\|_2^2 - \|\mathbf{f} - \mathbf{r}\|_2^2. \end{aligned}$$

Thus

$$\|\mathbf{f}\|_2 = \|\mathbf{r}\|_2 \ \Rightarrow \ \mathbf{f} = \mathbf{r} \ \Rightarrow \ (\text{by Eq. (32)}) \ \nabla F = 0 \ \Rightarrow \ \|\mathbf{f}\|_2 = \|\mathbf{r}\|_2.$$

Remark 5 *This result is important because it replaces the condition for a stationary point by a single scalar condition that is easy to test.*

4.2. The Gauss Newton algorithm

Newton's method computes a correction $\mathbf{h}$ to a current estimate $\mathbf{x}$ by solving a linear approximation to Eq. (31)

$$\nabla F^T + \nabla^2 F\mathbf{h} = 0. \tag{34}$$

It is well known to have a second order convergence rate, but also to have rather poor global convergence properties. Other disadvantages include the requirement to evaluate second derivatives of F, and the difficulty in finding a well scaled monitor function to use in the line search to improve the global behaviour.

In the data analysis application most of these problems are overcome in the Gauss–Newton method. In outline this has the form:
Until converged do {

(1) Solve for $\mathbf{h}$ the linear subproblem Eq. (32);

(2) Use $\mathbf{h}$ as a direction of search for a lower value of F. This is done by finding λ such that $F(\mathbf{x}+\lambda\mathbf{h})$ is significantly less than $F(\mathbf{x})$;

(3) Update $\mathbf{x}$ by setting $\mathbf{x} \leftarrow \mathbf{x}+\lambda\mathbf{h}$ and test for convergence}.

It turns out that this algorithm looks good in comparison with Newton's method on almost all the points raised [7]. Points of importance include:

(a) The algorithm requires only first derivative information on the components of $\mathbf{f}$;

(b) The algorithm is transformation invariant provided step (2) is;

(c) The algorithm can make use of parallel vector processors for solving Eq. (32). If $\mathbf{f}$ is distributed in banded fashion (compare Figure 2) then the evaluation of $\mathbf{f}$, A can proceed in parallel,

4.3. Implementation questions I – the line search

The Gauss–Newton direction is necessarily downhill for minimizing F when A has full rank. This is in contrast to the Newton direction where a corresponding result cannot be guaranteed. Thus F can be reduced always, at least in theory, by moving a small enough amount in the direction of $\mathbf{h}$. Two major approaches are employed in determining the distance to move. These are:

(1) *Armijo*: Let

$$\lambda_i = \theta^i,\ 0 < \theta < 1,\ i = 1, 2, \cdots.$$

A suitable λ is the largest value (corresponding to smallest i) that gives

$$F(\mathbf{x}+\lambda_i\mathbf{h}) < F(\mathbf{x});$$

(2) *Goldstein*: Let

$$\psi(\lambda, \mathbf{h}) = \frac{F(\mathbf{x}+\lambda\mathbf{h}) - F(\mathbf{x})}{\lambda\nabla F(\mathbf{x})\mathbf{h}}, \tag{35}$$

and choose λ to satisfy $0 < \rho < \psi(\lambda, \mathbf{h}) < 1 - \rho$.

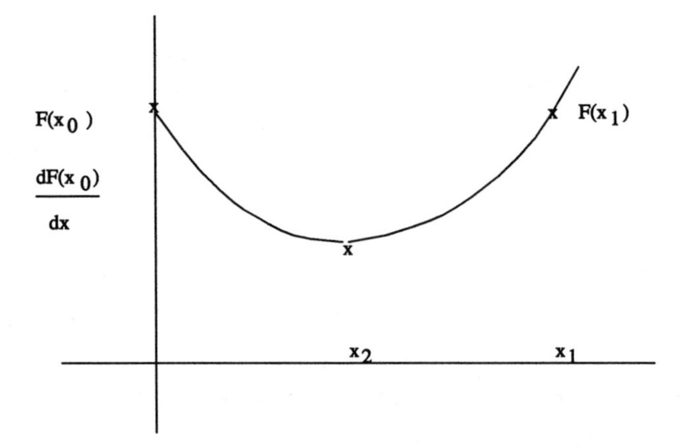

Fig. 6. Finding a suitable step

Remark 6 *The condition $\lambda < \psi$ ensures that F is reduced in the current step, while the condition $\psi < 1 - \rho$ ensures that a large enough step is taken. This is needed to avoid the possibility of premature termination. If the Gauss–Newton method is working well then $\psi \approx .5$.*

To evaluate ψ note that

$$\begin{aligned}\nabla F\mathbf{h} &= 2\mathbf{f}^T J\mathbf{h}, \\ &= -2\|\mathbf{f}-\mathbf{r}\|_2^2, \\ &= -2(\|\mathbf{f}\|_2^2 - \|\mathbf{r}\|_2^2). \end{aligned} \tag{36}$$

In this form it requires only function values. A convenient method for finding a suitable λ is illustrated in Figure 6. The data values $(F(\mathbf{x}),\ \frac{dF}{d\lambda}(\mathbf{x}),\ F(\mathbf{x}+\lambda\mathbf{h}))$ are sufficient to determine a quadratic approximation $G(\lambda)$ to F. Set

$$G(\lambda) = F(\mathbf{x}) + \lambda\frac{dF}{d\lambda}(\mathbf{x}) + \lambda^2 C.$$

Then C can be found by equating $G(\lambda_1) = F(\mathbf{x}+\lambda_1\mathbf{h})$. This gives

$$\begin{aligned} C &= \frac{1}{\lambda_1^2}\{F(\mathbf{x}+\lambda_1\mathbf{h}) - F(\mathbf{x}) - \lambda_1\frac{dF}{d\lambda}\}, \\ &= \frac{1}{\lambda_1}\frac{dF}{d\lambda}\{\psi(\lambda_1) - 1\}. \end{aligned}$$

At a minimum

$$\frac{dF}{d\lambda} + 2\lambda C = 0,$$

so that

$$\lambda = -\frac{\frac{dF}{d\lambda}}{2C} = \frac{\lambda_1}{2(1-\psi)}.$$

The procedure implemented updates λ_1 by

$$\lambda_2 = \max(\theta\lambda_1, \frac{\lambda_1}{2(1-\psi)}). \tag{37}$$

It represents a convenient combination of the two approaches.

4.4. Implementation questions II – termination

Several possibilities are available for testing for termination. These include the following possibilities which all relate to the conditions for a stationary point of the nonlinear sum of squares:

(1) $\|\mathbf{h}\|_2$ small. This test on the magnitude of $\|\mathbf{h}\|$ is transformation dependent in the sense that if $T\mathbf{x} \to \mathbf{u}$ as new dependent variables then $\mathbf{h} \to T\mathbf{h}$ so that the magnitude of $\mathbf{h}$ is invariant only if the transformation is orthogonal. But it is independent of scale as the magnitude of $\mathbf{h}$ is unchanged under the transformation $\mathbf{f} \to \alpha\mathbf{f}$.

(2) $\|\mathbf{f}-\mathbf{r}\|_2$ small. This measure is transformation invariant as

$$\|\mathbf{f}-\mathbf{r}\|_2 = \|\nabla_u\mathbf{f}\mathbf{h}_u\|_2 = \|\nabla_x\mathbf{f}T^{-1}T\mathbf{h}\|_2,$$

but it is not scale invariant.

(3) $\{1 - \|\mathbf{r}\|_2^2/\|\mathbf{f}\|_2^2\}^{1/2}$ small. This provides a test which both transformation and scale invariant.

(4) $\|\nabla_x F\|_2$ small. This is neither scale nor transformation invariant.

The invariance properties suggest that test (3) would be most suitable. Here the quantity required enters already into the linesearch calculation in a natural fashion.

4.5. Implementation questions III – rate of convergence

The Gauss–Newton algorithm has attractive convergence properties in the data analysis application if n is large enough [7]. The argument is based on the condition for a simple iteration to have a fixed point. The result required is that if the iteration

$$\mathbf{x}_{i+1} = \mathbf{H}(\mathbf{x}_i)$$

is locally convergent to $\mathbf{x} = \hat{\mathbf{x}}$ from any close enough starting point then

$$\varpi\{\mathbf{H}'(\hat{\mathbf{x}})\} < 1, \tag{38}$$

where ϖ denotes the spectral radius. This is applied by noting that if $\lambda = 1$ is acceptable (this is discussed in Osborne [7]) then the Gauss–Newton algorithm can be written in this fixed point form as

$$\mathbf{x}_{i+1} = \mathbf{x}_i - (A^T A)^{-1} A^T \mathbf{f}.$$

Using the condition that F is stationary at $\hat{\mathbf{x}}$, and assuming the consistency of the estimation procedure which holds under very general conditions, it follows that

$$\begin{aligned} \mathbf{H}'(\hat{\mathbf{x}}) &= I - (A^T A)^{-1}\{A^T A + \sum_{i=1}^{n} f_i \nabla^2 f_i\}, \\ &= -(A^T A)^{-1} \sum_{i=1}^{n} f_i \nabla^2 f_i. \end{aligned} \tag{39}$$

Now define $\mathbf{K}(\mathbf{x})$ to have the functional dependence specified by the right hand side of Eq. (39). Then

$$\mathbf{K}(\mathbf{x}) = -(\frac{1}{n} A^T A)^{-1} \frac{1}{n} \sum_{i=1}^{n} f_i \nabla^2 f_i. \tag{40}$$

It follows that

$$\mathbf{H}'(\hat{\mathbf{x}}) = \mathbf{K}(\mathbf{x}^*) + O(\|\hat{\mathbf{x}} - \mathbf{x}^*\|), \tag{41}$$

and that the law of large numbers Eq. (19) can be applied to show almost surely that $\mathbf{K}(\mathbf{x}^*) \to 0,\ n \to \infty$. This shows that

$$\mathbf{H}'(\hat{\mathbf{x}}) \to 0, \quad n \to \infty.$$

Thus the iteration can be expected to be very well behaved given sufficient data. Also the mean value theorem gives

$$\begin{aligned} \mathbf{h}_{i+1} &= \mathbf{H}(\mathbf{x}_i) - \mathbf{H}(\mathbf{x}_{i-1}), \\ &= \mathbf{H}'(\mathbf{x}_{MV})\mathbf{h}_i + O(\|\mathbf{h}_i\|^2), \\ &= \mathbf{H}'(\hat{\mathbf{x}})\mathbf{h}_i + \max(O(\|\mathbf{h}_i\|\|\hat{\mathbf{x}} - \mathbf{x}_{MV}\|), O(\|\mathbf{h}_i\|^2)). \end{aligned} \tag{42}$$

This suggests that the power method could be used to estimate ϖ. It fails most often when convergence is very rapid so it is of little interest anyway. It is a useful diagnostic tool as slow convergence of the iteration indicates either a poor model or insufficient data. The iteration is usually well defined in these cases. At the other extreme of gross violation of the modelling assumptions it is again hardly needed.

4.6. Implementation questions IV – Data distribution

Use of the MGS algorithm to solve Eq. (32) requires that compatible data structures be used (see Figure 2). Communication is required in forming the nonlinear

```
!XOCL SPREAD DO /(PP)
      do 998 L=1,nproc
         if (irge.GT.0) then
!XOCL SPREAD MOVE
c     collect the results of the MGS calculation
            do 20 i=1,irge
 20            gh(i+ip,1)=A(i+vrge,pp1)
!XOCL END SPREAD (W1)
!XOCL MOVEWAIT (W1)
         endif
         if (L.EQ.1) then
            flag=.TRUE.
         else
            flag=.FALSE.
         endif
 998  continue
!XOCL END SPREAD
!XOCL BROADCAST (h)(flag)
```

Fig. 7. Annotated Fortran to spread the computed correction

sum of squares and sending the result to each processor. The logic is similar to that used in evaluating and distributing the length of the pivotal column using the SUM function in the MGS algorithm. It is required also that the pieces of the correction vector **h** be retrieved, assembled, and then distributed to each processor. The form of annotation used here for this purpose is indicated in Figure 7. One feature is that the line search decisions can be carried out on each processor with minimal overhead. The computation is dominated by the function value evaluations required, and this does involve communication.

4.7. Numerical examples

A small simulation has been carried out to test the performance of the Gauss–Newton algorithm. The first idea has been to try and develop a family of models that would maintain its credibility as the number of parameters was increased. To this end the family of models

$$y(t) = x_1 \exp(-x_2 t) \sin(x_3 \pi t + x_4) + \sum_{i=1}^{k} x_{3i+2} \exp\left(-(t - x_{3i+3})^2 / x_{3i+5}\right) \qquad (43)$$

has been studied. It could be thought of as an attempt to model peaks in the presence of the background expressed by the trigonometric term. If peak widths are narrow,

Table 4. Fitting Gaussian peaks plus background, k=1

n		$\sigma = .01$	$\sigma = .03$	$\sigma = .1$
	$\|\mathbf{h}\|$	$.68e-5$	$.63e-3$	$.53e-5$
	ss	$.8075e-2$	$.7270e-1$	$.8077$
100	$cond$	$.17e1$	$.17e1$	$.16e1$
	its	6	7	10
	ϖ	**	.07	.22
	$\|\mathbf{h}\|$	$.33e-6$	$.17e-5$	$.99e-5$
	ss	$.1013$	$.9118$	$.1013e2$
1000	$cond$	$.17e1$	$.17e1$	$.17e1$
	its	6	6	6
	ϖ	**	.01	.03
	$\|\mathbf{h}\|$	$.90e-4$	$.13e-3$	$.39e-5$
	ss	$.1018e1$	$.9163e1$	$.1018e3$
10000	$cond$	$.17e1$	$.17e1$	$.17e1$
	its	5	5	6
	ϖ	**	**	.01

and peak locations well spaced then it might be hoped that well behaved fitting problems would be obtained. Here k is the number of peaks.

Example 1 *The simplest case corresponds to $k = 1$. The form of the model considered is*

$$y(t) = \exp(-3t)\sin(5\pi t + .25) + 18\exp(-(t-.5)^2/.05).$$

The initial $\mathbf{x}$ *is given by*

$$\mathbf{x} = \{1.1, 2.9, 5.1, .27,\ 17.8, .52, .049\}^T.$$

The numerical results are summarised in Table 4.

Example 2 *An increment of degree of difficulty occurs if another peak is added. In this case the model has the form*

$$y(t) = \exp(-3t)\sin(5\pi t + .25) + 18\exp(-(t-.25)^2/.1) + 6\exp(-(t-.75)^2/.05).$$

The initial $\mathbf{x}$ *in this case is*

$$\mathbf{x} = \{1.1, 2.9, 5.1, .27,\ 17.8, .26, .12,\ 5.9, .73, .049\}^T.$$

The numerical results are summarised in Table 5.

Table 5. Fitting Gaussian peaks plus background, k=2

n		$\sigma = .01$	$\sigma = .03$	$\sigma = .1$
	$\|\mathbf{h}\|$	$.12e-4$	$.12e-2$	$.66e-4$
	ss	$.7619e-2$	$.6861e-1$	.7618
100	$cond$	.90e2	.12e3	.67e2
	its	10	15	28
	ϖ	.24	**	.57
	$\|\mathbf{h}\|$	$.10e-4$	$.22e-4$	$.22e-3$
	ss	.1012	.9106	.1011e2
1000	$cond$	.68e2	.64e2	.93e2
	its	8	11	10
	ϖ	.13	.32	**
	$\|\mathbf{h}\|$	$.72e-5$	$.18e-4$	$.55e-4$
	ss	.1018e1	.9161e1	.1018e3
10000	$cond$	.70e2	.68e2	.62e2
	its	7	9	19
	ϖ	.08	.22	.63

Example 3 *The problem becomes distinctly harder if another peak is added. In this case the model is taken to have the form*

$$\begin{aligned} y(t) &= \exp(-3t)\sin(5\pi t + .25) + 18\exp(-(t-1/6)^2/.15) \\ &\quad +6\exp(-(t-.5)^2/.1) + 3.6\exp(-(t-5/6)^2/.05). \end{aligned}$$

The initial $\mathbf{x}$ *in this case is*

$$\mathbf{x} = \{1.1, 2.9, 5.1, .27,\ 17.8, .16, .16,\ 6.0, .52, .095,\ 3.55, .82, .049\}^T.$$

The numerical results are summarised in Table 6.

The case of no convergence is indicated by '**'. But the other results evidence a fair amount of variation, and are not immune from question. The variation can be explained at least in part by noting that the initial conditions are not by any means uniformly good for fixed n as σ varies. However. the general trend appears to be compatible with the theory. The problems considered prove harder than expected suggesting that significantly narrower peaks will be needed if problems with more variables are to be attempted. The detailed results showed that the phase constant x_4 in the background term was poorly determined relative to the others.

Table 6. Fitting Gaussian peaks plus background, k=3

n		$\sigma = .01$	$\sigma = .03$	$\sigma = .1$
100		**	**	**
	$\|\|\mathbf{h}\|\|$	$.41e-3$	$.11e2$	$.23e-2$
	ss	.1010	.9089	$.1010e2$
1000	$cond$	$.58e5$	$.69e4$	$.22e4$
	its	18	25	49
	ϖ	.27	.68	**
	$\|\|\mathbf{h}\|\|$	**	$.52e-3$	$.59e-3$
	ss	**	$.9155e1$	$1017e3$
10000	$cond$	**	$.29e4$	$.11e4$
	its	**	16	21
	ϖ	**	.27	**

5. References

1. Å. Björck, 'Solving linear least squares problems by Gram–Schmidt orthogonalisation', *BIT*, **7** (1967) 1–21.
2. Å. Björck, 'Numerics of Gram–Schmidt Orthogonalisation', *Lin.Alg. and Applic.*, **194** (1993) 1–19.
3. G.H. Golub and C.F. Van Loan, *Matrix Computations*, (Johns Hopkins University Press, 1989).
4. G.H. Golub and J.H. Wilkinson, 'Iterative refinement of least squares solutions', *Num. Math.*, **9** (1967) 139–148.
5. C.L. Lawson and R.J. Hanson, *Solving Least Squares Problems*, (Prentice Hall, 1974).
6. J.W. Longley, *Least Squares Computations Using Orthogonal Methods*, (Marcel Decker, 1984).
7. M.R. Osborne, 'Fisher's method of scoring', *Int. Stat. Rev.*, **60** (1992) 99–117.

ASPECTS OF M-ESTIMATION AND l_1 FITTING PROBLEMS

M. R. OSBORNE
Centre for Mathematics and its Applications,
School of Mathematical Sciences, Australian National University,
Canberra, ACT 0200, Australia
E-mail: mike@thrain.anu.edu.au

and

G. A. WATSON
Department of Mathematics and Computer Science,
University of Dundee, Dundee DD1 4HN, Scotland
E-mail: gawatson@mcs.dundee.ac.uk

ABSTRACT

There has been a recent flurry of interest in the properties of the Huber M-estimator and its relation to l_1 fitting. In particular, it has been shown that the dual of the M–estimation problem is an interval constrained quadratic program. Thus methods for solving such quadratic programs can be applied directly to the M–estimation problem, and existing methods such as Clark's partitioning algorithm can be interpreted in the dual framework. Also, the duality result has been applied to study the limiting behavior of the M–estimator as the scale parameter $\gamma \to 0$, and results have been reported which support the claim that the l_1 fitting problem should be solved as the limiting case of M–estimation as $\gamma \to 0$. This claim is reviewed, and it is concluded that its basis is by no means as clear cut as has been suggested. Simplicial methods for the l_1 problem involve a descent step, and the manner in which this is carried out can be important for the overall effectiveness of the procedure. Some variants are discussed.

1. Introduction

Fitting a linear model containing p parameters to $n > p$ observed data points is a fundamental problem in data analysis. The typical formulation of the problem involves the minimization of some loss function $\psi(\mathbf{r})$, where

$$\mathbf{r} = A\mathbf{x} - \mathbf{b} \tag{1}$$

is defined for all $\mathbf{x} \in R^p$ and depends on the given data $A \in R^{n \times p}$, $\mathbf{b} \in R^n$. The components of $\mathbf{r} \in R^n$ can be interpreted as errors in the dependent variable values, and they are required to be made small by suitable choice of $\mathbf{x}$. The best known and most widely used loss function is the sum of squares. However, robust estimation has been a source of considerable activity in recent years. Here the aim is to compute parameter estimates with the property that any outliers corresponding to wild observations in the data have little influence on the estimation of $\mathbf{x}$. The minimization of the sum of absolute deviations or l_1 norm provides one possible approach. The solution normally fixes p components of $\mathbf{r}$ at zero, and is unaffected by (possibly large) changes in other components.

Another possibility which has been widely investigated is the Huber M-estimator. The loss function is defined by

$$\psi(\mathbf{r}) = \sum_{i=1}^{n} \rho(r_i), \tag{2}$$

where

$$\rho(t) = \begin{cases} t^2/2, & |t| \leq \gamma \\ \gamma(|t| - \gamma/2), & |t| > \gamma, \end{cases} \tag{3}$$

and γ is a scale factor or tuning constant. The function (2) is convex and once continuously differentiable, but has discontinuous second derivatives at points where $|r_i| = \gamma$ (the tight residuals). Clearly if γ is chosen large enough, then the loss function is just the least squares function, and if γ tends to zero, then limit points of the set of solutions minimize the l_1 function[5], so in a sense the M–estimator represents a compromise between these two. Frequently, the scale factor γ has to be obtained by satisfying an auxilliary equation as part of the estimation process. The ability to do this is an important feature of the continuation algorithm of Clark and Osborne[6], but most algorithms have been developed for the case when γ is assumed given. Then, as observed by Huber, the search for the minimum is the search for the correct partition of the components of $\mathbf{r}$, and the correct signs θ_i of the $|r_i| > \gamma$ (see (5) below). Once these are established, then the solution can be obtained by solving the necessary conditions. Let the partition be defined by an index set σ and its complement σ^c as follows:

$$\sigma = \{i : |r_i| \leq \gamma\}, \quad \sigma \cup \sigma^c = \{1, 2, \cdots, n\}. \tag{4}$$

Then the system of equations determined by the necessary conditions is

$$A_\sigma^T A_\sigma \mathbf{x} = A_\sigma^T \mathbf{b}_\sigma - \gamma \sum_{i \in \sigma^c} \theta_i \mathbf{a}_i, \tag{5}$$

where $\mathbf{a}_i^T$ denotes the ith row of A, where A_σ is obtained from A by deleting rows corresponding to indices $i \in \sigma^c$, and $\mathbf{b}_\sigma$ is defined similarly. There are two recent developments which are of particular interest here.

1. It has been suggested[10,9] that the preferred method for solving the l_1 problem is via a sequence of Huber problems for a sequence of scale values $\gamma \to 0$. This algorithmic development has lead to increased interest in the relationship between the Huber M-estimator and the l_1 problem[11,9].

2. Some duality results have appeared for the Huber problem: in particular, it has been shown that it has a dual which is a bounded variable, quadratic programming problem[13].

This gives rise to some further questions about whether there may be computational advantages in solving the Huber problem by attacking the dual directly. Also it is of interest to see if any current methods are dual methods implicitly.

In the next section, we examine the primal-dual relationship, and make some observations about uniqueness. In the following section, we look at some algorithmic developments, and, in particular, at an algorithm due to Clark for solving the Huber problem. A MATLAB implementation has been used to assess its potential performance, and its relation to methods for the solution of the dual problem is discussed. In particular we see that the steps of Clark's method correspond to solutions of the dual problem with changing subsets of bounded variables. Finally the recommendation[9,10] of M–estimation methods for solving the l_1–fitting problem is discussed. It is suggested that a more careful evaluation also of the usual simplicial methods is required before these recommendations can be accepted. Aspects of the linesearch relevant to the implementation of the simplicial methods are discussed. Tentative conclusions are drawn, which include the importance of using realistic test problems in evaluating algorithms.

2. Aspects of duality

It is shown by Michelot and Bougeard[13] that there is a simple interval bound quadratic program dual to the M–estimation problem. Applications of the dual have been considered by Li and Swetits[9]. The dual problem is

$$\min_{\mathbf{y}\in Y} \frac{\gamma}{2}\mathbf{y}^T\mathbf{y} - \mathbf{b}^T\mathbf{y}; \quad Y = \{\mathbf{y} : A^T\mathbf{y} = 0, \ -\mathbf{e} \leq \mathbf{y} \leq \mathbf{e}\}. \tag{6}$$

This result is readily verified by computing the Kuhn–Tucker conditions for (6). These give

$$\begin{aligned} \gamma\mathbf{y} - \mathbf{b} &= A\mathbf{u} + \mathbf{v} - \mathbf{w}, \quad \mathbf{v},\ \mathbf{w} \geq 0, \\ \mathbf{v}^T(\mathbf{y} + \mathbf{e}) &= \mathbf{w}^T(\mathbf{e} - \mathbf{y}) = 0, \end{aligned}$$

where $\mathbf{u},\ \mathbf{v},\ \mathbf{w}$ are the Kuhn–Tucker multipliers. It follows that

$$-\gamma\mathbf{y} = A(-\mathbf{u}) - \mathbf{b} - \mathbf{v} + \mathbf{w}.$$

The necessary conditions for a minimum of the M-estimation problem (2) give

$$A_\sigma^T\mathbf{r}_\sigma + \gamma\sum_{i\in\sigma^c}\theta_i\mathbf{a}_i = 0,$$

where

$$\mathbf{r}_\sigma = A_\sigma\mathbf{x} - \mathbf{b}_\sigma.$$

Identifying the M–estimation solution $\mathbf{x}$ with $-\mathbf{u}$ permits the further correspondences

$$\begin{aligned} v_i = w_i &= 0 && \Rightarrow i \in \sigma, \\ -\gamma y_i &= r_i = \mathbf{a}_i^T(-\mathbf{u}) - b_i, && i \in \sigma, \ (\Rightarrow |r_i| \leq \gamma), \\ -\gamma y_i &= r_i - \mathbf{v}_i + \mathbf{w}_i, && i \in \sigma^c \ (\Rightarrow |r_i| > \gamma). \end{aligned}$$

Thus the necessary conditions for a solution of the quadratic programming problem provide a consistent solution of the necessary conditions for the M–estimation problem, and a solution of the M–estimation problem permits a solution of the dual quadratic program to be written down directly. This result confirms a conjecture made by Clark in his thesis[4] that a natural dual should exist to the M–estimation problem. He was lead to this observation after developing his partitioning algorithm[6] by noting that it had many of the characteristic features of a dual algorithm. This point is discussed further in the next section.

An interesting motivation for this form of the dual is given by Li and Swetits[9]. Here the authors are interested in the relationship between the uniqueness of the solution of the l_1 fitting problem and solutions of the M–estimation problem as $\gamma \to 0$. This question has been considered by Clark[5]. He showed that limit points of the sequence of M–estimation problems as $\gamma \to 0$ solve the l_1 fitting problem, and gave an example to show that the M–estimation problem need not be unique for $\gamma > 0$ yet the limiting l_1 fitting problem can have an unique solution. Effectively the same questions are considered by Li and Swetits[9], but the methods are of independent interest. They employ the dual of the M–estimation problem to connect solutions of this problem with minimum norm solutions of the l_1 fitting problem by means of a result due to Mangasarian and Meyer[12]. Recall that the l_1 problem has the dual formulation as the interval linear program[14]

$$\min_{\mathbf{y} \in Y} -\mathbf{b}^T\mathbf{y}; \quad Y = \{\mathbf{y} : A^T\mathbf{y} = 0, \ -\mathbf{e} \leq \mathbf{y} \leq \mathbf{e}\}. \tag{7}$$

Let the set of optimal solutions to (7) be S. This is a nonempty, polyhedral convex set in R^n, and it may be a singleton. Let $\hat{\mathbf{y}}$ be the element of minimum Euclidean norm in S. It solves the problem

$$\min_{\mathbf{y} \in S} \frac{\gamma}{2}\mathbf{y}^T\mathbf{y}. \tag{8}$$

Mangasarian and Meyer show that there exist Kuhn–Tucker multipliers ($\mathbf{u}(\gamma)$, $\mathbf{v}(\gamma)$, $\mathbf{w}(\gamma)$) such that $\hat{\mathbf{y}}$, ($\mathbf{u}(\gamma)$, $\mathbf{v}(\gamma)$, $\mathbf{w}(\gamma)$) is a Kuhn–Tucker point for the M–estimation dual (which is a perturbation of the l_1 dual (7) in an obvious sense) for sufficiently small $\gamma > 0$. This result is readily verified in the same manner as before by calculating the Kuhn–Tucker conditions.

3. Algorithms

Here we consider algorithms for the Huber problem, and their relationship. Methods for minimising (2) have so far been broadly of two types:

1. methods based on applying Newton's method (or variants) to the zero derivative conditions characterizing a minimum[10];

2. methods which work towards the correct partition by explicit interchanges of components of $\mathbf{r}$ from one group to another in a systematic way[6].

In the light of the developments in duality theory, the question arises as to whether there are advantages to be gained in solving the dual. It is of interest also to consider the relationship of partitioning methods with the solution of the dual. We argue that there is a sense in which the Newton class of methods may be regarded as primal methods and the partitioning class as dual methods. To do this we will look at a particular method of the latter class and show how it can be interpreted as a method for solving the dual. But first we consider the Newton methods, which set out to solve the M–estimation problem for a fixed value of γ. For given $\mathbf{x} \in R^p$, define W as a diagonal matrix with elements 1 if $|r_i| \leq \gamma$ and 0 otherwise. Then, assuming that no value of $|r_i|$ is equal to γ, and letting $\mathbf{s} \in R^n$ be defined by

$$s_i = \begin{cases} 0, & i \in \sigma \\ \text{sign}(r_i), & i \in \sigma^c, \end{cases}$$

it is easily seen by differentiating the Huber function that ψ is minimized if and only if

$$A^T \left[\frac{1}{\gamma} W \mathbf{r} + \mathbf{s} \right] = 0.$$

The formal Newton step $\mathbf{d}$ for solving this system of equations ignores the discontinuity in derivative. It satisfies

$$\frac{1}{\gamma} A^T W A \mathbf{d} = -A^T \left[\frac{1}{\gamma} W \mathbf{r} + \mathbf{s} \right]. \tag{9}$$

If A has full rank p then the rank of W can always be taken as $\geq p$ at the solution of the M–estimation problem[14]. However, this does not ensure that W has rank $\geq p$ in a step of the Newton iteration. Problems with singularity of the linear system can be avoided by inserting additional 1's into the diagonal positions of W or indeed the unit matrix can be used. A line search may be necessary to ensure descent. For further details, see Madsen and Neilsen[10].

Now consider the alternative of solving the dual formulation described in the previous section. Since the solution of the dual problem is essentially a search for which dual variables attain their bounds, one might ask to what extent the idea of solving the dual problem is already foreshadowed in existing methods which arrive at the correct partition by suitable swapping. It is in fact possible to interpret the method of Clark[6] as a strategy for solving the dual by a method which itself is a dual procedure in the sense that feasibility and optimality are achieved together. In this sense it is a solution strategy similar to that proposed by Goldfarb and Idnani[8]. We now describe Clark's method, and begin with some definitions and notation.

Any choice of the index set σ defines a partition, P say, of $\{1, 2, \cdots, n\}$. Let $\mathbf{x}_\sigma$ minimise

$$\frac{1}{2}\sum_{i\in\sigma} r_i^2 + \sum_{i\in\sigma^c}(\gamma|r_i| - \frac{1}{2}\gamma^2). \tag{10}$$

Then the partition P is called σ–feasible if

$$i \in \sigma \Rightarrow |r_i(\mathbf{x}_\sigma)| \leq \gamma.$$

If

$$i \in \sigma^c \Rightarrow |r_i(\mathbf{x}_\sigma)| > \gamma,$$

then P is said to be σ^c–feasible. Note that these feasibility tests require the minimization of (10).

Clearly the search for the minimum of the Huber function is the search for a partition which is both σ–feasible and σ^c–feasible. Such a partition is called feasible. This search is performed in Clark's algorithm in the following way.

Step 1: Starting from an initial partition P_1, find a σ_1^c feasible partition.
Set $i = 1$.
While P_i is σ_i^c–feasible and P_i is σ_i–infeasible do:

Step 2: While P_i is σ_i–infeasible (so that $\exists k \in \sigma_i,\ |r_k| > \gamma$) do:

$$\begin{aligned} \sigma_{i+1} &= \sigma_i \setminus k, \\ i &= i + 1. \end{aligned}$$

end
Set $\mathbf{z}_{i-1} = \mathbf{x}_{i-1}$.

Step 3: While P_i is σ_i^c–infeasible do:
Find $\alpha,\ 0 < \alpha < 1$, such that

$$\begin{aligned} \mathbf{z}_i &= \alpha\mathbf{x}_i + (1-\alpha)\mathbf{z}_{i-1}, \\ |r_j(\mathbf{z}_i)| &\geq \gamma,\ \forall j \in \sigma_i^c, \\ \exists k &\in \sigma_i^c,\ |r_k(\mathbf{z}_i)| = \gamma, \end{aligned}$$

and set

$$\begin{aligned} \sigma_{i+1} &= \sigma_i \cup k, \\ i &= i + 1. \end{aligned}$$

end
end

n	p	q_1	q_2	q_3
100	10	80	0	0
160	10	140	1	1
160	20	131	2	2

Table 1: steps in the partitioning algorithm.

It is shown by Clark and Osborne[6] that this algorithm gives a sequence of partitions with the property that the minima of (10) are strictly decreasing. It terminates in a finite number of steps at a feasible partition. The first step of this algorithm (finding an initial σ^c–feasible partition) can be achieved easily by taking σ^c to be empty. The corresponding minimum of (10) is the least squares solution.

The dual to the problem of minimizing (10) is

$$\min_{\mathbf{y}\in Y} \frac{\gamma}{2}\mathbf{y}^T\mathbf{y} - \mathbf{b}^T\mathbf{y}; \quad Y = \{\mathbf{y} : A^T\mathbf{y} = 0, \ -1 \leq y_i \leq 1, \ i \in \sigma^c\}. \tag{11}$$

Clark's algorithm makes a systematic change of σ (and therefore of σ^c) by one index at a time until the correct partition is identified. It has been noted this can be interpreted as a particular dual strategy for solving the dual of the Huber problem.

A MATLAB implementation of Clark's algorithm has been used in order to study its performance. Interest is primarily in establishing numbers of interchanges, and there are potential efficiency gains which have not been exploited. Initially $\sigma_1^c = \emptyset$ was used to define the initial partition P_1, so that the least squares solution is generated as a first approximation, and small values of γ were chosen so that the final solution is an approximation to the l_1 solution. Most of the effort was required in Step 2, with very few iterations from Step 3 ever being performed. To illustrate, Table 1 shows results for computations performed using randomly generated data. The number of exchanges from $\sigma \to \sigma^c$ in Step 2 is denoted by q_1, the number of exchanges from $\sigma^c \to \sigma$ in Step 3 is denoted by q_2, and the number of times Step 3 is entered is denoted by q_3. The iteration counts are typical. The value of γ was 0.05 for these examples, and the final σ contai ned between p and $2p$ indices.

These and other examples suggest that the complexity of the method increases essentially linearly with n. The dual problem (11) was solved using the quadratic programming algorithm from the MATLAB optimisation toolbox. This is a primal active set method and so has the disadvantage in this context that a warm start is possible following an interchange between σ and σ^c only if the current solution $\mathbf{y}$ is feasible for the updated partition. This requirement is incompatible with the selection criterion used in Step 2. Also, little monitoring on the progress of the algorithm is available.

Our main aim here has been twofold: firstly, to obtain some information about iteration counts for the Clarke algorithm, and secondly to show how the method can be interpreted in the dual framework. Whether Clarke's method (or indeed any method applied directly or indirectly to the dual) can give significant computational advantages over primal methods remains to be established, and we will not pursue that question further here. Certainly, one advantage to be gained in using the dual formulation (11) is that standard software can be used.

4. l_1 problem: complexity of algorithms

We turn next to the solution of the l_1 problem. Traditionally, the favoured methods are simplicial (direct) methods, such as the reduced gradient algorithm[14]. Effective implementations use techniques which are closely related to those used in the implementation of linear programming methods, but with the additional feature that an efficient line search is required in each descent step[2,14]. The dual of the l_1 problem (7) follows directly from (6). It is an interval linear program, and the analogue of the line search is a succession of moves from one bound to its opposite as this leads to a trivial modification to the basis matrix factorization and hence to a cheap computation of a tentative solution estimate. This observation is used in the popular algorithm by Barrodale and Roberts[3]. However, recently it has been suggested that better results can be achieved by solving a sequence of Huber problem with γ allowed to decrease to zero systematically[10,9]. This proposal is based on the observations that the l_1 problem is the limiting case of the M–estimation problem as γ tends to zero, and that the l_1 solution can be detected for γ sufficiently small but non-zero. The justification claimed for this approach is that it avoids the use of simplicial methods for solving the l_1 problem.

The fact that γ need not reach zero is claimed to be the key to the efficiency and numerical stability of the algorithm. In particular, it is shown by Madsen and Neilsen[10] by numerical examples that starting from the l_2 solution only a small number of M-estimation problems need to be solved. Comparisons with results for the solution of the l_1 problem by the algorithm of Barrodale and Roberts[3] are given, and it is shown that greater efficiency can be obtained through the use of the Huber function. A variant of this Huber based method is proposed by Li and Swetits[9]. They claim that duality can be used to improve the selection of the sequence of values of $\gamma \to 0$, and that the assumption that both the l_1 and Huber estimation problems have unique solutions can be avoided. Both use the Newton class of methods for each fixed γ. Warm starts for Newton's method are based on the use of low rank updating for the exchange of single indices between σ and σ^c. Refactorisation of the matrix on the left hand side of (9) is done only when exchange of blocks of indices proves necessary, and iterative refinement is used to help overcome any problems with poor conditioning of the Hessian matrix.

There are reasons for caution in accepting wider implications of the comparisons given by Madsen and Neilsen[10].

- Firstly, more efficient implementations of the simplicial methods, with careful attention to line search performance and scaling, are available for solving the l_1 problem[2].

- Secondly, the larger scale calculations using randomly generated data reported by Madsen and Neilsen[10] are carried out only for $n/p = 2$. This does not seem appropriate for many problems of practical interest.

- Thirdly, it is legitimate to question conclusions as to which of the two approaches to solving the l_1 problem is superior based on the results of random simulations: such problems are easy to generate, but they are unlikely to have a great deal of relevance to actual practice.

It is important to stress that the data analytic origins of the M–estimation and l_1–fitting problems make it plausible that the chief interest is in a comparison of the methods for fixed p as n gets large ($\rightarrow \infty$). If the target is the l_1 solution corresponding to γ sufficiently small, then the work required in using an M–estimation algorithm, and starting with the least squares solution, involves swapping $O(n-p) = O(n)$ indices from σ to σ^c. As this reflects in the structure of A^TWA, a minimum of $O(n)$ update steps are needed in addition to the initial least squares solution. Typically these steps would be mostly rank one updates. However, if, at some point, a block of indices is switched between σ and σ^c simultaneously then a more substantial refactorization may be required. The result is a complexity estimate which is sum of the three components:

1. $\sim np^2$ for the initial least squares solution,

2. $\sim Jp^2$ for the rank one updates for interchanges between σ and σ^c, and

3. $\sim Knp^2$ for refactorizations (possibly corresponding to changes in γ).

Here J depends both on n (probably linearly) and on the cost of a rank one up or down dating step. The latter can be estimated by the work involved in a back substitution followed by p Givens rotations to sweep out a row appended to a triangular matrix ($\sim 2.5p^2$ multiplications). The reported evidence suggests that K is small, and it will be ignored here as will the cost of any linesearch (Newton step accepted).

The corresponding estimate for the l_1–fitting problem solved by the reduced gradient algorithm has the form

$$I\{np + \zeta(n)\},$$

where I is the number of tableau updating steps in the reduced gradient algorithm, $\sim np$ is the work done in a tableau update, and $\zeta(n)$ is an estimate of the linesearch complexity. Here $I \geq p$ which is the minimum number of steps needed to establish an initial basis, but it probably also grows with n. The significance of this growth is not too easy to ascertain. The work of Anderson and Steiger[1] shows that a normalised steepest edge test[14] significantly reduces the rate of growth on their random test problems (perhaps by as much as 25%). The unnormalised steepest edge computations reported in Osborne[14] are applied to similar problems and show a rate of growth $I \approx (1.6 + .25 \log \frac{n}{20})p$ for $p = 10$, and $20 \leq n \leq 320$. Also, the line search cost $\zeta(n)$ can be important as $n \to \infty$ for p fixed. It is of lower order than the tableau updating cost if the fast median methods are used, but it can be significant if analogues of a comparison sort are used. This problem occurs in the original Barrodale and Roberts program[2].

5. How good are the fast median methods?

In this section we report some results which compare the fast median algorithm suggested by Bloomfield and Steiger[2] with an alternative approach based on the use of the secant algorithm[7] for rank regression problems and applied to separable piecewise linear problems with complex structure[15]. Here 'complex structure' means that some at least of the component functions may have enormous numbers of linear pieces. In the l_1 case the circumstances are different because here it is n, the number of functions, that is large, not the number of linear pieces per function. The idea behind the partitioning methods is that of cutting down computations with complexity proportional to n. The Bloomfield and Steiger algorithm uses the heuristic from Hoare's quicksort method. They compute the set S of positive distances to zero of the current residuals, and then take the median of the first, middle, and last members of this set to define a partition bound pb. One pass through the set using Hoare's partitioning algorithm suffices to sort it into sets S_1, S_2 such that

$$\lambda \in S_1 \Rightarrow \lambda < pb, \quad \lambda \in S_2 \Rightarrow \lambda \geq pb.$$

The knowledge of S_1 permits the change in directional derivative in stepping to pb to be computed without additional sorting. If this quantity is non-negative then the crossing point is in S_1, else it is in S_2. Thus one pass through the set has the potential to significantly reduce the size of the set containing the crossing point. The effectiveness of this method depends on the effectiveness of the method for estimating the crossing point. It is here that the secant algorithm applied to the directional derivative of the l_1 function can be used. This is a piecewise constant, increasing function of λ, and the secant is fitted to the values corresponding to the largest and smallest elements in S. To set up its computation at the next step requires not only the contribution to the directional derivative from the points in S_1, but also the largest and smallest elements in each set together with the associated index information. This does involve a further comparison and possibly two further assignments – possibly a doubling in the computation required compared with that required by the median of three technique. This is illustrated in the code fragment below which represents the inner (partitioning) loop of the linesearch. Here σ, σ^C are held in `ib`, `dl` is the set of positive distances, **kp** the corresponding references to **ib**, and **da** the increments in directional derivative.

```
do
  {
   while (dl[i]<pb)
      {
       Fl+=fabs(da[ib[kp[i]]]);
       if (xl<dl[i])              //this block is not needed
          {                       //in the median of three
            xl=dl[i];             //computation of the
            imax=i;               //partition bound
          }
       i++;
      }
    while(dl[j]>=pb)
      {
        Fu-=fabs(da[ib[kp[j]]]);//can be avoided
        if (xu>dl[j])             //this block is not needed
           {                      //in the median of three
            xu=dl[j];             //computation of the
            jmin=j;               //partition bound
           }
       j--;
      }
   if (j>i)
      {
        s=dl[i]; dl[i]=dl[j];dl[j]=s;
        l=kp[i]; kp[i]=kp[j];kp[j]=l;
      }
  }while (j>i);
```

Two different series of experiments have been carried out to test the secant algorithm on l_1 problems. The aim here has been to test the use of the secant algorithm to select the partition bound, and to compare this with the use of the median of three heuristic. The problems used are:

1. Random problems generated in the manner described in Osborne[14], and with the dimension values $n = 20(*2)1280,\ p = 10(*2)40$.

2. Two structured l_1 problems with

$$b_i = \phi(t_i) + \epsilon_i, i = 1, 2, \cdots, n,$$

where the ϵ_i are independent, negative exponential distributed random variables

	n=20	40	80	160	320	640	1280
	17	23	29	33	38	40	41
p=10	2.67	3.40	4.12	4.85	5.50	6.32	6.95
	2.31	3.35	4.48	5.47	6.13	7.39	8.19
		41	63	78	92	104	113
p=20		3.36	3.95	4.43	4.91	5.56	6.00
		3.06	4.23	5.29	6.29	7.32	8.24
			106	172	222	263	295
p=40			3.82	4.11	4.35	4.70	5.12
			3.77	4.97	6.00	7.04	8.00

Table 2: Secant algorithm and median of 3 - randomly generated problems.

with common variance τ^2. In the first case

$$\phi(t) = exp(t), \ 0 < t \leq 1,$$

and ϕ is approximated by a polynomial in t of degree $p-1$. In the second case

$$\phi(t) = t^2, \ 0 < t \leq 1,$$

and ϕ is approximated by p terms of a Fourier sine series.

In the first of these structured problems it is easy to get high accuracy in the approximation to ϕ so that when the fitting problem is nearly converged then what it sees is dominated by the random error unless τ^2 is very small. On the other hand the Fourier series cannot be expected to give a good estimate of t^2 so that systematic error will compete with random error at each stage of the computation.

Results are given in Tables 2 and 3. The first gives median numbers of iterates for the l_1 solver, and medians of line search averages per iteration for both the secant and median of three methods for choosing the partition bound for 100 randomly generated problems. The results show the secant algorithm shapes up quite well, but, because it is somewhat more complex, there is certainly no clear case in its favour. The situation is rather different for the structured problems in Table 3. where results are given for $p = 5$ and $\tau^2 = 1(*.01).0001$, and $n = 100(*10)10000$. Here the secant algorithm gains a very significant advantage for small τ^2, especially in the case of the good approximation (small ϵ). The difference in the numbers of iterations may be noted. Apart from the line search the only difference between the programs was in the calculation of the initial directional derivative. This was found from the Kuhn-Tucker estimate in the method that used the secant algorithm, and was calculated directly in the median of three program.

p=5			median of three		secant	
τ^2	n	ϵ	its	aver.	its	aver.
1.	100	9.06-1	15	5.00	8	3.38
	1000	1.11+0	22	8.82	16	4.78
	10000	1.03+0	24	12.13	25	6.04
.01	100	9.05-3	16	4.75	20	3.50
	1000	1.11-2	21	8.57	19	7.11
	10000	1.03-2	27	12.70	27	8.56
.001	100	8.81-5	15	7.33	12	5.67
	1000	1.12-4	16	21.63	18	6.78
	10000	1.03-4	29	19.07	36	6.83
.0001	100	1.93-5	11	10.73	11	4.36
	1000	1.86-5	12	80.33	12	5.25
	10000	1.86-5	12	781.67	12	7.25
1.	100	1.13+0	10	6.10	9	5.44
	1000	1.15+0	21	9.33	13	5.15
	10000	1.11+0	20	14.0	21	7.29
.01	100	3.33-1	14	6.57	11	8.00
	1000	2.20-1	23	9.22	16	10.38
	10000	2.11-1	26	17.60	24	11.63
.001	100	3.28-1	12	10.25	12	8.25
	1000	2.20-1	12	71.92	13	13.69
	10000	2.10-1	15	264.87	15.	18.13
.0001	100	3.28-1	12	10.25	12	8.50
	1000	2.20-1	13	82.77	12	17.75
	10000	2.10-1	15	813.87	15	34.13

Table 3: Secant algorithm and median of 3 - structured problems.

6. Conclusion

Dual algorithms for the M–estimation problem have been discussed, and they are shown to widen the scope of computational methods in a significant manner. In particular, Clark's partitioning algorithm is interpreted as a dual method. This method has been suggested for removing degeneracy in the continuation approach. The significance of this approach in solving real data estimation problems is discussed by Clark and Osborne[6].

The conclusion from the complexity discussion would seem to be that simplicial methods using normalised steepest edge tests could be superior in l_1 fitting problems for the small p, moderate n values considered here. The behaviour for very large n should not be prejudged, and nothing is offered in the case n/p small as it would appear to have no practical interest in the classic data fitting paradigm, although it may be relevant in other circumstances. The comparison between the two linesearch algorithms reinforces the point about drawing conclusions based on randomly generated problems. The generally satisfactory behaviour of the secant algorithm suggests that it could be an important component of linesearch computations in this class of problems.

Stability considerations may also be relevant. The Newton class of methods use solutions found by solving normal equations where principally downdating style Hessian updates are used for economy at each value of γ. This is not ideal from a stability perspective as computing the normal matrix involves squaring the scale of the design matrix, and a priori this is not an ideal setting for using downdating. The implementation of the reduced gradient algorithm follows the description of the tableau algorithm given in by Osborne[14]. It uses a form of Bartels–Golub updating with partial pivoting being employed for stability. The strong dependence in the number of iterations required is on p. It is argued that this is typically not large in the data analytic context.

Acknowledgement The work of the second author was carried out while a Departmental Visitor in the Centre for Mathematics and its Applications, School of Mathematical Sciences, Australian National University, and was supported by a grant from the Centre. This visit was assisted by a grant from the Carnegie Trust for the Universities of Scotland.

7. References

1. D.H. Anderson and W.L. Steiger, A comparison of methods for discrete L_1 curve–fitting, (Technical Report 96, Department of Computer Science, Rutgers University, 1982).

2. P. Bloomfield and W.L. Steiger, *Least Absolute Deviations* (Birkhaüser, 1983).
3. I. Barrodale and F.D.K. Roberts, An improved algorithm for discrete l_1 linear approximation, *SIAM J. Num. Anal.* **10** (1973) 839–848.
4. D.I. Clark, *Finite Algorithms for Linear Optimisation Algorithms*, (Ph.D thesis, Australian National University, 1981).
5. D.I. Clark, The mathematical structure of Huber's M–estimator, *SIAM J. Sci. Stat. Comp.* **6** (1985) 209–219.
6. D.I. Clark and M.R. Osborne, Finite algorithms for Huber's M–estimator, *SIAM J. Sci. Stat. Comp.* **7** (1986) 72–85.
7. Karen George and M.R. Osborne, The efficient computation of linear rank statistics, *J.Statist. Comput. Simul.* **35** (1990) 227–237.
8. D. Goldfarb and A. Idnani, A numerically stable dual method for solving strictly convex quadratic programs, *Math. Programming* **27** (1983) 1–33.
9. W. Li and J.J. Swetits, Linear l_1 estimator and Huber M–estimator, (preprint, 1995).
10. K. Madsen and H.B. Nielsen, A finite smoothing algorithm for linear l_1 estimation, *SIAM J. Optim.* **3** (1993) 223–235.
11. K. Madsen, H.B. Nielson and M.C. Pinar, New characterizations of l_1 solutions to overdetermined systems of linear equations, (Institute for Numerical Analysis, Technical University of Denmark Report NI-93-09, 1993).
12. O.L. Mangasarian and R.R. Meyer, Nonlinear perturbations of linear programs, *SIAM J. Control Optim.* **17** (1979) 745–752.
13. C. Michelot and M.L. Bougeard, Duality results and proximal solutions of the Huber M–estimator problem , *Appl. Math. Optim.* **30** (1994) 203–221.
14. M.R. Osborne, *Finite Algorithms in Optimisation and Data Analysis*, (Wiley, 1985).
15. M.R. Osborne, On simplicial methods for minimizing piecewise linear functions, (C.M.A. Tech. Report MRR 076-95, Australian National University, 1995).

ON GENERALIZED BERNSTEIN POLYNOMIALS

GEORGE M PHILLIPS
Mathematical Institute, University of St Andrews
North Haugh, St Andrews, Fife KY16 9SS, Scotland
E-mail: gmp@st-andrews.ac.uk

ABSTRACT

The classical Bernstein polynomials require the approximated function to be evaluated at equal intervals. This paper is concerned with a generalization where the function is evaluated at intervals which are in geometric progression.

1. Introduction

It seems appropriate to begin with some family history. This paper and its sibling, a paper by Koçak and Phillips [4] concerning B-splines with q-integer knots, are offspring of a paper by Lee and Phillips [5] which discusses interpolation on the triangle at nodes based on the q-integers. The latter paper is, in turn, the offspring of a paper by Mitchell and Phillips [7] concerned with interpolation on the triangle and of a paper by Schoenberg [10] dealing with univariate interpolation on the q-integers. There is a pleasing symmetry in the parenthood of the Lee–Phillips paper, in that A. R. Mitchell and I. J. Schoenberg were Ph.D examiners for G. M. Phillips and S. L. Lee respectively. Finally, I mention in confidence something not usually discussed outside the family, that I was a referee for one of the above papers.

Recently [8] I proposed the following generalization of the Bernstein polynomials, based on the q-integers. For each integer n, we define

$$B_n(f;x) = \sum_{r=0}^{n} f_r \begin{bmatrix} n \\ r \end{bmatrix} x^r \prod_{s=0}^{n-r-1} (1-q^s x), \tag{1}$$

where an empty product denotes 1 and $f_r = f([r]/[n])$. The notation requires some explanation. The function f is evaluated at ratios of the q-integers $[r]$ and $[n]$, where q is a positive real number and

$$[r] = \begin{cases} (1-q^r)/(1-q), & q \neq 1, \\ r, & q = 1. \end{cases} \tag{2}$$

The expression $\begin{bmatrix} n \\ r \end{bmatrix}$, called a q-binomial coefficient, is defined by

$$\begin{bmatrix} n \\ r \end{bmatrix} = \frac{[n]!}{[r]![n-r]!} \tag{3}$$

for integers $n \geq r \geq 0$, where$[r]!$ has the obvious definition

$$[r]! = \begin{cases} [r].[r-1]\ldots[1], & r = 1,2,\ldots, \\ 1, & r = 0. \end{cases} \tag{4}$$

It is easily verified by induction on k that

$$\begin{bmatrix} k+1 \\ r \end{bmatrix} = q^{k-r+1} \begin{bmatrix} k \\ r-1 \end{bmatrix} + \begin{bmatrix} k \\ r \end{bmatrix} \tag{5}$$

and

$$\begin{bmatrix} k+1 \\ r \end{bmatrix} = \begin{bmatrix} k \\ r-1 \end{bmatrix} + q^{r} \begin{bmatrix} k \\ r \end{bmatrix}. \tag{6}$$

The q-binomial coefficients are also called Gaussian polynomials. (See Andrews. [1]) Indeed, it follows from either Eq. (5) or (6) that $\begin{bmatrix} n \\ r \end{bmatrix}$ is a polynomial of degree $r(n-r)$ in q with positive integral coefficients and is thus a monotonic increasing function of q. When $q = 1$, $\begin{bmatrix} n \\ r \end{bmatrix}$ reduces to the ordinary binomial coefficient and Eq. (1) gives the classical Bernstein polynomial. (See, for example, Cheney, [2] Davis, [3] Rivlin. [9])

The generalized Bernstein polynomial defined by Eq. (1) can be expressed in terms of q-differences. For any real function f we define

$$\Delta^0 f_i = f_i$$

for $i = 0, 1, \ldots, n$ and, recursively,

$$\Delta^{k+1} f_i = \Delta^k f_{i+1} - q^k \Delta^k f_i \tag{7}$$

for $k = 0, 1, \ldots, n-i-1$, where f_i denotes $f([i]/[n])$. See Schoenberg, [10] Lee and Phillips. [5] When $q = 1$, these q-differences reduce to ordinary forward differences and it is easily established by induction that

$$\Delta^k f_i = \sum_{r=0}^{k} (-1)^r q^{r(r-1)/2} \begin{bmatrix} k \\ r \end{bmatrix} f_{i+k-r}. \tag{8}$$

Then we may write (see Phillips [8])

$$B_n(f;x) = \sum_{r=0}^{n} \begin{bmatrix} n \\ r \end{bmatrix} \Delta^r f_0 \, x^r \tag{9}$$

which generalizes the well known forward difference form (see, for example, Davis [3]) of the classical Bernstein polynomial.

2. Results on convergence

For the generalized Bernstein operator defined by Eq. (1), we readily find from Eq. (9) that, for $n \geq 0, 1$ and 2 respectively,

$$B_n(1;x) = 1, \quad B_n(x;x) = x, \quad B_n(x^2;x) = x^2 + \frac{x(1-x)}{[n]}. \tag{10}$$

For the classical Bernstein operator, the uniform convergence of the sequence of polynomials $B_n(f;x)$ to $f \in C[0,1]$ follows as a special case of the Bohman–Korovkin theorem. (See Cheney, [2] Lorentz. [6]) Convergence is assured by the following two properties:

1. B_n is a monotone operator,
2. $B_n(f;x)$ converges uniformly to $f \in C[0,1]$ for $f(x) = 1, x$ and x^2.

Recall that if a linear operator L maps an element $f \in C[0,1]$ to $Lf \in C[0,1]$, then L is said to be *monotone* if $f(x) \geq 0$ on $[0,1]$ implies that $Lf(x) \geq 0$ on $[0,1]$. The generalized Bernstein operator defined by Eq. (1) is monotone for $0 < q \leq 1$. Yet if $0 < q < 1$ it is clear from Eqs. (10) and (2) that $B_n(x^2;x)$ does not converge to x^2. The following theorem, which is concerned with convergent sequences of Bernstein polynomials (other than the classical case with $q = 1$), is a special case of the Bohman–Korovkin theorem.

Theorem 1 *Let $q = q_n$ satisfy $0 < q_n < 1$ and let $q_n \to 1$ as $n \to \infty$. Then, for any $f \in C[0,1]$, the sequence of generalized Bernstein polynomials defined by*

$$B_n(f;x) = \sum_{r=0}^{n} f_r \begin{bmatrix} n \\ r \end{bmatrix} x^r \prod_{s=0}^{n-r-1} (1 - q_n^s x),$$

where $f_r = f([r]/[n])$, converges uniformly to $f(x)$ on $[0,1]$. □

The above result is discussed in Phillips [8], where there is also a proof of the following generalization of Voronovskaya's theorem. (This proof in Phillips [8] closely follows that presented in Davis [3] for the classical Bernstein polynomials.)

Theorem 2 *Let f be bounded on $[0,1]$ and let x_0 be a point of $[0,1]$ at which $f''(x_0)$ exists. Further, let $q = q_n$ satisfy $0 < q_n < 1$ and let $q_n \to 1$ as $n \to \infty$. Then the rate of convergence of the sequence of generalized Bernstein polynomials is governed by*

$$\lim_{n\to\infty} [n](B_n(f;x_0) - f(x_0)) = \frac{1}{2} x_0(1 - x_0) f''(x_0). \quad \square \tag{11}$$

3. Convergence of first derivative

For the classical Bernstein polynomials we have (see Davis [3]) that if $f \in C^p[0,1]$ the pth derivative of $B_n(f;x)$ converges uniformly on $[0,1]$ to the pth derivative of f. In view of the results of Sec. 2, it appears worthwhile to explore whether these results may be generalized, for suitable sequences (q_n). For the remainder of this paper, we will assume that $0 < q_n \leq 1$ and that $q_n \to 1$ as $n \to \infty$.

In what follows, we need to consider a modified form of the generalized Bernstein operator, namely

$$\tilde{B}_{n-1}(f;x) = \sum_{r=0}^{n-1} f_r \begin{bmatrix} n-1 \\ r \end{bmatrix} x^r \prod_{s=0}^{n-r-2} (1 - q_n^s x), \tag{12}$$

where $f_r = f([r]/[n])$, as above. Note that $\tilde{B}_{n-1}(f;x)$ is not the same as $B_{n-1}(f;x)$, since in the latter polynomial the function f is evaluated at the points $[r]/[n-1]$ and the q-integers are based on $q = q_{n-1}$. Then, just as we can write Eq. (1) in the form of Eq. (9), we can express Eq. (12) in the q-difference form

$$\tilde{B}_{n-1}(f;x) = \sum_{r=0}^{n-1} \begin{bmatrix} n-1 \\ r \end{bmatrix} \Delta^r f_0 \, x^r. \tag{13}$$

Using Eq. (13), a small calculation shows that

$$\tilde{B}_{n-1}(1;x) - 1 = 0, \quad \tilde{B}_{n-1}(x;x) - x = -\frac{q_n^{n-1}}{[n]} x$$

for $n \geq 1$ and 2 respectively and

$$\tilde{B}_{n-1}(x^2;x) - x^2 = \frac{[n-1]}{[n]^2} x + \left(\frac{q_n[n-1][n-2]}{[n]^2} - 1 \right) x^2$$

for $n \geq 3$. The latter equation may be recast in the form

$$\tilde{B}_{n-1}(x^2;x) - x^2 = \frac{1}{q_n[n]} \left(1 - \frac{1}{[n]} \right) x + \frac{1}{q_n^2} \left(1 - q_n^2 - \frac{2+q_n}{[n]} + \frac{1+q_n}{[n]^2} \right) x^2$$

and thus it is clear that $\tilde{B}_{n-1}(f;x)$ converges uniformly to f on $[0,1]$ for each of the three functions $1, x$ and x^2. Since $\tilde{B}_{n-1}$ is also a monotone operator on $[0,1]$, it follows from the Bohman–Korovkin theorem that $\tilde{B}_{n-1}(f;x)$ converges uniformly on $[0,1]$ to f, for each $f \in C[0,1]$.

We now express $B_n(f;x)$ in the q-difference form of Eq. (9) and differentiate to give

$$B_n'(f;x) = \sum_{r=1}^{n} r \begin{bmatrix} n \\ r \end{bmatrix} \Delta^r f_0 \, x^{r-1}.$$

This may be written in the form

$$B_n'(f;x) = \sum_{r=0}^{n-1} \frac{r+1}{[r+1]} \begin{bmatrix} n-1 \\ r \end{bmatrix} \Delta^r([n]\Delta f_0) \, x^r. \tag{14}$$

We note that, if f' exists, $[n]\Delta f_0$ is close to $f'(0)$ for large n. If we omit the quotient $(r+1)/([r+1])$, we change Eq. (14) into

$$\sum_{r=0}^{n-1} \begin{bmatrix} n-1 \\ r \end{bmatrix} \Delta^r([n]\Delta f_0) \, x^r = \tilde{B}_{n-1}([n]\Delta f;x), \tag{15}$$

on using Eq. (13). Since the operators concerned are linear we can easily compare the term $\tilde{B}_{n-1}([n]\Delta f;x)$ in Eq. (15) with $\tilde{B}_{n-1}(f';x)$, by writing

$$\tilde{B}_{n-1}(f';x) - \tilde{B}_{n-1}([n]\Delta f;x) = \sum_{r=0}^{n-1} (f_r' - [n]\Delta f_r) \begin{bmatrix} n-1 \\ r \end{bmatrix} x^r \prod_{s=0}^{n-r-2} (1 - q_n^s x). \tag{16}$$

We now assume that $f \in C^1[0,1]$. Then, by the mean value theorem, there exists a real θ, with $0 < \theta < 1$, such that

$$f'_r - [n]\triangle f_r = f'\left(\frac{[r]}{[n]}\right) - q_n^r f'\left(\frac{[r]+\theta q_n^r}{[n]}\right).$$

Since $q_n \to 1$ as $n \to \infty$ and $f \in C^1[0,1]$, given any $\epsilon > 0$, there exists a positive integer $N = N(\epsilon)$ such that

$$| f'_r - [n]\triangle f_r | < \epsilon, \quad 0 \leq r \leq n-1,$$

forall $n \geq N$. From Eqs. (16) and (10) it follows that

$$| \tilde{B}_{n-1}(f';x) - \tilde{B}_{n-1}([n]\triangle f;x) | < \epsilon \sum_{r=0}^{n-1} \begin{bmatrix} n-1 \\ r \end{bmatrix} x^r \prod_{s=0}^{n-r-2} (1-q_n^s x) = \epsilon.$$

This shows that, given any $\epsilon > 0$, there exists $N = N(\epsilon)$ such that

$$\| \tilde{B}_{n-1}(f';x) - \tilde{B}_{n-1}([n]\triangle f;x) \|_\infty < \epsilon \tag{17}$$

for all $n \geq N$.

In order to examine how close Eq. (14) is to Eq. (15), we will require the following inequality.

Lemma 1 *For* $0 < q \leq 1$,

$$\sum_{r=0}^{n} \begin{bmatrix} n \\ r \end{bmatrix} \left(\sum_{s=0}^{r} q^{s(s-1)/2} \begin{bmatrix} r \\ s \end{bmatrix} \right) \leq 3^n, \tag{18}$$

with equality when $q = 1$.

Proof Since the q-binomial coefficients are monotonic increasing functions of q, it suffices to consider the case where $q = 1$. We then have

$$\sum_{r=0}^{n} \binom{n}{r} \left(\sum_{s=0}^{r} \binom{r}{s} \right) = \sum_{r=0}^{n} \binom{n}{r} 2^r = (1+2)^n = 3^n. \quad \square$$

We now write

$$e_r^{(n)} = \frac{r+1}{[r+1]} - 1, \quad 0 \leq r \leq n-1,$$

where the superfix n emphasizes the dependence of $e_r^{(n)}$ on $q = q_n$. Then $e_r^{(n)} \geq 0$ for all r and $e_r^{(n)}$ is monotonic increasing in r. From Eqs. (14) and (15) we then obtain

$$B'_n(f;x) - \tilde{B}_{n-1}([n]\triangle f;x) = \sum_{r=0}^{n-1} e_r^{(n)} \begin{bmatrix} n-1 \\ r \end{bmatrix} \triangle^r([n]\triangle f_0)\, x^r$$

and, using Eq. (8) to expand the q-differences, we obtain

$$B_n'(f;x) - \tilde{B}_{n-1}([n]\Delta f;x) = \sum_{r=0}^{n-1} e_r^{(n)} \begin{bmatrix} n-1 \\ r \end{bmatrix} x^r \sum_{s=0}^{r} (-1)^s q^{s(s-1)/2} \begin{bmatrix} r \\ s \end{bmatrix} ([n]\,\Delta f_{r-s}).$$

On putting $t = r - s$, we derive

$$B_n'(f;x) - \tilde{B}_{n-1}([n]\Delta f;x) = \sum_{t=0}^{n-1} \begin{bmatrix} n-1 \\ t \end{bmatrix} ([n]\,\Delta f_t)\, x^t \sum_{s=0}^{n-1-t} (-1)^s q^{s(s-1)/2} \begin{bmatrix} n-1-t \\ s \end{bmatrix} e_{s+t}^{(n)} x^s.$$

Since $|\, [n]\Delta f_t \,| \leq \| f' \|_\infty$ and $0 \leq e_r^{(n)} \leq e_{n-1}^{(n)}$ for $0 \leq r \leq n-1$ it follows that

$$\| B_n'(f;x) - \tilde{B}_{n-1}([n]\Delta f;x) \|_\infty \leq \epsilon_n \| f' \|_\infty \sum_{t=0}^{n-1} \begin{bmatrix} n-1 \\ t \end{bmatrix} \sum_{s=0}^{n-1-t} q^{s(s-1)/2} \begin{bmatrix} n-1-t \\ s \end{bmatrix},$$

where ϵ_n denotes the number $e_{n-1}^{(n)}$. From the latter inequality and the above Lemma, we deduce that

$$\| B_n'(f;x) - \tilde{B}_{n-1}([n]\Delta f;x) \|_\infty \leq 3^{n-1} \epsilon_n \| f' \|_\infty. \tag{19}$$

From the inequalities (17) and(19) and the uniform convergence of $\tilde{B}_{n-1}(f';x)$ to f' on $[0,1]$ for $f \in C^1[0,1]$, we obtain the following result.

Theorem 3 *Let $f \in C^1[0,1]$ and let the sequence (q_n) be chosen so that the sequence (ϵ_n) converges to zero from above faster than $(1/3^n)$, where*

$$\epsilon_n = \frac{n}{1 + q_n + q_n{}^2 + \cdots + q_n{}^{n-1}} - 1. \tag{20}$$

Then the sequence of derivatives of the generalized Bernstein polynomials $B_n(f;x)$ converges uniformly on $[0,1]$ to $f'(x)$ □.

We note that the above condition in Theorem 3 can be satisfied by choosing (q_n) such that

$$1 - \frac{a^{-n}}{n} \leq q_n < 1,$$

where $a > 3$. For then

$$q_n{}^r \geq 1 - \frac{ra^{-n}}{n}, \quad 0 \leq r \leq n-1,$$

so that

$$[n] \geq n - \frac{1}{2}(n-1)a^{-n} > n\left(1 - \frac{1}{2}a^{-n}\right).$$

Thus

$$\epsilon_n = \frac{n}{[n]} - 1 < \frac{1}{1-\frac{1}{2}a^{-n}} - 1 < a^{-n}$$

and so (ϵ_n) converges to zero faster than $(1/3^n)$.

4. Higher derivatives

The analysis of Sec. 3 can be adapted to discuss the convergence of higher derivatives of generalized Bernstein polynomials. Since this is rather messy we will give merely an outline. First we extend Eq. (14) to give

$$B_n^{(p)}(f;x) = \sum_{r=0}^{n-p} c_{p,r} \begin{bmatrix} n-p \\ r \end{bmatrix} \Delta^r \left([n][n-1]\cdots[n-p+1]\,\Delta^p f_0\right) x^r, \qquad (21)$$

where

$$c_{p,r} = \frac{(r+p)(r+p-1)\cdots(r+1)}{[r+p][r+p-1]\cdots[r+1]}.$$

We delete the term $c_{p,r}$ in Eq. (21) to define a function $\tilde{B}_{n-p}$ which generalizes the function $\tilde{B}_{n-1}$ defined in Eq. (15) for the case where $p = 1$. We may then compare this function $\tilde{B}_{n-p}$ with $\tilde{B}_{n-p}(f^{(p)};x)$, thus generalizing Eq. (17). Hence we generalize Theorem 3 to pth derivatives, replacing ϵ_n in Eq. (20) by

$$\epsilon_n = \frac{n(n-1)\cdots(n-p+1)}{[n][n-1]\cdots[n-p+1]} - 1.$$

5. References

1. G. E. Andrews, *The Theory of Partitions* (Addison-Wesley, Reading, Mass., 1976).
2. E. W. Cheney, *Introduction to Approximation Theory* (McGraw-Hill, New York,1966).
3. P. J. Davis, *Interpolation and Approximation* (Dover, New York, 1976).
4. Z. F. Koçak and G. M. Phillips, *BIT* **34** (1994) 388.
5. S. L. Lee and G. M. Phillips, *Proc. Roy. Soc. Edin.* **108A** (1988) 75.
6. G. G. Lorentz, *Approximation of Functions* (Holt, Rinehart and Winston, New York, 1966).
7. A. R. Mitchell and G. M. Phillips, *BIT* **12** (1972) 81.
8. G. M. Phillips, in *Festschrift for T. J. Rivlin* (in the press).
9. T. J. Rivlin, *An Introduction to the Approximation of Functions* (Dover, New York, 1981).
10. I. J. Schoenberg, *Proc. Roy. Soc. Edin.* **90A** (1981) 195.

THE FINITE VOLUME ELEMENT METHOD FOR ELLIPTIC AND PARABOLIC EQUATIONS

THOMAS F. RUSSELL
Department of Mathematics, University of Colorado at Denver, Campus Box 170
Denver, CO 80217-3364, USA
E-mail: trussell@carbon.cudenver.edu

and

RICK V. TRUJILLO
Ecodynamics Research Associates, P.O. Box 9229, Albuquerque, NM 87119, USA

ABSTRACT

An error analysis of the finite volume element method for elliptic and parabolic partial differential equations is presented. Existing results apply to discretizations of steady diffusion equations by linear finite elements. These results are extended to advection–reaction–diffusion equations and are generalized to polynomial finite elements of arbitrary order. Optimal-order error estimates are derived in a discrete H^1 norm, under minimal regularity assumptions for the exact solution, the finite element triangulation, and the finite volume construction. With additional uniformity assumptions for the finite volumes, H^1 superconvergence results are obtained for linear finite elements.

1. Introduction

This paper presents an error analysis of the finite volume element method (FVE) for elliptic and parabolic partial differential equations. Full details of this analysis are given in the thesis of Trujillo[18]. As introduced by Baliga and Patankar[1] and elaborated by McCormick[14] and in the references cited therein, we can view FVE as a combination of the standard finite volume method (FV), also known as cell-centered finite differences (see Mitchell and Griffiths[12] for details), and the standard Galerkin finite element method (FE) (see Ciarlet[6] for details). Although a large body of theoretical infrastructure and of results for elliptic equations exists for both FV and FE, analysis for FVE is limited to the foundation laid by Cai and McCormick[3], Cai *et al.*[4], and culminating in Cai[5] for the numerical solution of steady diffusion equations by linear finite elements.

Starting from this foundation, we extend and generalize FVE error analysis to advection–reaction–diffusion equations and to polynomial finite elements of arbitrary order. To construct an error analysis for FVE, we are guided by the large body of FE theory as contained in Ciarlet[6] and in references cited therein, and by modern FV error analysis based on FE-style arguments: e.g., Bank and Rose[2], Ewing *et al.*[7], Hackbusch[8], Heinrich[9], Herbin[10], Lazarov *et al.*[11], Morton and Süli[13], Samarskii *et al.*[16], Süli[17], and Weiser and Wheeler[19].

By adapting FV and FE arguments to fit the context of FVE and developing some new arguments unique to FVE, we derive optimal-order error estimates for FVE in a discrete H^1 norm for polynomial finite elements of arbitrary order—under minimal regularity assumptions for the exact solution, the finite element triangulation, and the finite volume construction. With additional uniformity assumptions for the finite volumes, H^1 superconvergence results are obtained for linear finite elements.

The remainder of the paper is organized as follows. In Section 2, we outline how FVE arises from the approximation of certain integral conservation laws that lead to elliptic partial differential equations. In Section 3, we define and discuss the computational meshes (i.e., the FE triangulation $\mathcal{T}^h$, the FVE primal mesh $\mathcal{T}^{h/k}$, and the FVE dual mesh $\mathcal{V}^h$) that are fundamental to the implementation of FVE. In Section 4, we briefly outline an FE analysis for the steady diffusion equation to set a reference point for an FVE analysis in Sections 5–10. In Section 11, we note modifications for an FVE analysis for general elliptic equations. Finally, in Section 12 we briefly outline an extension of FVE analysis to parabolic equations.

2. FVE Background

The finite volume element method for elliptic equations is based on the fact that these equations arise from integral conservation laws. For definiteness and simplicity, we work in two spatial dimensions and consider the following model elliptic equation for an unknown distribution u:

$$\nabla \cdot (\mathbf{a}\, u - D\nabla u) + r\, u = f \quad \text{in } \Omega, \tag{1}$$

where $\Omega \subset \Re^2$ is the spatial domain, f is a source term, and $\mathbf{a} = (a_1, a_2)$, D, and r are advection, diffusion and reaction coefficients. For simplicity, we assume here that the coefficients are smooth functions of space alone to avoid difficulties caused by discontinuities and nonlinearities in our analysis, though these cases can be handled (see Section 12).

For (1), we consider Dirichlet or flux boundary conditions:

$$u = g \quad \text{or} \quad (\mathbf{a}\, u - D\nabla u) \cdot \mathbf{n} = g \quad \text{on } \partial\Omega, \tag{2}$$

where $\partial\Omega$ is the boundary of Ω, g is boundary data on $\partial\Omega$, and $\mathbf{n}$ is the outward unit normal on $\partial\Omega$.

Although we stated at the outset that we are investigating numerical solutions of the elliptic partial differential equation (1), we are more precisely studying integral conservation laws on subdomains $V \subseteq \Omega$ that lead to equations such as (1):

$$\int_{\partial V} (\mathbf{a}\, u - D\nabla u) \cdot \mathbf{n}\, dS + \int_V r\, u\, d\mathbf{x} = \int_V f\, d\mathbf{x}, \ \forall\, V \subseteq \Omega, \tag{3}$$

where ∂V is the boundary of V, $\mathbf{n}$ is the outward unit normal on the boundary ∂V, and dS is a spatial boundary measure. The conservation law (3) states that the advective-diffusive flux across volume boundary ∂V is counter-balanced by reactions and sources within the volume V. Equation (3) is called the primitive form of the elliptic equation (1) because it contains the essence of the physical model within the least restrictive mathematical model. Finally, it is (3)—and *not* (1)—that we mimic discretely in FVE.

FVE approximates (3) using meshes described in the next section, by replacing u and f in (3) with finite element approximations u^h and f^h which are based on a finite element triangulation $\mathcal{T}^h$ that partitions Ω, and by posing equation (3) on a finite subset $\mathcal{V}^h$ of volumes that partitions Ω.

3. FVE Meshes

The most basic component in the implementation of the finite volume element method is the discretization of the domain Ω into computational meshes. The finite element triangulation $\mathcal{T}^h$ of Ω and the finite volume element volumization $\mathcal{V}^h$ of Ω are two different—yet interconnected—meshes or discretizations of Ω. The two discretizations are connected by a third mesh or discretization of Ω: $\mathcal{T}^{h/k}$, the k-fold refinement of $\mathcal{T}^h$. The nodes or vertices of $\mathcal{T}^{h/k}$ are the locations of the degrees of freedom (DOF) for $u^h \in P^k(\mathcal{T}^h)$: u^h is a C^0 function on Ω that is a polynomial of degree $k \geq 1$ when restricted to each element T of $\mathcal{T}^h$. In FVE terminology, the triangulation $\mathcal{T}^{h/k}$ is the *primal* mesh and the volumization $\mathcal{V}^h$ is the *dual* mesh. Next, we describe these three meshes and their relationships.

3.1. FE Triangulation

Let $\mathcal{T}^h$ be a non-overlapping triangulation of $\overline{\Omega}$, the closure of Ω, into a finite number of elements T. To simplify the discussion, we assume the elements of the triangulation are triangles. Other types of elements could be considered: in particular, rectangular elements for a rectangular domain Ω.

For each T of $\mathcal{T}^h$, we define the mesh parameters h_T, h, and ρ_T as follows: h_T is the diameter of the circumscribing circle for T; h is the maximum value of h_T; and ρ_T is the diameter of the inscribed circle in T.

We assume that the triangulation is *regular:* there exists a positive constant σ such that

$$\frac{h_T}{\rho_T} \leq \sigma, \quad \forall T \in \mathcal{T}^h; \tag{4}$$

the family (h_T) is bounded and 0 is its unique accumulation point—i.e., h approaches zero. The assumption of regularity is used to simplify error estimates and to avoid degenerate triangulations (i.e., a minimum angle condition is satisfied).

3.2. FVE Primal Mesh

Once a nodal or base triangulation $\mathcal{T}^h$ is defined, we can define elements of $P^k(\mathcal{T}^h)$. The degrees of freedom (DOF) for $u^h \in P^k(\mathcal{T}^h)$ are located in a regular fashion in $\overline{T}$: on the vertices, along edges, and in the interior of T. Just as the FE triangulation $\mathcal{T}^h$ can be determined by connecting the vertices of T to their nearest neighbors, an alternative triangulation $\mathcal{T}^{h/k}$ can be determined by connecting the DOF in T to their nearest neighbors within T: let T_k denote the triangular elements of $\mathcal{T}^{h/k}$. The primal mesh $\mathcal{T}^{h/k}$ can be seen as the k-fold refinement of $\mathcal{T}^h$: if $\mathrm{diam}(T) = h$, then $\mathrm{diam}(T_k) = h/k$.

3.3. FVE Dual Mesh

The FVE dual mesh $\mathcal{V}^h$ or volumization of Ω partitions $\overline{\Omega}$ into a finite number of non-overlapping elements V_i—the index i refers to a one-to-one correspondence between volumes and DOF for $u^h \in P^k(\mathcal{T}^h)$. To define $\mathcal{V}^h$ from the primal mesh $\mathcal{T}^{h/k}$, we make frequent reference to $\mathcal{T}_i^{h/k}$: the union of all $T_k \in \mathcal{T}^{h/k}$ that have the location of the ith DOF, $\mathbf{x}_i$, as a vertex. To ensure a one-to-one correspondence between volumes and DOF, we require $V_i \subset \mathcal{T}_i^{h/k}$—further specifications for V_i are outlined below.

3.3.1. Volume Construction

Following Bank and Rose[2], we construct volumes as follows: (1) select a point $P \in \overline{T_k}, \quad \forall T_k \in \mathcal{T}_i^{h/k}$; (2) connect P by straight line segments to edge midpoints of T_k for the two edges of T_k adjacent to the vertex $\mathbf{x}_i$, $\forall T_k \in \mathcal{T}_i^{h/k}$; (3) for each $T_k \in \mathcal{T}_i^{h/k}$, define a *sub-volume*, $v_i(T_k)$, as the region bounded by the line segments formed in Step 2 and line segments connecting the edge midpoints of T_k with the vertex $\mathbf{x}_i$. Finally, define the volume V_i as the region enclosing $\mathbf{x}_i$ in Step 2:

$$V_i \equiv \bigcup_{T_k \in \mathcal{T}_i^{h/k}} v_i(T_k). \tag{5}$$

The choice of P in Step 1 is crucial in volume construction. In practice (see Cai[5]), we typically use the circumcenter volume: P is the center of the circle circumscribed about T_k, or equivalently, the intersection of the perpendicular bisectors of the edges of T_k. To ensure that $P \in \overline{T_k}$, this case requires that no interior angle of T_k exceed $\pi/2$—no obtuse triangles are permitted. In applications, advantages of circumcenter volumes are that they are always convex and geometrically simple, while other types of volumes (e.g., centroid, orthocenter, incenter, etc.) are usually non-convex and geometrically complex (see Bank and Rose[2] for details).

3.3.2. Volume Symmetry

In the analysis to follow, superconvergence results for linear finite elements can be demonstrated for "symmetric" circumcenter volumes. To define this symmetry precisely, we need additional notation: X_{ij} is the edge or line segment connecting nodes $\mathbf{x}_i$ and $\mathbf{x}_j$ of the triangulation $\mathcal{T}^{h/k}$; γ_{ij} is the interface or volume boundary between volumes V_i and V_j of the volumization $\mathcal{V}^h$—i.e., $\gamma_{ij} \equiv \overline{V_i} \cap \overline{V_j}$. Following Cai *et al.*[4] and Cai[5], the volumization $\mathcal{V}^h$ is *symmetric* to the triangulation $\mathcal{T}^{h/k}$ if the following two symmetries hold for all volumes in $\mathcal{V}^h$: (X-symmetry) γ_{ij} is a perpendicular bisector of X_{ij}; (γ-symmetry) X_{ij} is a perpendicular bisector of γ_{ij}. We remark that for rectangular elements, only the first condition of X-symmetry is required for $\mathcal{V}^h$ to be symmetric to $\mathcal{T}^{h/k}$.

3.3.3. Volume Regularity

Of more general significance is the notion of volume "regularity," which is assumed in all of our results. Form an auxiliary triangulation $\widetilde{\mathcal{T}^{h/k}}$ of triangular elements by connecting the endpoints of γ_{ij} with the endpoints of X_{ij} for every volume interface in $\mathcal{V}^h$. For circumcenter volumes, if this auxiliary triangulation $\widetilde{\mathcal{T}^{h/k}}$ is regular according to the definition in Section 3.1 (cf. (4)) and no interior angle of $\widetilde{T_k} \in \widetilde{\mathcal{T}^{h/k}}$ exceeds $\pi/2$, then the volumization $\mathcal{V}^h$ is *regular*.

4. Finite Element Analysis

Here we present the rudiments of FE analysis to establish a reference point for analogous developments to follow for FVE. For simplicity, consider the diffusion equation with homogeneous Dirichlet boundary condition:

$$-\nabla \cdot (D\nabla u) = f, \quad x \in \Omega, \tag{6}$$

$$u|_{\partial\Omega} = 0, \tag{7}$$

where D is a continuous, bounded, non-degenerate ($0 < D_m \le D \le D_M < \infty$) diffusion coefficient and $\Omega \subset \Re^2$ is a domain.

The weak solution u of (6) (see (8) below) lies in the Sobolev space $\mathcal{W} \equiv H^k(\Omega) \cap H_0^1(\Omega)$, where $k \ge 1$. The numerical solution u^h of (6) (see (11) below) will lie in $\mathcal{W}^h \equiv P_0^k(\mathcal{T}^h)$—the space of C^0 functions that are polynomials of order $\le k$ on each element T of the triangulation $\mathcal{T}^h$ and vanish on $\partial\Omega$. Thus, $\mathcal{W}^h \subset \mathcal{W}$. Similarly, $\mathcal{W}_- = H^{k-1}(\Omega) \cap L^2(\Omega)$ and $\mathcal{W}_-^h = P^{k-1}(\mathcal{T}^h)$ are the lower-order spaces used to represent the continuous and numerical source terms f and f^h, respectively.

A weak solution $u \in \mathcal{W}$ to (6), corresponding to $f \in \mathcal{W}_-$, satisfies:

$$A(u,w) = (f,w), \quad \forall\, w \in \mathcal{W}, \tag{8}$$

where

$$A(u,w) = \int_\Omega D\nabla u \cdot \nabla w \, d\mathbf{x}, \tag{9}$$

$$(f,w) = \int_\Omega f\, w \, d\mathbf{x}. \tag{10}$$

Although it is clear from (8)–(10) that $f \in H^{k-1}(\Omega)$ implies $u \in H^k(\Omega)$, it is well-known that for a domain Ω with sufficiently smooth boundary and for sufficiently smooth D that we have an elliptic regularity result: $u \in H^{k+1}(\Omega)$. In our error estimate (19) below, we assume elliptic regularity; later, in Section 4.5, we discuss modifications to the analysis in the absence of elliptic regularity.

In direct analogy with (8), the (finite element) numerical solution $u^h \in \mathcal{W}^h$ to (6), corresponding to $f^h \in \mathcal{W}_-^h$, satisfies:

$$A(u^h,w) = (f^h,w), \quad \forall\, w \in \mathcal{W}^h, \tag{11}$$

where $f^h = \Pi_{k-1}^h f$ is an interpolant of the source term f from $\mathcal{W}_-$ into $\mathcal{W}_-^h$.

The key components of finite element error analysis are ellipticity, boundedness, and approximation theory results: we introduce and discuss each below.

4.1. Ellipticity

An *ellipticity* (or lower bound) condition that there exists a positive constant α such that

$$A(u,u) \geq \alpha\, |u|_{1,\Omega}^2, \quad \forall u \in \mathcal{W}, \tag{12}$$

is perhaps the most important component of FE error analysis: it implies that the bilinear form $A(\cdot,\cdot)$ of (9)—hereafter called the "A–form"—is positive definite.

4.2. Boundedness

Upper bounds (or *continuity*) conditions for the A-form (9), that there exists a positive constant M such that

$$A(u,w) \leq M\, |u|_{1,\Omega} |w|_{1,\Omega}, \quad \forall\, u,w \in \mathcal{W}, \tag{13}$$

and for the source term (10),

$$(f,w) \leq |f|_{-1,\Omega} |w|_{1,\Omega}, \quad \forall\, f \in \mathcal{W}_-, w \in \mathcal{W}, \tag{14}$$

are necessary to demonstrate convergence of the numerical method as seen below.

4.3. Approximation Theory

For the $H^m(\Omega)$ semi-norms $(-1 \le m \le 1)$ employed in (12), (13), and (14), we cite standard approximation theory results for polynomial interpolation: there exists a constant C such that

$$|\eta_k(u)|_{m,\Omega} \le C\, h^{k+1-m} \|u\|_{k+1,\Omega}, \quad \forall\, u \in H^{k+1}(\Omega), \tag{15}$$

where $\eta_k(u) \equiv \Pi_k^h u - u$ is the interpolation error for u in the polynomial space $P^k(\mathcal{T}^h)$. In the remainder of the text, C will represent a (generic) positive constant.

4.4. Error Analysis

In the standard FE error analysis to follow, we apply the results (12) and (13)–(15); for (6), (12) and (13) hold with $\alpha = D_m$ and $M = D_M$, respectively.

4.4.1. Error Splitting

Let $e \equiv u - u^h \in \mathcal{W}$ be the error in the FE approximation to (6). For theoretical purposes, define $\zeta \equiv \Pi_k^h e$ as an interpolation of the error from $\mathcal{W}$ into $\mathcal{W}^h$, so that $\zeta \equiv \Pi_k^h u - u^h \in \mathcal{W}^h$. We refer to ζ as the *representation error* in $\mathcal{W}^h$: i.e., the difference between interpolated and computed approximations in $\mathcal{W}^h$ of the exact solution u. Then, we employ the *error splitting* $e = \zeta - \eta_k(u)$, in which the FE error is split into its representation error and interpolation error components: $|u - u^h|_{1,\Omega} \le |\zeta|_{1,\Omega} + |\eta_k(u)|_{1,\Omega}$. With the estimate of $|\eta_k(u)|_{1,\Omega}$ in (15), estimating $|e|_{1,\Omega}$ is reduced to estimating $|\zeta|_{1,\Omega}$. Below, we develop an estimate for $|\zeta|_{1,\Omega}$ in terms of estimates for $|\eta_k(u)|_{1,\Omega}$ and $|\eta_{k-1}(f)|_{1,\Omega}$—the interpolation errors for u and f, respectively.

4.4.2. Zero Property

For all $w \in \mathcal{W}^h$, the exact solution satisfies (8) and the numerical solution satisfies (11). Subtracting, we have the *zero property* or *orthogonality* of the error,

$$A(u - u^h, w) - (f - f^h, w) = 0, \quad \forall\, w \in \mathcal{W}^h, \tag{16}$$

which in our context is more appropriately expressed as

$$A(\zeta, w) = A(\eta_k(u), w) + (\eta_{k-1}(f), w), \quad \forall\, w \in \mathcal{W}^h. \tag{17}$$

4.4.3. Error Estimate

Since the representation error, ζ, is in $\mathcal{W}^h$, we take $w = \zeta$ in (17) to obtain

$$\alpha\, |\zeta|_{1,\Omega}^2 \;\le\; A(\zeta, \zeta)$$

$$\begin{aligned} &= A(\eta_k(u), \zeta) + (\eta_{k-1}(f), \zeta) \\ &\leq M\, |\eta_k(u)|_{1,\Omega} |\zeta|_{1,\Omega} + |\eta_{k-1}(f)|_{-1,\Omega} |\zeta|_{1,\Omega}. \end{aligned} \tag{18}$$

The three lines of (18) use (12), (17), and (13) and (14), respectively. By (15), (18) implies the following optimal-order *error estimate* for the finite element method (under the assumption of elliptic regularity):

$$|u - u^h|_{1,\Omega} \leq C\, h^k \,(\|u\|_{k+1,\Omega} + \|f\|_{k-1,\Omega}). \tag{19}$$

4.5. Modified Error Estimate

In the absence of elliptic regularity, a different proof of the optimal-order estimate is needed, one that is more typical of FVE analysis. Recalling the weak form (8), we assume that $f \in H^k(\Omega)$, which immediately implies that $u \in \mathcal{W}^+ \equiv H^{k+1}(\Omega) \cap H_0^1(\Omega)$; however, in the numerical scheme, we retain the same discrete space structure (i.e., $u^h \in \mathcal{W}^h$ and $f^h \in \mathcal{W}_-^h$) as before. Under these assumptions, we may effectively replace (14) with the less elegant bound

$$(f, w) \leq |f|_{0,\Omega} |w|_{0,\Omega}, \quad \forall\, f, w \in \mathcal{W}. \tag{20}$$

Then the third line of (18) is replaced by

$$\alpha\, |\zeta|_{1,\Omega}^2 \leq M\, |\eta_k(u)|_{1,\Omega} |\zeta|_{1,\Omega} + |\eta_{k-1}(f)|_{0,\Omega} |\zeta|_{0,\Omega}. \tag{21}$$

Finally, after applying the Poincaré inequality to $|\zeta|_{0,\Omega}$, we obtain:

$$|u - u^h|_{1,\Omega} \leq C\, h^k \,(\|u\|_{k+1,\Omega} + \|f\|_{k,\Omega}). \tag{22}$$

Regarding elliptic regularity in FE error analysis, its presence is easily incorporated and its absence causes complications—this is reversed in the FVE error analysis below.

5. FVE Preliminaries

We employ the same continuous and discrete space notations, definitions, and structures as in Section 4.5 for the FVE continuous and discrete trial spaces that represent the solution and the source term data: for $k \geq 1$, $u \in \mathcal{W}^+ \equiv H^{k+1}(\Omega) \cap H_0^1(\Omega)$, $u^h \in \mathcal{W}^h \equiv P^k(\mathcal{T}^h)$, $f \in \mathcal{W} \equiv H^k(\Omega) \cap H^1(\Omega)$, and $f^h \in \mathcal{W}_-^h \equiv P^{k-1}(\mathcal{T}^h)$. Furthermore, let $\mathcal{X}^h$ denote the FVE test space $P_0^0(\mathcal{V}^h)$—the polynomials that are constant on each volume V of $\mathcal{V}^h$ and vanish on $\partial\Omega$. Thus, test functions corresponding to volumes that intersect $\partial\Omega$ are identically zero. Also, let χ_V denote the characteristic function associated with volume $V \subseteq \Omega$ and let $\mathcal{V}$ denote the collection of all such V.

Recalling the integral equation (3), (in the absence of elliptic regularity) a weak solution u of (6) satisfies: given $f \in \mathcal{W}$, find $u \in \mathcal{W}^+$ such that

$$B(u, \chi_V) = (f, \chi_V), \quad \forall\, V \in \mathcal{V}, \tag{23}$$

where

$$B(u, \chi_V) = -\int_{\partial V} D\nabla u \cdot \mathbf{n}\, dS, \tag{24}$$

$$(f, \chi_V) = \int_V f\, d\mathbf{x}. \tag{25}$$

FVE poses the weak form (23) on the volumization $\mathcal{V}^h$—a finite subset of $\mathcal{V}$ with non-overlapping volumes—and employs standard FE representations of u and f. Thus, FVE replaces (23) with: given $f^h \in \mathcal{W}^h_-$, find $u^h \in \mathcal{W}^h$ such that

$$B(u^h, \chi_V) = (f^h, \chi_V), \quad \forall\, V \in \mathcal{V}^h. \tag{26}$$

Since $\mathcal{V}^h$ partitions Ω and the volumes in $\mathcal{V}^h$ are related to the triangulation $\mathcal{T}^h$ as discussed in Section 3.3, (26) is more amenable than (23) to an FE-style analysis. Below, we point out differences between FVE and FE from the perspective of an FE-style framework for analysis.

6. FVE Motivation

To motivate our FVE analysis and its relation to previous FE work, reconsider (6)–(7) and the FE analysis of Section 4. We saw that the crucial ellipticity result (12) for the A-form and boundedness results (13), (14), and (20) are stated in terms of the continuum space $\mathcal{W}$, but since the method is *conforming* (i.e., $\mathcal{W}^h \subset \mathcal{W}$), the result also held for the discrete space $\mathcal{W}^h$. It could appear that the focus was on the continuum problem (8) rather than on the FE discrete problem (11). However, in Section 4.4, the error analysis emphasized results for the discrete FE problem (11) and the associated discrete spaces. Recall that the exact weak solution u was invoked only in the zero property (17); furthermore, recall that the key observation in (17) was that u conformed to the FE discrete equation (11). Therefore, to begin an FE-style analysis for FVE, we need to generate results analogous to those of Sections 4.1, 4.2, and 4.3 for the FVE discrete problem (26) rather than for (23).

The continuum problems (8) and (23) are different in physical character (minimization vs. conservation) as well as in mathematical content, discouraging a single, unifying framework for their analysis. For the discrete problems (11) and (26), the situation is more favorable for this. We now manipulate (26) to be more reminiscent of (11).

If $w \in \mathcal{X}^h$, put

$$w \equiv \sum_{i \in I} w_i \chi_i, \tag{27}$$

where $\chi_i = \chi_{V_i}$ and $\mathcal{V}^h = \{V_i : i \in I\}$. Now the FVE solution $u^h \in \mathcal{W}^h$ of (26) also satisfies:

$$B(u^h, w) = (f^h, w), \quad \forall\, w \in \mathcal{X}^h, \tag{28}$$

where

$$B(u^h, w) = \sum_{i \in I} B(u^h, w_i \chi_i), \tag{29}$$

$$(f^h, w) = \sum_{i \in I} (f^h, w_i \chi_i). \tag{30}$$

That is, if u^h satisfies the each of the local equations of (26), then by (28) u^h also satisfies any linear combination of those equations. Notice that (29) and (30) are now global in extent, like their FE counterparts (9) and (10), and (28) with $w \equiv 1$ represents a global conservation law on the union or aggregate of all volumes in $\mathcal{V}^h$. Like the FE A-form, we refer to (29) as the B-form.

As in FE error analysis, the key components of FVE analysis are ellipticity, boundedness, and approximation theory results: we introduce and discuss each result below. All follow the basic framework introduced in the previous section after appropriate modifications for FVE. The proofs of these results are in the thesis of Trujillo[18]; the intent here is to summarize them and to compare and contrast FE and FVE analysis.

7. FVE Ellipticity

In Section 4, the FE ellipticity result required that the trial space representation error be used as a test function. In FE, test and trial spaces coincide so this was straightforward. In FVE, test and trial spaces differ, and we must describe the test function representation of a trial function.

Let $u \in \mathcal{W}^h$ with nodal representation $u \equiv \sum_{i \in I} u_i \phi_i$, where $\{\phi_i\}_{i \in I}$ is the DOF basis for $\mathcal{W}^h$. Then u also has a corresponding *lumped* or test space representation, $\overline{u}$, in $\mathcal{X}^h$:

$$\overline{u} \equiv \sum_{i \in I} u_i \chi_i. \tag{31}$$

In other words, if $u \in P^k(\mathcal{T}^h)$, $\overline{u} \in P^0(\mathcal{V}^h)$ is the piecewise constant interpolant of u. Our ellipticity result for FVE will use the lumped representation defined by (31) to characterize test functions.

As in FE, a discrete ellipticity result (for h sufficiently small when $k \geq 2$),

$$B(u, \overline{u}) \geq \alpha \, |u|^2_{1,\omega}, \quad \forall u \in \mathcal{W}^h, \tag{32}$$

is perhaps the most important component of FVE analysis. On the right-hand side of (32), $|\cdot|_{1,\omega}$ is a discrete $H^1_0(\Omega)$ norm; this can be understood as a restriction of the continuous $H^1_0(\Omega)$ norm to $P^1_0(\mathcal{T}^{h/k})$: i.e., $|u|_{1,\omega} \equiv |\Pi_1^{h/k} u|_{1,\Omega}$. Therefore, (32) implies that the B-form is positive definite.

In addition, we see that $B(u, \overline{u}) \equiv \mathbf{u}^T \mathbf{B} \mathbf{u} \equiv Q_{\mathbf{B}}(\mathbf{u}), \quad \forall u \in \mathcal{W}^h$, where $\mathbf{u}$ is the vector of nodal values for $u \in \mathcal{W}^h$, $\mathbf{B}$ is the FVE matrix generated by the B-form, and $Q_{\mathbf{B}}(\cdot)$ is the associated quadratic form. Therefore, (32) simply states that $\mathbf{B}$ is positive definite: i.e.,

$$Q_{\mathbf{B}}(\mathbf{u}) \geq 0, \quad \forall \mathbf{u}, \tag{33}$$

with equality holding in (33) only when $\mathbf{u}$ (i.e., $u \in \mathcal{W}^h$) is identically zero. Hence, (32) guarantees the existence and uniqueness of the numerical solution $u^h \in \mathcal{W}^h$.

The FVE ellipticity result (32) clearly has the same form as the corresponding FE result (12) (when restricted to $\mathcal{W}^h$). In fact, the two results are identical for (6) when D is constant and $\mathcal{W}^h = P_0^1(\mathcal{T}^h)$; that is, the FVE (and FV) matrix $\mathbf{B}$ and the FE matrix $\mathbf{A}$ are identical for piecewise linear polynomials on a general triangulation as demonstrated by Bank and Rose.[2] This equivalence between FV and FE for a piecewise linear trial space is the key observation that drives much of the FV analysis in the references cited in Section 1. Such arguments may seem to imply dependence of FV (or FVE) analysis on the corresponding FE formulation of the same problem. As we will see in the following sections, this is not the case: FVE (or FV) operates independently of FE and its analysis need not refer to FE or be confined to the cases when FVE and FE are equivalent in some sense. Our FVE analysis is similar to FE in style but not in substance.

8. FVE Boundedness

In Sections 4.2 and 4.5, the upper bounds (13), (14), and (20) for the A-form and the source term were given in terms of products of certain bounded functionals or norms (to be specified according to the context of a problem) of a continuous trial function and a discrete test function. In the case of FE, these functionals were standard integer-order Sobolev norms (or sub-linear functionals). Since these bounds are to be used in conjunction with the ellipticity result previously discussed, the only theoretical constraint on the bound is that the functional or norm applied to the test function must match (or can be made to match via an auxiliary result like the Poincaré inequality) the functional or norm in the ellipticity result—i.e., $|\cdot|_{1,\Omega}$ for FE and $|\cdot|_{1,\omega}$ for FVE.

After this requirement is satisfied, there is great freedom of choice in the functional applied to the continuous trial function: this can be problem-dependent and can adapt to any additional constraints or exploit any additional information in a problem. Practically speaking, we want this functional to admit optimal-order or even superconvergent approximation theorems, since the continuous trial functions in our analysis are the interpolation errors for the exact solution (i.e., $\eta_k(u) \in \mathcal{W}$) and for the source term (i.e., $\eta_{k-1}(f) \in \mathcal{W}_-$).

In FE analysis, the $H^p(\Omega)$ norm constraint (i.e., $p \in \{0,1\}$) on the test function leads to the choice of $H^m(\Omega)$ Sobolev norms ($m \in \{-1,0,1\}$) for the trial functions. However, in FVE analysis, the $H^p(\omega)$ constraint on the test function leads to more general choices of problem-dependent bounded linear functionals for the trial functions; these functionals correspond to the diffusion term and to the source term (and in Section 11.2 to reaction and advection terms).

For (6), the focus is on diffusion and source terms. First, the boundedness result for the B-form (29) is (cf. (13) of Section 4.2):

$$B(u,\overline{w}) \leq \mathcal{D}(u)|w|_{1,\omega}, \quad \forall\, u \in \mathcal{W}, w \in \mathcal{W}^h\,(\overline{w} \in \mathcal{X}^h). \tag{34}$$

Here $\mathcal{D}(\cdot)$ is a bounded functional defined as

$$\mathcal{D}(u) \equiv \Big(\sum_{\{i,j\}} \mathcal{D}_{ij}^2(u)\,\frac{|X_{ij}|}{|\gamma_{ij}|}\Big)^{1/2}, \tag{35}$$

where $|\cdot|$ is the (local) Lebesgue measure, X_{ij} and γ_{ij} are defined in Section 3.3.2, and $\mathcal{D}_{ij}(\cdot)$ is a bounded linear functional involving the FVE diffusion term,

$$\mathcal{D}_{ij}(u) \equiv -\int_{\gamma_{ij}} D\nabla u \cdot \mathbf{n}_{ij}\, dS, \tag{36}$$

and $\mathbf{n}_{ij}$ is a normal pointing outward from V_i into V_j. Second, the boundedness result for the source term (30) is (cf. (20) of Section 4.5):

$$(f,\overline{w}) \leq \mathcal{F}(f)|w|_{0,\omega}, \quad \forall\, f \in \mathcal{W}_{-}, w \in \mathcal{W}^h\,(\overline{w} \in \mathcal{X}^h). \tag{37}$$

In (37), $\mathcal{F}(\cdot)$ is a bounded functional defined as

$$\mathcal{F}(f) \equiv \Big(\sum_{i\in I} \frac{1}{|V_i|}\,\mathcal{F}_i^2(f)\Big)^{1/2}, \tag{38}$$

where $\mathcal{F}_i(\cdot)$ is a bounded linear functional corresponding to the FVE source term:

$$\mathcal{F}_i(f) = \int_{V_i} f\, d\mathbf{x}. \tag{39}$$

Also, $|\cdot|_{0,\omega}$ is a discrete L^2 norm (restricting the usual continuous L^2 norm to $P^0(\mathcal{V}^h)$) to be related to the discrete H^1 semi-norm by a discrete Poincaré inequality.

8.1. Elliptic Regularity

To take advantage of elliptic regularity (i.e., if $(k-1)$-regularity of f implies $(k+1)$-regularity of u), we must modify the source term bound (37). Similar to (14), we have the alternative source term bound (40):

$$(f,\overline{w}) \leq (|f|_{-1,\Omega} + Ch|f|_{0,\Omega})|w|_{1,\omega}, \quad \forall\, f \in \mathcal{W}_{-}, w \in \mathcal{W}^h\,(\overline{w} \in \mathcal{X}^h). \tag{40}$$

With upper bounds for the B-form and source term in terms of non-standard, even problem-dependent, bounded linear functionals, we need approximation theorems for these functionals: the results are summarized and discussed below.

9. FVE Approximation Theory

By applying the linear and bilinear Bramble-Hilbert lemmas[6] to the bounded linear functionals (36) and (39), we can demonstrate approximation theorems for the diffusion and source term functionals $\mathcal{D}(\cdot)$ and $\mathcal{F}(\cdot)$. Our criterion is that the Bramble-Hilbert results be equivalent to the standard optimal-order estimates for polynomial interpolation (in terms of the integer-order Sobolev norms) as demonstrated by (15) of Section 4.3. Furthermore, in the case of the diffusion functional, we can prove superconvergence (cf. Cai[5]) that actually surpasses the standard results.

For $\mathcal{D}(\cdot)$, we have an (H^1-equivalent) optimal-order result for interpolation into $P^k(\mathcal{T}^h)$ when the triangulation $\mathcal{T}^h$ and volumization $\mathcal{V}^h$ are regular (defined in Sections 3.1 and 3.3.3):

$$\mathcal{D}(\eta_k(u)) \leq C\, h^S \|u\|_{S+1,\Omega}, \quad S \in (1/2, k]. \tag{41}$$

Notice that (41) holds across a scale of regularities for the generalized function u. However, for interpolation into $P_0^1(\mathcal{T}^h)$, we have a choice between an (H^1-equivalent) optimal-order result and a superconvergence result that depends on whether the volumization $\mathcal{V}^h$ is symmetric to the triangulation $\mathcal{T}^h$ (see Section 3.3.2):

$$\mathcal{D}(\eta_1(u)) \leq C\, h^S \|u\|_{S+1,\Omega}, \quad S \in (1/2, K], \tag{42}$$

where $K = 1$ (optimal-order) or 2 (superconvergence) in the absence or presence of volume symmetry, respectively. Notice (cf. (41)) that the superconvergence is consistent with the use of a quadratic, rather than linear, trial space; indeed, the linear FVE on a symmetric volumization is identical to a quadratic FVE on a special "degenerate" volumization (see Trujillo[18] for details).

Unlike standard FE superconvergence results that are usually local and one-dimensional, the FVE superconvergence result (42) is always global and multidimensional. Since superconvergence is of great importance in both theory and practice, the focus of FVE analysis here is on establishing superconvergence results for the linear trial space before moving on to optimal-order results for higher-order trial spaces.

For $\mathcal{F}(\cdot)$, we have an (L^2-equivalent) optimal-order result for interpolation into $P^{k-1}(\mathcal{T}^h)$ when the triangulation $\mathcal{T}^h$ and volumization $\mathcal{V}^h$ are regular:

$$\mathcal{F}(\eta_{k-1}(f)) \leq C\, h^S \|f\|_{S,\Omega}, \quad S \in (1/2, k]. \tag{43}$$

To be consistent with the superconvergence result (42) and its interpretation as a special quadratic (not linear) FVE, we modify the implementation of (43) when $k = 1$:

$$\mathcal{F}(\eta_{K-1}(f)) \leq C\, h^S \|f\|_{S,\Omega}, \quad S \in (1/2, K], \tag{44}$$

where K is defined as in the discussion following (42).

10. Error Analysis

With the components developed in the preceding sections, expressed in (32), (34), (37), and (40)–(44), we can outline the standard FVE analysis. To combine optimal-order and superconvergence results, define

$$\kappa \equiv \max\{k, K\}, \tag{45}$$

where k and K are the optimal-order and superconvergence parameters employed in the preceding section. The pathway to our error estimate is detailed below.

10.1. Error Splitting

As in the FE analysis of Section 4.4, we define the error in the FVE approximation to (6) as $e \equiv u - u^h \in \mathcal{W}$ and write $e = \zeta - \eta_k(u)$, so that the FVE error is split into its representation error ($\zeta \equiv \Pi_k^h u - u^h \in \mathcal{W}^h$) and interpolation error ($\eta_k(u) \equiv \Pi_k^h u - u \in \mathcal{W}$) components. Since $|u - u^h|_{1,\omega} \equiv |\zeta|_{1,\omega}$, estimating $|e|_{1,\omega}$ is reduced to estimating $|\zeta|_{1,\omega}$. The latter estimate can be developed as below in terms of estimates (41)–(44) for the interpolation errors $\eta_k(u)$ and $\eta_{\kappa-1}(f)$.

10.2. Zero Property

Knowing that (for all $w \in \mathcal{X}^h$) the exact solution of (23) satisfies $B(u, w) = (f, w)$ and the numerical solution satisfies (28), we have the *zero property*,

$$B(u - u^h, w) - (f - f^h, w) = 0, \quad \forall\, w \in \mathcal{X}^h, \tag{46}$$

which in our context is more appropriately expressed as

$$B(\zeta, w) = B(\eta_k(u), w) + (\eta_{\kappa-1}(f), w), \quad \forall\, w \in \mathcal{X}^h. \tag{47}$$

10.3. Error Estimate

Since $\overline{\zeta}$, the lumped representation (or $P_0^0(\mathcal{V}^h)$-interpolant) of ζ, is in $\mathcal{X}^h$, take $w = \overline{\zeta}$ in (47) and obtain

$$\begin{aligned} \alpha |\zeta|_{1,\omega}^2 &\le D(\zeta, \overline{\zeta}) \\ &= B(\eta_k(u), \overline{\zeta}) + (\eta_{\kappa-1}(f), \overline{\zeta}) \\ &\le \mathcal{D}(\eta_k(u))|\zeta|_{1,\omega} + \mathcal{F}(\eta_{\kappa-1}(f))|\zeta|_{0,\omega}. \end{aligned} \tag{48}$$

The three lines of (48) use (32), (47), and (34) and (37), respectively. Finally, after applying a discrete Poincaré inequality to $|\zeta|_{0,\omega}$, by (41)–(44) we have the following combined optimal-order and superconvergent error estimate (cf. (22)) for the finite volume element method (in the absence of elliptic regularity):

$$|u - u^h|_{1,\omega} \leq C\, h^S\, (\|u\|_{S+1,\Omega} + \|f\|_{S,\Omega}), \quad S \in (1/2, \kappa], \tag{49}$$

where κ is defined in (45).

10.4. Elliptic Regularity

With elliptic regularity, we modify the preceding argument by replacing (37) with the alternative bound (40). Then the third line of (48) is replaced by

$$\alpha|\zeta|^2_{1,\omega} \leq \mathcal{D}(\eta_k(u))|\zeta|_{1,\omega} + (|\eta_{\kappa-1}(f)|_{-1,\Omega} + Ch|\eta_{\kappa-1}(f)|_{0,\Omega})|\zeta|_{1,\omega}. \tag{50}$$

Finally, after an application of (41), (42), and the standard approximation theory, we have the following combined optimal-order and superconvergent error estimate (cf. (19)) for the finite volume element method (in the presence of elliptic regularity):

$$|u - u^h|_{1,\omega} \leq C\, h^\kappa\, (\|u\|_{\kappa+1,\Omega} + \|f\|_{\kappa-1,\Omega}), \tag{51}$$

where κ is defined in (45).

Regarding the role of elliptic regularity in an error analysis, the theoretical picture for FVE complements that of FE. That is, in FE, elliptic regularity is easily incorporated (Section 4.4) and its absence causes some complications (Section 4.5). In contrast, in FVE, the absence of elliptic regularity is easily incorporated and its presence causes some complications.

This concludes our presentation for diffusion equations. In the next section, we outline the necessary additions and modifications for the inclusion of reaction and advection terms, as in the general elliptic equation (1) that arises from the general integral conservation law (3) of Section 2.

11. General Elliptic Equations

In general, the FVE B-form is given by

$$B(u,w) \equiv \mathcal{A}(u,w) + \mathcal{D}(u,w) + \mathcal{R}(u,w), \quad u \in \mathcal{W}, w \in \mathcal{X}^h, \tag{52}$$

where the bilinear forms corresponding to advection, diffusion, and reaction are

$$\mathcal{A}(u,w) \equiv \sum_{i\in I} \int_{\partial V_i} \mathbf{a}u \cdot \mathbf{n}\, w_i\, dS, \tag{53}$$

$$\mathcal{D}(u,w) \equiv \sum_{i\in I} - \int_{\partial V_i} D\nabla u \cdot \mathbf{n}\, w_i\, dS, \tag{54}$$

$$\mathcal{R}(u,w) \equiv \sum_{i\in I} \int_{V_i} ru\, w_i\, d\mathbf{x}. \tag{55}$$

Then (53)–(55) correspond to the left-hand side of the general integral equation (3), so that (52) is the starting point for an FVE analysis for the general elliptic equation (1). Sections 7–9 analyzed the diffusion bilinear form (54); here we outline the corresponding results for the advection and reaction forms (53) and (55).

11.1. Ellipticity

For reaction problems, we assume continuous and bounded ($0 < r_m \leq r \leq r_M < \infty$) reaction coefficients. If u in (55) is replaced by the lumped representation $\overline{u} \in \mathcal{X}^h$ (recall (31)), a discrete ellipticity condition analogous to (32) is:

$$\mathcal{R}(\overline{u}, \overline{u}) \geq \beta \, |u|_{0,\omega}^2, \quad \forall u \in \mathcal{W}^h, \tag{56}$$

where $\beta \equiv r_m > 0$ is the lower bound on the reaction coefficient. Using (32) and (56), we find that for the general reaction term (55) we have

$$\mathcal{R}(u, \overline{u}) \geq \beta \, |u|_{0,\omega}^2 - \alpha \, C \, h^2 |u|_{1,\omega}^2, \quad \forall u \in \mathcal{W}^h. \tag{57}$$

Then (57) and (32) imply discrete ellipticity for a reaction-diffusion problem when h is sufficiently small (i.e., we must have $C \, h^2 \leq \epsilon < 1$). For advection problems, we assume continuously differentiable and bounded ($||\mathbf{a}||, ||\nabla \cdot \mathbf{a}|| \leq M < \infty$) velocity vectors. After some algebra to split the general advection term (53) into L^2 and H^1-equivalent components, we find that

$$\mathcal{A}(u, \overline{u}) \geq -\frac{M^2}{2\epsilon} \, |u|_{0,\omega}^2 - \epsilon \, |u|_{1,\omega}^2, \quad \forall u \in \mathcal{W}^h. \tag{58}$$

Then (58), with (32) and (57), implies discrete ellipticity for an advection-reaction-diffusion problem when ϵ is sufficiently small and β is sufficiently large.

11.2. Boundedness

For the advection bilinear form (53), similar to the diffusion bound (34) we have

$$\mathcal{A}(u, w) \leq \mathcal{A}(u) |w|_{1,\omega}, \quad \forall \, u \in \mathcal{W}, w \in \mathcal{X}^h. \tag{59}$$

Here $\mathcal{A}(\cdot)$ is a bounded functional defined as

$$\mathcal{A}(u) \equiv \Big(\sum_{\{i,j\}} \mathcal{A}_{ij}^2(u) \frac{|X_{ij}|}{|\gamma_{ij}|} \Big)^{1/2}, \tag{60}$$

where $\mathcal{A}_{ij}(\cdot)$ is a bounded linear functional relating to the FVE advection term:

$$\mathcal{A}_{ij}(u) \equiv \int_{\gamma_{ij}} \mathbf{a} u \cdot \mathbf{n}_{ij} \, dS. \tag{61}$$

For the reaction bilinear form (55), we derive a bound nearly identical to (37):

$$\mathcal{R}(u,w) \leq \mathcal{R}(u)|w|_{0,\omega}, \quad \forall\, u \in \mathcal{W}, w \in \mathcal{X}^h. \tag{62}$$

$\mathcal{R}(\cdot)$ is a bounded functional defined as

$$\mathcal{R}(u) \equiv \Big(\sum_{i\in I} \frac{1}{|V_i|} \mathcal{R}_i^2(u)\Big)^{1/2}, \tag{63}$$

where $\mathcal{R}_i(\cdot)$ is a bounded linear functional relating to the FVE reaction term:

$$\mathcal{R}_i(u) = \int_{V_i} ru\, d\mathbf{x}. \tag{64}$$

Therefore, the reaction bound (62) is simply a weighted generalization of the source term bound (37) with the reaction coefficient r as a weighting function.

11.3. Approximation Theory

For the advection functional (60), we apply the Bramble-Hilbert lemmas to the bounded linear functional (61) to obtain the analogue of (41):

$$\mathcal{A}(\eta_k(u)) \leq C\, h^{S+1}\|f\|_{S+1,\Omega}, \quad S \in (1/2, k], \tag{65}$$

which is an (L^2-equivalent) optimal-order result for interpolation into $P^k(\mathcal{T}^h)$ when the triangulation $\mathcal{T}^h$ and volumization $\mathcal{V}^h$ are regular. Bramble-Hilbert arguments applied to (64) show that the reaction functional (63) shares the same estimate as (65).

11.4. Error Analysis

With the results of the previous subsections, the basic structure of the error analysis (48) presented in Section 10 is relatively unchanged. Combining the ellipticity results of Sections 7 and 11.1, we find that the first line of (18) is modified to read

$$\alpha_0\, |\zeta|_{1,\omega}^2 \leq \mathcal{A}(\zeta,\overline{\zeta}) + \mathcal{D}(\zeta,\overline{\zeta}) + \mathcal{R}(\zeta,\overline{\zeta}), \tag{66}$$

where α_0 is a positive constant equal to the minimum value of the positive constants formed after combining (32), (57), and (58)—a combination of ellipticity results for h and ϵ sufficiently small and β sufficiently large, respectively. Combining the upper bound results of Sections 8 and 11.2, the third line of (48) becomes

$$B(\eta_k(u),\overline{\zeta}) \leq (\mathcal{A}(\eta_k(u)) + \mathcal{D}(\eta_k(u)))|\zeta|_{1,\omega} + \mathcal{R}(\eta_k(u))|\zeta|_{0,\omega}. \tag{67}$$

With these modifications and the approximation results of Sections 9 and 11.3, the optimal-order and superconvergence results of (49) and (51) are maintained.

12. Conclusions

A summary of FVE error analysis for elliptic partial differential equations with smooth coefficients has been presented: H^1-equivalent optimal-order and superconvergence results have been obtained. These FVE results can be extended to problems with discontinuous and nonlinear coefficients: the modifications for discontinuous coefficients are analogous to those of Samarskii *et al.*[16] for FV; the modifications for nonlinear coefficients are analogous to those of Russell[15] for FE.

Furthermore, by an FVE variant of the elliptic projection argument of Wheeler[20], the FVE results presented here can be extended to parabolic equations in a straightforward manner. That is, an analysis of a method that uses finite differences in time and FVE in space yields error estimates of the form

$$\max_{0\le N\le NT} |(u-u^h)(\cdot,t^N)|_{1,\omega} \le C\left(\Delta t^R ||\partial_t^{R+1}u||_{L^2(L^2)} + h^S\left(||u_t||_{L^2(H^{S+1})} + ||u||_{L^\infty(H^{S+1})}\right)\right), \quad (68)$$

where $S \in (1/2, \kappa]$ and $R = 1$ or 2 depending on whether backward Euler ($R = 1$) or Crank-Nicolson ($R = 2$) time differencing is used.

Acknowledgments The first author was supported in part by National Science Foundation Grant No. DMS-9312752, by U.S. Army Research Office Grant No. 34593-GS, and by Ecodynamics Research Associates, Inc.

13. References

1. B. R. Baliga and S. V. Patankar, "A new finite element formulation for convection-diffusion problems," *Numer. Heat Transfer*, **3** (1980), pp. 393–409.
2. R. E. Bank and D. J. Rose, "Some error estimates for the box method," *SIAM J. Numer. Anal.*, **24** (1987), pp. 777–787.
3. Z. Cai and S. F. McCormick, "On the accuracy of the finite volume element method for diffusion equations on composite grids," *SIAM J. Numer. Anal.*, **27** (1990), pp. 636–655.
4. Z. Cai, J. Mandel, and S. F. McCormick, "The finite volume element method for diffusion equations on general triangulations," *SIAM J. Numer. Anal.*, **28** (1991), pp. 392–402.
5. Z. Cai, "On the finite volume element method," *Numer. Math.*, **58** (1991), pp. 713–735.

6. P. G. Ciarlet, *Handbook of Numerical Analysis: Vol. II, Finite Element Methods (Part 1)*, P. G. Ciarlet and J. L. Lions, ed., Elsevier (North-Holland), Amsterdam, 1991.
7. R. E. Ewing, R. D. Lazarov, and P. S. Vassilevski, "Local refinement techniques for elliptic problems on cell-centered grids: I. error analysis," *Math. Comp.*, **56** (1991), pp. 437–461.
8. W. Hackbusch, "On first and second order box schemes," *Computing*, **41** (1989), pp. 277–296.
9. B. Heinrich, *Finite Difference Methods on Irregular Networks: generalized approach to second order elliptic problems*, International Series in Numerical Mathematics, 82, Birkhäuser, Stuttgart, 1987.
10. R. Herbin, "An error estimate for a finite volume scheme for a diffusion-convection problem on a triangular mesh," *Numer. Meth. PDE*, **11** (1995), pp. 165–173.
11. R. D. Lazarov, I. D. Mishev, and P. S. Vassilevski, "Finite volume methods with local refinement for convection-diffusion problems," *Computing*, **53** (1994), pp. 33–57.
12. A. R. Mitchell and D. F. Griffiths, *The Finite Difference Method in Partial Differential Equations*, Wiley, New York, 1980.
13. K. W. Morton and E. Süli, "Finite volume methods and their analysis," *IMA J. Numer. Anal.*, **11** (1991), pp. 241–260.
14. S. F. McCormick, *Multilevel Adaptive Methods for Partial Differential Equations*, Chapter 2: The Finite Volume Element Method, Frontiers in Applied Mathematics, Vol. 6, SIAM, Philadelphia, 1989.
15. T. F. Russell, "Time-stepping along characteristics with incomplete iteration for a Galerkin approximation of miscible displacement in porous media," *SIAM J. Numer. Anal.*, **22** (1985), pp. 970–1013.
16. A. A. Samarskii, R. D. Lazarov, and V. L. Makarov, *Finite Difference Schemes for Differential Equations with Weak Solutions*, Vysshaya Shkola Publishers, Moscow, 1987. (In Russian.)
17. E. Süli, "Convergence of finite volume schemes for Poisson's equation on nonuniform meshes," *SIAM J. Numer. Anal.*, **28** (1991), pp. 1419-1430.
18. R. V. Trujillo, "Analysis of the Finite Volume Element Method for Elliptic and Parabolic Partial Differential Equations," Ph.D. Thesis, University of Colorado, Denver, CO, May 1996.
19. A. Weiser and M. F. Wheeler, "On convergence of block-centered finite differences for elliptic problems," *SIAM J. Numer. Anal.*, **25** (1988), pp. 351–357.
20. M. F. Wheeler, "*A priori* L^2 error estimates for Galerkin approximations to parabolic partial differential equations," *SIAM J. Numer. Anal.*, **10** (1973), pp. 723–759.

A GENERALIZED DISCRETE MULTIPLE SCALES ANALYSIS TECHNIQUE

S.W. SCHOOMBIE AND E. MARE
Department of Applied Mathematics, University of the Orange Free State, P.O. Box 339, Bloemfontein 9300, South Africa
E-mail: schalk@wwg3.uovs.ac.za

ABSTRACT

We derive an extension of the discrete multiple scales analysis technique developed by Schoombie [1]. We show the similarity between the generalized technique and results obtained by a continuous multiple scales analysis for the Korteweg-de Vries equation.

1. Introduction

Recently one of the authors [1] developed a discrete multiple scales technique mainly to analyze modulational properties of dispersive finite difference schemes. In particular the Zabusky–Kruskal discretization [3] of the Korteweg–de Vries (KdV) equation was analyzed in this way.

Schoombie [1] also showed that the technique can be used to identify carrier modes of small amplitude modulated harmonic waves for which the solution of the Zabusky–Kruskal scheme deviates from that of the KdV equation.

However, the Zabusky–Kruskal scheme is not by far the best numerical scheme for the KdV equation. Higher order finite difference or finite element methods [4,5] and spectral methods [6] have been developed since then, and are usually much more efficient.

Thus there is a need to extend the multiple scales analysis for second order finite difference schemes to higher order finite difference, and eventually also spectral and pseudospectral methods. In this paper we take the first step: We extend the previous work to cover finite central difference stencils of arbitrary finite order.

In section 2 a brief review of the well established theory of the continuous equation is given. Section 3 defines the notation we will be using, and quotes an important lemma of Schoombie [1] which is essential for the multiple scales analysis. Section 4 contains our main results, and we conclude the paper with a few final remarks in section 5.

2. Multiple Scales Analysis of the KdV Equation

In this section we review the type of multiple scales analysis of the KdV equation which was used, among others, by Kawahara [7] in 1975 and Zakharov and Kusnetsov [8] in 1986 to investigate various aspects of the modulational theory of the KdV equation. Equations in this section are then conveniently available for comparison with their discrete analogs in section 4.

Consider the KdV equation in the form

$$u_t + \eta u_x + \zeta u u_x + \gamma u_{xxx} = 0, \tag{1}$$

where the subscripts denote partial differentiation as usual and η,ζ and γ are constants, with $\gamma \neq 0$. Furthermore we assume that suitable initial data

$$u(x,0) = \epsilon f(x), \;\; f(x) = O(1), \tag{2}$$

be prescribed, where ϵ is a small, real, positive number. We also enforce the following periodicity conditions

$$u(x \pm L, t) = u(x,t), \;\; f(x \pm L) = f(x), \;\; t > 0, \; x \in \Re. \tag{3}$$

A multiple scales analysis of Eq. (1) can now be performed if we assume that its solution $u(x,t)$ can be expanded in the form

$$u(x,t) = \sum_{r=-\infty}^{\infty} u_r(X_1, T_1, T_2, \epsilon) e^{ir\theta}, \tag{4}$$

where

$$X_k = \epsilon^k x, k = 0, 1, \tag{5}$$

and

$$T_k = \epsilon^k t, k = 0, 1, 2. \tag{6}$$

Following Kawahara [7] and Zakharov and Kusnetsov [8], the phase variable, θ, is given by

$$\theta = kX_0 - \omega T_0, \tag{7}$$

with the carrier wave number k related to ω by the linear dispersion relation

$$\omega = \eta k - \gamma k^3. \tag{8}$$

Furthermore, to satisfy the periodicity conditions Eq. (3), we have to restrict values of k to the following

$$k = k_m = 2\pi m / L, \;\; m = 0, 1, \ldots \tag{9}$$

Then (following Kawahara [7] and Zakharov and Kusnetsov [8]) we use

$$u_r = \epsilon^{\delta_r} v_r(X_1, T_1, T_2, \epsilon), \tag{10}$$

with

$$\delta_0 = 2, \delta_r = |r|, \tag{11}$$

and

$$v_0 = V_0(X_1, T_1, T_2) \tag{12}$$

$$v_1 = V_1(X_1, T_1, T_2). \tag{13}$$

When $r > 1$ we have

$$v_r = V_r(X_1, T_1, T_2) + \sum_{s=r}^{\infty} \epsilon^{s+1-r} W_{rs}(X_1, T_1, T_2). \tag{14}$$

To enable a multiple scales analysis of Eq. (1), we use the chain rule for derivatives on Eq. (5) and Eq. (6) to give

$$\partial_t = -ir\omega + \epsilon\partial_{T_1} + \epsilon^2\partial_{T_2}, \tag{15}$$

and

$$\partial_x = irk + \epsilon\partial_{X_1}. \tag{16}$$

Substituting Eq. (4), Eq. (15) and Eq. (16) into Eq. (1) yields

$$\begin{aligned} &-ir\omega u_r + \epsilon\partial_{T_1}u_r + \epsilon^2\partial_{T_2}u_r + ir\eta k u_r + \eta\epsilon\partial_{X_1}u_r \\ &+ \gamma[irk + \epsilon\partial_{X_1}]^3 u_r + \zeta\textstyle\sum_{s=-\infty}^{\infty}[iksu_s + \partial_{X_1}\epsilon u_s]u_{r-s} = 0, \end{aligned} \tag{17}$$

an infinite set of equations for u_r, upon equating coefficients of $e^{ir\theta}$ to zero. We now wish to have Eq. (1) and therefore Eq. (17) satisfied for each r up to terms $O(\epsilon^3)$. We commence by putting $r = 0$ in Eq. (17). By equating the $O(\epsilon^3)$ terms to zero we obtain the following equation

$$\partial_{T_1}V_0 + \eta\partial_{X_1}V_0 + \zeta\partial_{X_1}|V_1^2| = 0. \tag{18}$$

Next, put $r = 1$ in Eq. (17). Equating the $O(\epsilon)$ terms to zero reproduces the linear dispersion relation Eq. (8). Similarly the $O(\epsilon^2)$ and $O(\epsilon^3)$ terms yield the following equations respectively

$$\partial_{T_1}V_1 + c_g\partial_{X_1}V_1 = 0 \tag{19}$$

and

$$\partial_{T_2}V_1 + 3i\gamma k\partial_{X_1}^2 V_1 + ik\zeta[V_1V_0 + V_1^*V_2] = 0, \tag{20}$$

where

$$c_g = \frac{d\omega}{dk} = \eta - 3\gamma k^2 \tag{21}$$

is the linear group velocity. When we use $r = 2$ we obtain from the $O(\epsilon^2)$ terms

$$V_2 = \zeta V_1^2/(6\gamma k^2), \tag{22}$$

by making use of the linear dispersion relation Eq. (8).

By making the physically meaningful assumption that V_0 also satisfies Eq. (19), we obtain from Eq. (18) that

$$V_0 = -(\zeta/3\gamma k^2)|V_1|^2. \tag{23}$$

We can now rewrite Eq. (20) by making use of Eq. (23) and Eq. (22). The result

$$\partial_{T_2} V_1 - i[\zeta^2/(6\gamma k)]V_1(|V_1|^2) + 3i\gamma k\, \partial^2_{X_1} V_1 = 0, \tag{24}$$

is the nonlinear Schrödinger equation in the variables T_2 and X_1.

It should be noted at this point that, since ϵ is small,

$$u(x,t) \approx \epsilon(V_1^* e^{-i\theta} + V_1 e^{i\theta}). \tag{25}$$

Thus V_1 can be considered to be a small, variable amplitude of a monochromatic wave. On the timescale T_1, Eq. (19) tells us that the modulation envelope (V_1) moves at linear group velocity without changing its shape. On the timescale T_2, however, the envelope does change its shape, according to Eq. (24). Thus Eq. (19) describes the linear, and Eq. (24) describes the nonlinear modulation properties of the KdV equation.

3. Preliminaries and Notation

We adopt the notation used by Schoombie [1]. We divide the interval $[0, L]$ into N subintervals and obtain the following grid length

$$h = L/N. \tag{26}$$

(In this paper we shall confine ourselves to discretisations in space only.) We will use the symbol $u_j(t)$ to denote the solution of a difference scheme at $x = hj$, where j is a given integer, i.e.,

$$u_j(t) = u(hj, t). \tag{27}$$

Analogous to Eq. (5) we will use the following discrete multiple scales coordinates in space, to wit

$$X_p = \epsilon^p hj, p = 0, 1. \tag{28}$$

We now define the following shift operators

$$Ef(x) = f(x + h), \tag{29}$$

as well as

$$E_p f(\ldots, X_p, \ldots) = f(\ldots, X_p + \epsilon^p h, \ldots). \tag{30}$$

We also define the divided difference operators

$$\begin{aligned} \Delta &= (E-1)/h & (31)\\ \nabla &= (1-E^{-1})/h & (32)\\ \delta &= (\Delta+\nabla)/2. & (33) \end{aligned}$$

and similarly the partial divided difference operators

$$\begin{aligned} \Delta_p &= (E_p-1)/h & (34)\\ \nabla_p &= (1-E_p^{-1})/h & (35)\\ \delta_p &= (\Delta_p+\nabla_p)/2. & (36) \end{aligned}$$

We are now in a position to state the following lemma proved by Schoombie [1], which forms the keystone to the discrete multiple scales analysis:

Lemma 1 *In terms of the divided difference operators defined above,*

$$\begin{aligned} \Delta &= \Delta_0+\epsilon\Delta_1 E_0 & (37)\\ & \quad \textit{and} & (38)\\ \nabla &= \nabla_0+\epsilon\nabla_1 E_0^{-1} & (39) \end{aligned}$$

Lemma 1 now provides us with the discrete counterparts of the chain rules Eq. (15) and Eq. (16).

Schoombie [1] showed how Eq. (37) and Eq. (39) may be used to perform a multiple scales analysis on second order central difference equations. In the next section we will show how these same equations can also be used in the case of higher order finite difference schemes.

4. Generalized Discrete Multiple Scales Analysis

In what follows we extend the approach outlined above to make provision for central finite difference approximations of higher order than two.

Consider the grid introduced in the previous section. Suppose we wish to approximate an m-th order derivative at the grid point $x_j = jh$, using the stencil points

$$x_j+\alpha h, \quad \alpha = 0, \pm 1, \pm 2, \cdots \pm p,$$

with p a positive integer indicating the width of the stencil. Following Fornberg [2], we write our finite difference scheme in the form

$$\frac{d^m f}{dx^m}\Big|_{x=x_j} \approx D_p^m f(x_j) = \frac{1}{h^m}\sum_{\alpha=-p}^{p} \delta_{2p,\alpha}^m f(x_j+\alpha h), \qquad (40)$$

where, as in Fornberg's paper [2],

$$\frac{1}{h^m}\delta^m_{2p,\alpha} = \left[\frac{d^m}{dx^m}F_{2p,\alpha}(x)\right]_{x=x_j}, \tag{41}$$

with

$$F_{2p,\alpha}(x) = \frac{\omega_{2p}(x)}{\omega'_{2p}(x_j+\alpha h)(x-x_j-\alpha h)}, \tag{42}$$

and

$$\omega_{2p}(x) = \prod_{\beta=-p}^{p}(x-x_j-\beta h). \tag{43}$$

Moreover,

$$p(x) = \sum_{\alpha=-p}^{p} F_{2p,\alpha}(x) f(x_j+\alpha h), \tag{44}$$

is the Lagrange interpolation polynomial of degree $2p$ on our finite difference stencil.

Note that, from Eq. (41), an explicit expression for the difference weights is

$$\delta^m_{2p,\alpha} = \sum_{l_1,l_2,\cdots l_m=-p}^{p} \prod_{s=1}^{m}\frac{1}{\alpha-l_s} \prod_{\substack{\beta=-p \\ \beta\neq\alpha,l_1,\ldots,l_m}}^{p} \frac{(-\beta)}{(\alpha-\beta)}. \tag{45}$$

This expression is not very useful for the explicit calculation of the weights for especially high order schemes (high values of p). For that, Fornberg [2] provided an efficient algorithm. From Eq. (45), however, it is easy to see that

$$\delta^m_{2p,-\alpha} = (-1)^m\delta^m_{2p,\alpha}. \tag{46}$$

We shall only be concerned with odd values of m in this paper, so henceforth we shall assume m to be an odd positive integer. Then

$$\delta^m_{2p,-\alpha} = -\delta^m_{2p,\alpha}. \tag{47}$$

Using the shift operator E defined in Eq. (29), we are now in a position to write down the following expression for the finite difference operator D^m_p:

$$D^m_p = \frac{1}{h^m}\sum_{\alpha=1}^{p}\delta^m_{2p,\alpha}(E^\alpha - E^{-\alpha}). \tag{48}$$

The truncation error of this divided difference operator is in general of order $2p+1-m$, although it can be higher in special cases [2].

Thus, for the second order scheme for a first derivative we have $p = m = 1$ and $\delta^1_{2,1} = 1/2$. For the fourth order scheme, $p = 2, m = 1$ and $\delta^1_{4,1} = 2/3$, $\delta^1_{4,2} = -1/12$.

Further explicit values of the $\delta^m_{2p,\alpha}$ can be found in Fornberg's paper [2].

Later on we shall also have need of the following identities which we formulate in the following lemma:

Lemma 2 *Let m be an odd integer, with $1 \leq m < 2p$, and let $\delta^m_{2p,\alpha}$ be defined as in Eq. (41). Then*

$$\sum_{\alpha=1}^{p} \alpha^k \delta^m_{2p,\alpha} = 0 \quad for \ 0 < k < m, \tag{49}$$

with k an integer, and

$$\sum_{\alpha=1}^{p} \alpha^m \delta^m_{2p,\alpha} = \frac{m!}{2}. \tag{50}$$

Proof: Since in Eq. (44) the interpolation approximation is exact for $f(x)$ a polynomial of degree $\leq 2p$,

$$x^k = \sum_{\alpha=-p}^{p} F_{2p,\alpha}(x)(x_j + \alpha h)^k, \tag{51}$$

with $0 < k \leq m$ an odd integer.

Differentiate m times with respect to x, and put $x = x_j = jh$, then

$$\begin{aligned} \sum_{\alpha=-p}^{p} \delta^m_{2p,\alpha}(j+\alpha)^j &= \sum_{\alpha=1}^{p} \delta^m_{2p,\alpha}[(j+\alpha)^k - (j-\alpha)^k] \\ &= \begin{cases} m! & \text{if } k = m \\ 0 & \text{if } k < m. \end{cases} \end{aligned} \tag{52}$$

Thus, for $m = 1$, the identity Eq. (50) follows immediately. For $m > 1$, first let $k = 1$ in Eq. (52). Then the identity Eq. (49) follows for $k = 1$. For $k = 3 < m$, Eq. (52) becomes

$$6j^2 \sum_{\alpha=1}^{p} \alpha \delta^m_{2p,\alpha} + 2 \sum_{\alpha=-p}^{p} \alpha^3 \delta^m_{2p,\alpha} = 0. \tag{53}$$

The first term vanishes by virtue of Eq. (49) for $k = 1$, and what remains proves Eq. (49) for $k = 3$. Proceeding in this fashion, Eq. (49) is proved for any odd k such that $0 < k < m$.

The identity Eq. (50) is proved similarly, putting $k = m$ in Eq. (52), and removing superfluous terms by means of Eq. (49). □

We can now proceed to the following lemma:

Lemma 3 *The difference approximation to any derivative of odd order m of a suitable function can be expanded into the following discrete scales summation*

$$D^m_p = D^m_{p,0} + \sum_{r=1}^{p} \epsilon^r d^m_r, \tag{54}$$

where

$$D^m_{p,0} = \frac{1}{h^m} \sum_{\alpha=1}^{p} \delta^m_{2p,\alpha}(E^\alpha_0 - E^{-\alpha}_0), \tag{55}$$

and where the d_r^m *are given by*

$$d_r^m = h^{r-m} \sum_{\alpha=1}^{p} \binom{\alpha}{r} \delta_{2p,\alpha}^m \{\Delta_1^r E_0^\alpha - \nabla_1^r E_0^{-\alpha}\}. \tag{56}$$

Proof: Consider the generalized central difference approximation given by Eq. (48)

$$D_p^m = \frac{1}{h^m} \sum_{\alpha=1}^{p} \delta_{2p,\alpha}^1 [E^\alpha - E^{-\alpha}]. \tag{57}$$

By making use of Eq. (31) and Eq. (32) we can rewrite Eq. (57) in the following form

$$D_p^m = \frac{1}{h^m} \sum_{\alpha=1}^{p} \delta_{2p,\alpha}^m [(1 + h\Delta)^\alpha - (1 - h\nabla)^\alpha]. \tag{58}$$

By making use of the Binomial theorem we see that

$$D_p^m = \frac{1}{h^m} \sum_{\alpha=1}^{p} \delta_{2p,\alpha}^m \sum_{j=1}^{\alpha} \binom{\alpha}{j} h^j [\Delta^j - (-1)^j \nabla^j]. \tag{59}$$

By making use of lemma 1 and the Binomial theorem we have

$$\Delta^j = (\Delta_0 + \epsilon \Delta_1 E_0)^j = \sum_{r=0}^{j} \binom{j}{r} \epsilon^r \Delta_0^{j-r} \Delta_1^r E_0^r. \tag{60}$$

and

$$\nabla^j = (\nabla_0 + \epsilon \nabla_1 E_0^{-1})^j = \sum_{r=0}^{j} \binom{j}{r} \epsilon^r \nabla_0^{j-r} \nabla_1^r E_0^{-r}. \tag{61}$$

By combining Eq. (60) and Eq. (61) with Eq. (59) we find that

$$D_p^m = \frac{1}{h^m} \sum_{\alpha=1}^{p} \delta_{2p,\alpha}^m \sum_{j=1}^{\alpha} \binom{\alpha}{j} h^j \sum_{r=0}^{j} \binom{j}{r} \epsilon^r [\Delta_0^{j-r} \Delta_1^r E_0^r - (-1)^j \nabla_0^{j-r} \nabla_1^r E_0^{-r}]. \tag{62}$$

Eq. (62) can now be rearranged in the form of Eq. (54), with

$$d_r^m = \frac{1}{h^m} \sum_{\alpha=1}^{p} \delta_{2p,\alpha}^m \sum_{j=r}^{\alpha} \binom{\alpha}{j} h^j \binom{j}{r} [\Delta_0^{j-r} \Delta_1^r E_0^r - (-1)^j \nabla_0^{j-r} \nabla_1^r E_0^{-r}]. \tag{63}$$

We can simplify Eq. (63) by noting that, since

$$\binom{\alpha}{j} \binom{j}{r} = \binom{\alpha}{r} \binom{\alpha - r}{j - r}, \tag{64}$$

$$\sum_{j=r}^{\alpha} \binom{\alpha}{j} h^j \binom{j}{r} \Delta_0^{j-r} E_0^r \tag{65}$$

$$= \binom{\alpha}{r} \sum_{j=r}^{\alpha} \binom{\alpha - r}{j - r} h^j \Delta_0^{j-r} E_0^r \tag{66}$$

$$= \binom{\alpha}{r} \sum_{j=0}^{\alpha-r} \binom{\alpha - r}{j} h^{j+r} \Delta_0^j E_0^r \tag{67}$$

$$= \binom{\alpha}{r} h^r (1 + h\Delta_0)^{\alpha - r} E_0^r \tag{68}$$

$$= \binom{\alpha}{r} h^r E_0^{\alpha}, \tag{69}$$

and likewise

$$\sum_{j=r}^{\alpha} \binom{\alpha}{j} h^j \binom{j}{r} (-1)^j \nabla_0^{j-r} E_0^{-r} = \binom{\alpha}{r} h^r (-1)^r E_0^{-\alpha}. \tag{70}$$

Hence Eq. (56) follows.

□

We obtain a discrete scales analysis by solving for an approximate solution $u_j(t)$ in the following form

$$u_j(t) = \sum_{r=-[l/2]}^{[l/2]} c_r u_r(X_1, t, \epsilon) e^{ir\theta}, \tag{71}$$

where θ is defined as

$$\theta = khj - \Omega t, \tag{72}$$

with k now restricted to the following finite set of values due to aliasing, namely

$$k = 2\pi m/L, m = -N/2 + 1, \ldots, N/2. \tag{73}$$

Following the details described in Schoombie [1], l is obtained by using the integers s and l with least absolute magnitude such that

$$\frac{m}{N} = \frac{s}{l}, \tag{74}$$

and

$$[l/2] = \begin{cases} l/2 & \text{if } l \text{ even} \\ (l-1)/2 & \text{if } l \text{ odd.} \end{cases} \tag{75}$$

We also use

$$c_r = \begin{cases} 1 & \text{if } |r| < l/2 \\ 1/2 & \text{if } |r| = l/2. \end{cases} \tag{76}$$

Then (in direct analogy with the continuous case) we use

$$u_r = \epsilon^{\delta_r} v_r(X_1, T_1, T_2, \epsilon), \tag{77}$$

with

$$\delta_0 = 2, \delta_r = |r|, |r| = 1, ..., [l/2], \tag{78}$$

and

$$v_0 = V_0(X_1, T_1, T_2) \tag{79}$$

$$v_1 = V_1(X_1, T_1, T_2). \tag{80}$$

When $r = 2, \ldots, [l/2]$ we have

$$v_r = V_r(X_1, T_1, T_2, \epsilon) + \sum_{s=r}^{\infty} \epsilon^{s+1-r} W_{rs}(X_1, T_1, T_2). \tag{81}$$

It now follows that we can write up to terms $O(\epsilon^2)$

$$D_p^m(u_r e^{ir\theta}) = Q_{r,p}^m(u_r e^{ir\theta}), \tag{82}$$

where

$$\begin{aligned} Q_{r,p}^m/2 = \quad & ih^{-m} \textstyle\sum_{\alpha=1}^p \delta_{2p,\alpha}^m \sin(rkh\alpha) + \epsilon h^{1-m} \sum_{\alpha=1}^p \alpha \delta_{2p,\alpha}^m \cos(rkh\alpha)\delta_1 \\ & + \epsilon^2 ih^{2-m} \textstyle\sum_{\alpha=1}^p \alpha^2 \delta_{2p,\alpha}^m \sin(rkh\alpha)\Delta_1\nabla_1. \end{aligned} \tag{83}$$

We now turn to the finite difference approximation of the KdV equation Eq. (1) in space alone, keeping the time variable continuous (i.e., a method of lines approach). Using Eq. (48) we can write

$$\partial_t u_j + \eta D_p^1 u_j + \gamma D_q^3 u_j + \zeta[\Theta D_p^1[(u_j)^2/2] + (1-\Theta)u_j D_p^1 u_j] = 0 \tag{84}$$

We use the variable $\Theta \in [0,1]$ to provide for general discretizations of the nonlinear term in Eq. (1). Usually, however, it is taken to be 2/3.

We introduced an integer q, since the third derivative would require a larger stencil than the first if both are to be of the same order. Usually $q = p + 1$.

Substitution of Eq. (71) into Eq. (84), and setting coefficients of $c_r e^{ir\theta}$ to zero (similar to the approach taken in the continuous case), we obtain

$$\begin{aligned} & \epsilon^{\delta_r}\partial_t v_r + \epsilon^{\delta_r}\eta Q_{r,p}^1 v_r + \gamma\epsilon^{\delta_r} Q_{r,q}^3 v_r \\ & \quad + \zeta \textstyle\sum_w [(1-\Theta)(Q_{r,p}^1 v_r)v_{r-w} + \Theta/2 Q_{r,p}^1(v_r v_{r-w})]\epsilon^{\delta_{r-w}+\delta_w} = 0, \end{aligned} \tag{85}$$

with ∂_t given by Eq. (15). Proceeding in a similar way as in the continuous case, we would now study Eq. (84) and therefore, similar as done by Schoombie [1], Eq. (85) up to terms $O(\epsilon^0)$. This leads us to consider the cases $r = 0, 1$, and 2 in Eq. (85).

If we put $r = 0$ in Eq. (85) find the lowest order term in ϵ to be $O(\epsilon^3)$, which yields the following equation when equated to zero

$$\partial_{T_1} V_0 + \eta\delta_1 V_0 + \zeta[2(1-\Theta)\sum_{\alpha=1}^{p} \alpha\delta_{2p,\alpha}^1 \cos(k\alpha h) + \Theta]\delta_1|V_1|^2 = 0. \tag{86}$$

In Eq. (86) above, we have made use of Eq. (50), with $m = 1$.

By putting $r = 1$ in Eq. (85) and equating coefficients of the $O(\epsilon)$ terms to zero we obtain

$$\Omega/2 = \eta/h \sum_{\alpha=1}^{p} \delta^1_{2p,\alpha} \sin(kh\alpha) + \gamma/h^3 \sum_{\alpha=1}^{p} \delta^3_{2p,\alpha} \sin(kh\alpha), \tag{87}$$

the discrete linear dispersion relation. Subsequently, for $r = 1$, by equating the $O(\epsilon^2)$ terms to zero, the resulting equation is found to be

$$\partial_{T_1} V_1 + V_g \delta_1 V_1 = 0. \tag{88}$$

where

$$V_g = \frac{d\Omega}{dk} = 2\eta \sum_{\alpha=1}^{p} \alpha\delta^1_{2p,\alpha} \cos(kh\alpha) + 2\gamma/h^2 \sum_{\alpha=1}^{p} \alpha\delta^3_{2p,\alpha} \cos(kh\alpha) \tag{89}$$

is the discrete linear group velocity. Finally, for $r = 1$ terms, the equation obtained by putting $O(\epsilon^3)$ terms equal to zero is found to be

$$\begin{aligned} &\partial_{T_2} V_1/2 + \eta i h \textstyle\sum_{\alpha=1}^{p} \delta^1_{2p,\alpha}\alpha^2 \sin(kh\alpha)\Delta_1\nabla_1 V_1 \\ &\quad + \gamma i/h \textstyle\sum_{\alpha=1}^{p} \delta^3_{2p,\alpha}\alpha^2 \sin(kh\alpha)\Delta_1\nabla_1 V_1 + \zeta \textstyle\sum_{\alpha=1}^{p} \delta^1_{2p,\alpha} \sin(kh\alpha) V_1 V_0 \\ &\quad + \zeta[(1-\Theta) \textstyle\sum_{\alpha=1}^{p} \delta^1_{2p,\alpha}(\sin(2kh\alpha) - \sin(kh\alpha)) \\ &\quad + \Theta \textstyle\sum_{\alpha=1}^{p} \delta^1_{2p,\alpha} \sin(kh\alpha)] V_1^* V_2 i/h = 0. \end{aligned} \tag{90}$$

Using $r = 2$ we obtain terms $O(\epsilon^2)$

$$\begin{aligned} &-\Omega V_2 i + \eta i/h \textstyle\sum_{\alpha=1}^{p} \delta_{2p,\alpha} \sin(2\alpha kh) V_2 + \gamma i/h^3 \textstyle\sum_{\alpha=1}^{p} \delta^3_{2p,\alpha} \sin(2\alpha kh) V_2 \\ &\quad + \zeta i/h[\Theta/2 \textstyle\sum_{\alpha=1}^{p} \delta_{2p,\alpha} \sin(2\alpha kh) V_1^2 + (1-\Theta) \textstyle\sum_{\alpha=1}^{p} \delta_{2p,\alpha} \sin(\alpha kh) V_1^2] = 0, \end{aligned} \tag{91}$$

which can be rewritten, using Ω given in Eq. (87), as

$$V_2 = \zeta\Lambda V_1^2, \tag{92}$$

where

$$\Lambda = i/h \frac{[\Theta/2 \sum_{\alpha=1}^{p} \delta_{2p,\alpha} \sin(2\alpha kh) + (1-\Theta) \sum_{\alpha=1}^{p} \delta_{2p,\alpha} \sin(\alpha kh)]}{g(h,k,t)}, \tag{93}$$

and

$$\begin{aligned} g(h,k,t) = \; &[-2\eta i/h \textstyle\sum_{\alpha=1}^{p} \delta_{2p,\alpha} \sin(\alpha kh) - 2\gamma i/h^3 \textstyle\sum_{\alpha=1}^{p} \delta^3_{2p,\alpha} \sin(\alpha kh) \\ &+ \eta i/h \textstyle\sum_{\alpha=1}^{p} \delta_{2p,\alpha} \sin(2\alpha kh) + \gamma i/h^3 \textstyle\sum_{\alpha=1}^{p} \delta^3_{2p,\alpha} \sin(2\alpha kh)]. \end{aligned} \tag{94}$$

Note that V_0 is not uniquely defined by Eq. (86). We make the physically reasonable assumption that V_0 must also satisfy Eq. (88), i.e.,

$$\partial_{T_1} V_0 + V_g \delta_1 V_0 = 0. \tag{95}$$

Hence, after substitution of Eq. (95) into Eq. (86) we find that

$$\partial_{X_1}\left(V_0 - \frac{\zeta[2(1-\Theta)\sum_{\alpha=1}^{p}\alpha\delta_{2p,\alpha}^{1}\cos(k\alpha h)+\Theta]|V_1|^2}{(\eta - V_g)}\right) = 0, \tag{96}$$

provided that

$$\eta - V_g \neq 0. \tag{97}$$

Solution of Eq. (96) leads to the following relation for V_0

$$V_0 = \frac{\zeta[2(1-\Theta)\sum_{\alpha=1}^{p}\alpha\delta_{2p,\alpha}^{1}\cos(k\alpha h)+\Theta]|V_1|^2}{(\eta - V_g)}. \tag{98}$$

Proceeding as in the continuous case, we can construct the discrete equivalent of the nonlinear Schrödinger equation, i.e., we replace the solutions for V_0 obtained in Eq. (96) and V_2 obtained in Eq. (91) in Eq. (90).

Note that

$$\lim_{h\to 0}\Lambda = \frac{1}{6\gamma k^2}, \tag{99}$$

and that

$$\lim_{h\to 0}\frac{d\Omega}{dk} = \eta - 3\gamma k^2. \tag{100}$$

We are led to conclude that in the limit $h \to 0$ Eq. (90) becomes the continuous nonlinear Schrödinger equation, to wit

$$\partial_{T_2}V_1 + 3i\gamma k\partial_{X_1}^2 V_1 - i\zeta^2/(6\gamma k)V_1|V_1|^2 = 0. \tag{101}$$

5. Conclusion

Using a generalized central difference approximation to the KdV, we have shown that the results obtained by Schoombie [1] hold for central finite difference schemes of arbitrary (finite) order.

In particular, we have shown how to obtain a discrete version of the nonlinear Schrödinger equation. This equation describes modulational properties of solutions of Eq. (84), the generalized central difference discrete approximation to Eq. (1).

Obviously the multiple scales analysis described in this paper can also be used to analyze other dispersive finite difference schemes than those designed to solve the KdV equation. This would include models which are formulated discretely in the first case, thus eliminating the necessity to first continuize the model before attempting an analysis.

What remains to be done is the development of a technique to perform a multiple scales analysis of a spectral or a pseudospectral scheme. Although Fornberg [9] has shown that a pseudospectral method can be considered to be a finite difference scheme of infinite order, the methods outlined in this paper are not applicable, since they are only valid for finite difference stencils. Hence another approach would be needed.

6. References

1. S.W. Schoombie, *J. Comput. Phys.* **101** (1992), 55–70.
2. Fornberg, B., Generation of Finite Difference formulas on Arbitrarily Spaced Grids, *Math. Comp.* **31** (1988), 699 – 706.
3. N.J. Zabusky and M.D. Kruskal, *Phys. Rev. Lett.* **15** (1965), 240-243.
4. S.W. Schoombie, *IMA J. Numer. Anal.* **2** (1982), 95-109.
5. J.M. Sanz-Serna and I. Christie, *J. Comput. Phys.* **39** (1981), 94-102.
6. B. Fornberg and G.B. Whitham, *Phil. Trans. R. Soc. London* **289** (1987), 373.
7. T. Kawahara, *Jnl. Phys. Soc. Japan* **38** (1975), 1200 – 1206.
8. V.E. Zakharov and E.A. Kusnetsov, *Phys. D* **18** (1986) 455-463.
9. B. Fornberg, *Geophys.* **52** (1987), 483-501.

NUMERICAL SOLUTION OF BURGERS' EQUATION USING A TWO-GRID METHOD

DAVID M. SLOAN
Department of Mathematics, University of Strathclyde
Livingstone Tower, 26 Richmond Street, Glasgow, G1 1XH, Scotland
E-mail: d.sloan@strath.ac.uk

and

RONNIE WALLACE*
Department of Mathematics, University of Strathclyde
Livingstone Tower, 26 Richmond Street, Glasgow, G1 1XH, Scotland
E-mail: r.wallace@strath.ac.uk

ABSTRACT

We describe a scheme for the numerical solution of nonlinear, dissipative partial differential equations that is based on the idea of approximate inertial manifolds. Spatial discretisation is effected on interlaced coarse and fine grids using finite differences, and a nonlinear mapping is employed to relate the solutions computed on these grids. The approach is illustrated by consideration of Burgers' equation in one space dimension, with Dirichlet boundary conditions. Numerical results are presented which demonstrate that the two-grid method is more accurate than the corresponding finite difference approximation on the coarse grid. Furthermore, the two-grid method is computationally more efficient, and the gain in efficiency is accentuated if a measure of adaptivity is incorporated.

1. Introduction

In a previous paper [7] we considered the numerical solution of nonlinear, dissipative partial differential equations (PDEs) by a pseudospectral method. Using the one-dimensional Kuramoto-Sivashinsky (KS) equation with spatial periodicity as an illustrative model, we presented solution algorithms based on the concept of approximate inertial manifolds (AIMs). An AIM is an approximation to a smooth invariant set in phase space called an inertial manifold, $\mathcal{M}$, which has the property that it attracts all trajectories at an exponential rate in time. The flow restricted to $\mathcal{M}$ is effectively described by a system of differential-algebraic equations (DAEs) of low dimension. Typically, the algebraic component of the DAE is a nonlinear relation that expresses the coefficients of the high order Fourier modes in terms of the coefficients of the lowest n Fourier modes, where n is small, and 'high' signifies a mode number greater than n. From a physical point of view the nonlinear relation may be regarded as an interaction law between small and large wavelengths such as those used in turbulence modelling. If p denotes the first n Fourier coefficients and q denotes the high

*Supported by the Engineering and Physical Sciences Research Council (EPSRC) of the United Kingdom.

Fourier coefficients, then the interaction law may be written as $q = \phi(p)$. An n-mode Galerkin approximation to a nonlinear, dissipative PDE, with q expressed in terms of p, leads to the *nonlinear Galerkin* (NLG) methods of Marion and Témam [6]. If $q = \phi(p)$ is replaced by the trivial equation $q \equiv 0$ the NLG method becomes a standard Galerkin method, usually referred to as a *flat Galerkin* method in this context. The reader is referred to [6] and to references therein for background information on AIMs based on approximate relations $q = \phi(p)$.

The feature that distinguishes the methods described in [7] from NLG methods is the replacement of the mapping $q = \phi(p)$ by a mapping that relates computed solutions on coarse and fine grids in physical space. Algorithms based on computation in physical space — rather than Fourier space — should have the advantage of being applicable in situations where boundary conditions are not restricted to spatial periodicity. Foias and Titi [2] introduced this potentially useful idea of characterising an AIM by nodal values of functions that are on the inertial manifold. In [7] we exploited this idea, albeit in situations where spatial periodicity pertained: the periodicity enabled us to use Fourier PS discretisation in space. Numerical experiments described in [7] show that the solution computed on a coarse grid enjoys a gain in accuracy when an AIM algorithm is used to incorporate a correction from a fine grid that is constructed by halving the mesh spacing of the coarse grid. It was noted, however, that this demanded a rather high cost in terms of computer usage time: indeed, to achieve a specified accuracy the cost of our AIM algorithms exceeds that of a standard PS method. In a study involving the two-dimensional KS equation, García-Archilla and de Frutos [3] observed analogous results with respect to accuracy and computational effort in comparing NLG methods to flat Galerkin methods.

In this paper we investigate the computational efficiency of our two-grid algorithm using Burgers' equation with Dirichlet boundary conditions. Discretisation in space is effected by means of a second-order finite difference approach, and time integration is carried out by a backward difference formulation (BDF) (see Section 3.12 of [4] for further information on BDF methods). Margolin and Jones [5] have used the ideas of AIMs to construct finite difference schemes for Burgers' equation in one space dimension. Their construction of a nonlinear relation between low and high modes was obtained by considering two sets of basis elements on a uniform grid. Elements in one set represent average values of the dependent variable within the difference cells, and elements in the second set represent first derivatives of the variable within cells. Despite their problematic formulation of boundary conditions for coefficients of basis elements in the second set, they obtain a simple and apparently effective approximation to the interaction law $q = \phi(p)$. Their results indicate that incorporation of the interaction law gives rise to improved accuracy.

The objective of this paper is to use the AIM approach to construct a simple finite difference scheme that is applicable for general boundary conditions. The formulation of our two-grid method is simpler than that adopted by Margolin and Jones [5], but it nevertheless offers an improvement in accuracy and computational efficiency relative to the standard finite difference method. In Section 2 we consider some features of the solution of the continuous problem. Section 3 deals with the discretisation in space, including the nonlinear stability of the semi-discrete system and the existence of an AIM for this system. Section 4 presents the two-grid algorithm, and numerical results are given in Section 5. This final section deals with accuracy and efficiency and it shows, inter alia, that computational efficiency is improved if the interaction between coarse and fine grids is limited to regions where the solution has a large spatial gradient.

2. Continuous Problem

As stated earlier, the problem we have chosen to study is the time-dependent Burgers' equation, namely

$$\frac{\partial u}{\partial t} = \lambda \frac{\partial^2 u}{\partial x^2} - u\frac{\partial u}{\partial x}, \quad \text{on } (0,1) \times (0,T], \tag{1}$$

where the boundary conditions are $u(0,t) = L > 0$ and $u(1,t) = 0$ for $0 \leq t \leq T$. Later we shall briefly consider the situation that arises when $u(0,t) = 0$ and $u(1,t) = R > 0$. Introducing the inner product $(u,v) = \int_0^1 u(x)v(x)dx$ and the corresponding L_2-norm it is readily shown that

$$\frac{1}{2}\frac{d}{dt}\|u(x,t)\|^2 = -\lambda\|u_x(x,t)\|^2 + \frac{L}{3}\left(L^2 - 3\lambda u_x(0,t)\right). \tag{2}$$

By setting $u_t = 0$ we can obtain the steady-state solution, u_0, of (1) in the form

$$u_0(x) = k \tanh \frac{k(1-x)}{2\lambda}, \tag{3a}$$

where

$$L = k \tanh \frac{k}{2\lambda}. \tag{3b}$$

Given a specific choice for λ, our first task is to solve (3b), a nonlinear equation in k. This can be done by using an appropriate iterative scheme such as Newton's method. We need only look for positive solutions of (3b) as the oddness of tanh implies that if $k = a$ is a solution then so is $k = -a$, and vice-versa. The range of tanh with a positive argument is $(0,1)$, so we can deduce from (3b) that $k > L$. We now go on to examine the asymptotics of the steady-state solution for small and large λ. We shall work with the function u_s, which represents u_0 in (3a) with λ treated as a parameter. To aid us in our search for asymptotic solutions we introduce the variable $z = k/2\lambda$. Considering the problem graphically, a solution occurs when the curves $y_1 = L/(2\lambda z)$ and $y_2 = \tanh z$ intersect. In the positive z half of the (y,z) plane they meet at only one point, which we shall refer to as (z_0, y_0). Indeed, due to the range of tanh, y_0

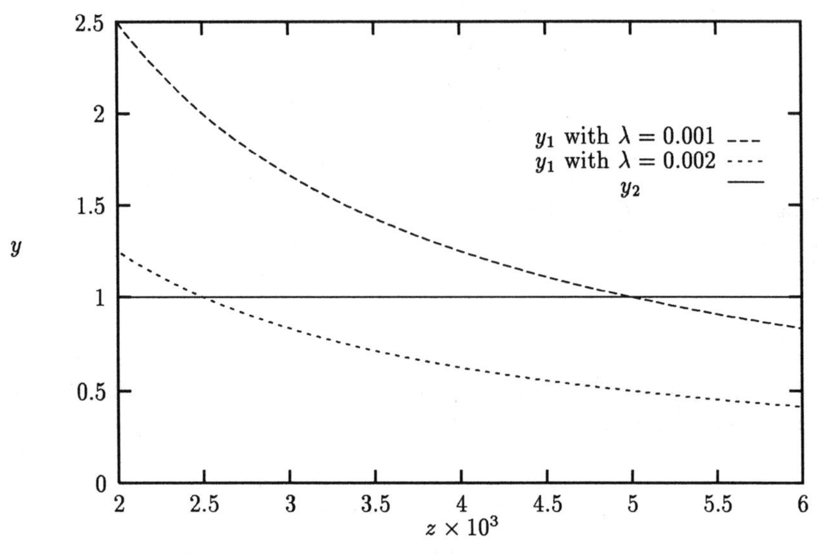

Fig. 1. Intersection of y_1 with y_2 as λ tends to zero.

must lie in $(0, 1)$. If λ is small then z must be sufficiently large for y_1 to fall within the range of y_2 (see Figure 1). More rigorously we may say that as $\lambda \downarrow 0$, $y_0 \to 1$, and as a consequence of this, $k \to L$. Applying this result for k and using our knowledge of tanh in (3a), we can say that

$$u_s(x, \lambda \to 0^+) = \begin{cases} L & \text{if } x \in [0, 1) \\ 0 & \text{if } x = 1. \end{cases} \tag{4}$$

For the case of large λ, as in Figure 2, the curve y_1 drops to zero very fast. When z is close to zero we are able to approximate y_2 by z: thus, we see that $z_0^2 = L/(2\lambda)$. As $\lambda \to \infty$, $k^2 \to 2\lambda L$, and so, by linearising the tanh function in (3a), we obtain the result

$$u_s(x, \lambda \to \infty) = L(1 - x). \tag{5}$$

Therefore we can say that for small λ the steady-state solution has a boundary layer at $x = 1$, while for large λ the solution develops into a straight line between the two boundary values (see Figure 3).

If, on the other hand, we wished to find a steady-state solution of (1) with the boundary conditions $u(0, t) = 0$ and $u(1, t) = R > 0$, the resulting calculation would proceed along a similar vein as that taken in the derivation of (3). This would yield the solution

$$u_1(x) = k \tan \frac{kx}{2\lambda}, \tag{6a}$$

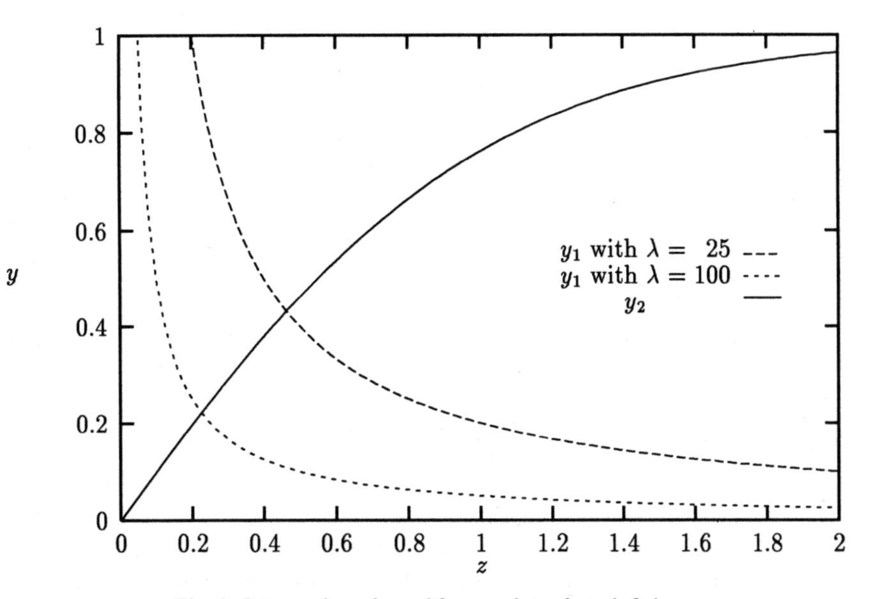

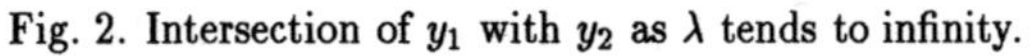
Fig. 2. Intersection of y_1 with y_2 as λ tends to infinity.

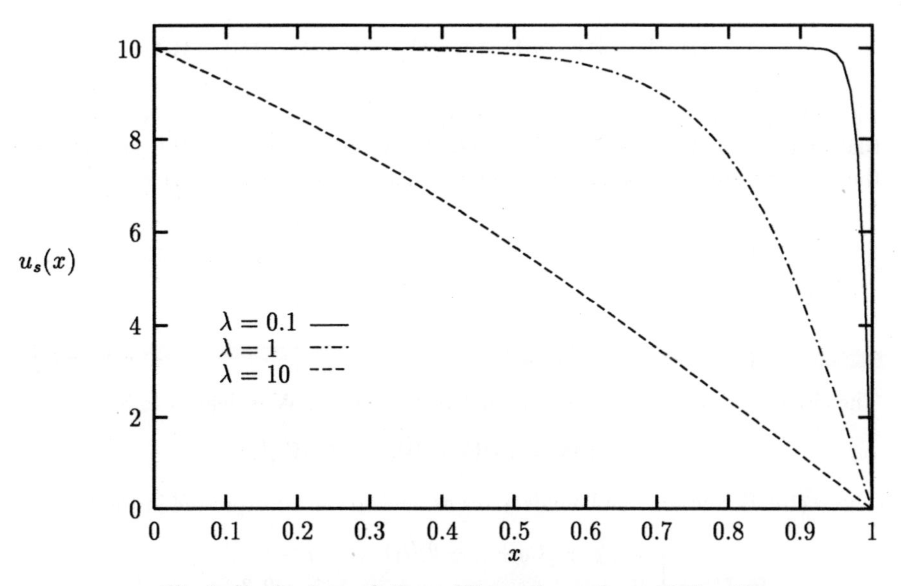

Fig. 3. Steady-state solutions of Burgers' equation, with $L = 10$.

where

$$R = k \tan \frac{k}{2\lambda}. \tag{6b}$$

This solution has a boundary layer at $x = 1$ for small λ.

3. Discrete Problem

3.1. Discretisation

We now go on to use a second-order central finite difference (FD) method to perform the discretisation of (1) on a uniform grid with n internal nodal points. The set of nodal values $\{x_j\}_{j=0}^{n+1}$ is such that $x_j = jh$, where the spatial step length $h = 1/(n+1)$. Note that we can express the nonlinear component of (1) in the form

$$uu_x = (1-\theta)uu_x + \frac{\theta}{2}(u^2)_x \text{ for } \theta \in [0,1]. \tag{7}$$

Since (7) is an identity, the term uu_x in (1) may be written equivalently by this expression involving θ. Note, however, that a spatial discretisation of uu_x is not necessarily equivalent to the analogous discretisation of the right-hand side of (7). Variation of θ may influence stability and accuracy of the discrete system. If (1) is written in semi-discrete form it yields

$$\dot{\mathbf{U}} = \lambda D_n^{(2)}\mathbf{U} - \left[(1-\theta)\mathbf{U} \otimes D_n^{(1)}\mathbf{U} + \frac{\theta}{2}D_n^{(1)}(\mathbf{U} \otimes \mathbf{U})\right] + \mathbf{B}(\mathbf{U}), \tag{8}$$

where $\mathbf{U} = [U_1, U_2, \ldots, U_n]^T$ and $U_j \approx u(x_j, t)$, $j = 0, 1, \ldots, n+1$. Note therefore that at the boundaries we have $U_0 = u(0,t)$ and $U_{n+1} = u(1,t)$. $D_n^{(1)}$ and $D_n^{(2)}$ are the regular n-dimensional second-order FD differentiation matrices given by

$$D_n^{(1)} = \frac{1}{2h}\begin{bmatrix} 0 & 1 & & & \\ -1 & 0 & 1 & & \\ & \ddots & \ddots & \ddots & \\ & & -1 & 0 & 1 \\ & & & -1 & 0 \end{bmatrix} \quad \text{and} \quad D_n^{(2)} = \frac{1}{h^2}\begin{bmatrix} -2 & 1 & & & \\ 1 & -2 & 1 & & \\ & \ddots & \ddots & \ddots & \\ & & 1 & -2 & 1 \\ & & & 1 & -2 \end{bmatrix}.$$

The circle product of two n-dimensional vectors $\mathbf{U}$ and $\mathbf{W}$ is defined to be

$$\mathbf{U} \otimes \mathbf{W} = [U_1W_1, U_2W_2, \ldots, U_nW_n]^T.$$

The vector $\mathbf{B}$ contains the boundary terms, and so can be expressed in the form

$$B_j(\mathbf{U}) = \begin{cases} \frac{U_0}{2h}(\frac{2\lambda}{h} + \frac{\theta}{2}U_0 + (1-\theta)U_1) & j = 1, \\ 0 & j = 2, 3, \ldots, n-1, \\ \frac{U_{n+1}}{2h}(\frac{2\lambda}{h} - \frac{\theta}{2}U_{n+1} - (1-\theta)U_n) & j = n. \end{cases}$$

Let us now define our discrete version of the inner product of two vectors in our solution space as

$$\langle \mathbf{U}, \mathbf{W} \rangle = h \sum_{j=1}^{n} U_j W_j, \tag{9}$$

with corresponding norm $\|\mathbf{U}\|^2 = \langle \mathbf{U}, \mathbf{U} \rangle$.

By using (9) to take the inner product of (8) with $\mathbf{U}$ we find that

$$\begin{aligned} \langle \mathbf{U}, \dot{\mathbf{U}} \rangle &= \frac{1}{2}\frac{d}{dt}\|\mathbf{U}\|^2 \\ &= \frac{\lambda}{h}\left(U_{n+1}(U_{n+1} - U_n) - U_1(U_1 - U_0) - h^2 \sum_{j=1}^{n} \left[\frac{U_{j+1} - U_j}{h}\right]^2 \right) \\ &\quad -\frac{\theta}{4}\left[U_{n+1}U_n(U_{n+1} + U_n) - U_0 U_1 (U_0 + U_1)\right] \\ &\quad +\frac{1}{4}(3\theta - 2) \sum_{j=1}^{n} U_j^2 (U_{j+1} - U_{j-1}). \end{aligned} \tag{10}$$

Using the boundary conditions and taking $\theta = \frac{2}{3}$, (10) can be written in the form

$$\frac{1}{2}\frac{d}{dt}\|\mathbf{U}\|^2 = -\lambda \|\delta\, \boldsymbol{U}\|^2 - \frac{\lambda}{h} U_1 (U_1 - L) + \frac{L}{6} U_1 (U_1 + L), \tag{11}$$

where $h\delta U_j = U_{j+1} - U_j$ for $j = 1, 2, \ldots, n$. To see how (11) compares with the continuous system (2) we replace $U_j(t)$ by the exact solution value $u(x_j, t)$ and make use of the vector $\mathbf{u} = [u(x_1, \cdot), u(x_2, \cdot), \ldots, u(x_n, \cdot)]^T$. Equation (11) then yields

$$\frac{1}{2}\frac{d}{dt}\|\mathbf{u}\|^2 = -\lambda \|\delta\, \boldsymbol{u}\|^2 - \frac{\lambda}{h} u(x_1, \cdot)[u(x_1, \cdot) - L] + \frac{L}{6} u(x_1, \cdot)[u(x_1, \cdot) + L] + \tau,$$

where τ represents the residual, a local truncation error. Making use of the differentiability of u it is readily shown that

$$\frac{1}{2}\frac{d}{dt}\|u\|^2 = -\lambda \|u_x\|^2 + \frac{L^3}{3} - \lambda L u_x(0, \cdot) + \tau + O(h) \tag{12}$$

as $h \to 0$. This shows that for $\theta = 2/3$, the residual satisfies $\tau = O(h)$, $h \to 0$, and we see that the discrete energy equation (12) mimics the continuous version (2).

3.2. Nonlinear stability

The procedure that we shall adopt here is identical to that which was used in Wallace and Sloan [7] for the KS equation. We begin by replacing $\mathbf{U}$ in (8) by

$$\mathbf{U}(t) := \mathbf{Z} + \mathbf{W}(t),$$

where $\mathbf{Z}$ is independent of t. By selecting $Z_0 = U_0$ and $Z_{n+1} = U_{n+1}$, the boundary conditions on $\mathbf{U}$ ensure that $W_0 = W_{n+1} = 0$. Consequently our substitution for $\mathbf{U}$ given above into (8) yields

$$\dot{\mathbf{W}} = \mathbf{g}(\mathbf{W}) + \mathbf{g}(\mathbf{Z}) + \mathbf{B}(\mathbf{Z}) + \mathbf{K}(\mathbf{Z},\mathbf{W}) - \mathbf{Q}(\mathbf{Z},\mathbf{W}), \tag{13}$$

where

$$\begin{aligned} \mathbf{g}(\mathbf{U}) &= \lambda D_n^{(2)}\mathbf{U} - (1-\theta)\mathbf{U}\otimes D_n^{(1)}\mathbf{U} - \frac{\theta}{2}D_n^{(1)}(\mathbf{U}\otimes\mathbf{U}), \\ \mathbf{Q}(\mathbf{Z},\mathbf{W}) &= (1-\theta)(\mathbf{Z}\otimes D_n^{(1)}\mathbf{W} + \mathbf{W}\otimes D_n^{(1)}\mathbf{Z}) + \theta D_n^{(1)}(\mathbf{Z}\otimes\mathbf{W}) \end{aligned}$$

and

$$2h\, K_j(\mathbf{Z},\mathbf{W}) = \begin{cases} (1-\theta)Z_0W_1 & j=1, \\ 0 & j=2,3,\ldots,n-1, \\ -(1-\theta)Z_{n+1}W_n & j=n. \end{cases}$$

By using the inner product (IP) as defined in the equation (9) we will show that $\|\mathbf{W}\|$ is bounded for sufficiently large t. Let us first consider the scalar term

$$\begin{aligned} 2h\,\mathbf{W}^{\mathrm{T}}\mathbf{Q} &= \sum_{i,j=1}^{n} d_{ij}^{(1)} W_i\,(\theta Z_jW_j + \theta_1 Z_iW_j + \theta_1 W_iZ_j) \\ &= \theta_2 \sum_{i=1}^{n-1} W_iW_{i+1}(Z_{i+1}-Z_i) + \theta_1\left(\sum_{i=1}^{n} W_i^2(Z_{i+1}-Z_{i-1}) + Z_0W_1^2 - Z_{n+1}W_n^2\right) \\ &= \theta_2 \sum_{i=1}^{n-1} W_iW_{i+1}(Z_{i+1}-Z_i) + \theta_1 \sum_{i=1}^{n} W_i^2(Z_{i+1}-Z_{i-1}) + 2h\mathbf{W}^{\mathrm{T}}\mathbf{K}, \end{aligned}$$

where $\theta_1 = 1-\theta$, $\theta_2 = 2\theta - 1$ and $d_{ij}^{(1)}$ is an element of the differentiation matrix $D_n^{(1)}$. Thus, by taking the IP of $\mathbf{W}$ with $\mathbf{Q}-\mathbf{K}$ and expressing the result in quadratic form we obtain

$$\langle \mathbf{W}, \mathbf{Q}-\mathbf{K}\rangle = \frac{1}{2}\mathbf{W}^{\mathrm{T}}B_n\mathbf{W}, \tag{14}$$

where B_n is the tridiagonal matrix

$$\begin{bmatrix} (Z_2-Z_0)\theta_1 & \frac{1}{2}(Z_2-Z_1)\theta_2 & & & \\ \frac{1}{2}(Z_2-Z_1)\theta_2 & (Z_3-Z_1)\theta_1 & \frac{1}{2}(Z_3-Z_2)\theta_2 & & \\ \ddots & \ddots & \ddots & & \\ & \frac{1}{2}(Z_{n-1}-Z_{n-2})\theta_2 & (Z_n-Z_{n-2})\theta_1 & \frac{1}{2}(Z_n-Z_{n-1})\theta_2 & \\ & & \frac{1}{2}(Z_n-Z_{n-1})\theta_2 & (Z_{n+1}-Z_{n-1})\theta_1 & \end{bmatrix}.$$

We now turn our attention to the vector $\mathbf{g}(\mathbf{W})$ that is in (13). If we define

$$\mathbf{W}^{\mathrm{T}}\mathbf{g}(\mathbf{W}) = \lambda\mathbf{W}^{\mathrm{T}}D_n^{(2)}\mathbf{W} + S,$$

then we must choose

$$\begin{aligned} S &= -\sum_{i,j=1}^{n} d_{ij}^{(1)} W_i \left(\frac{\theta}{2} W_j^2 + (1-\theta) W_i W_j\right) \\ &= \frac{1}{4h}(2-3\theta) \sum_{i=1}^{n-1} W_i W_{i+1}(W_{i+1} - W_i). \end{aligned}$$

This expression for S can also be written in quadratic form and so we can obtain

$$\langle \mathbf{W}, \mathbf{g}(\mathbf{W}) \rangle = h\lambda \mathbf{W}^{\mathrm{T}} D_n^{(2)} \mathbf{W} + \frac{1}{4}(2-3\theta) \mathbf{W}^{\mathrm{T}} \mathcal{W}_n \mathbf{W},$$

where $\mathcal{W}_n = \frac{1}{2}(C + C^{\mathrm{T}})$ and C is the n-dimensional matrix with non-zero elements $C_{i,i+1} = W_{i+1} - W_i$ for $i = 1, 2, \ldots, n-1$. It should now be obvious that by taking the IP of (13) with $\mathbf{W}$ we obtain

$$\frac{1}{2}\frac{\mathrm{d}}{\mathrm{dt}} \|\mathbf{W}\|^2 = h\lambda \mathbf{W}^{\mathrm{T}} D_n^{(2)} \mathbf{W} - \frac{1}{2} \mathbf{W}^{\mathrm{T}} B_n \mathbf{W} + \frac{1}{4}(2-3\theta) \mathbf{W}^{\mathrm{T}} \mathcal{W}_n \mathbf{W} + \langle \mathbf{W}, \mathbf{k} \rangle, \quad (15)$$

where for convenience we have set $\mathbf{k} = \mathbf{g}(\mathbf{Z}) + \mathbf{B}(\mathbf{Z})$.

Clearly the major obstacle to further progress is that the matrix $\mathcal{W}_n$ in (15) is dependent on elements of $\mathbf{W}$. In the ideal situation we would like to find some scalar ρ so that $\mathbf{W}^{\mathrm{T}} \mathcal{W}_n \mathbf{W} \leq \rho \|\mathbf{W}\|^2$. The equation (15) certainly illustrates the advantages to be gained by setting $\theta = 2/3$.

If we select $\theta = 2/3$ in (15) we may express it in the form

$$\frac{1}{2}\frac{\mathrm{d}}{\mathrm{dt}} \|\mathbf{W}\|^2 = h \mathbf{W}^{\mathrm{T}} \mathcal{D}_n \mathbf{W} + \langle \mathbf{W}, \mathbf{k} \rangle, \quad (16)$$

where

$$\mathcal{D}_n = \lambda D_n^{(2)} - \frac{1}{2h} B_n.$$

Note that for this particular choice of θ the constants θ_1 and θ_2 in B_n are both equal to 1/3.

We shall now look at various choices for $\mathbf{Z}$ in (16). Perhaps the most obvious to consider is the steady-state solution of (8), since $\mathbf{k}$ is then the zero vector. If this is the case we have

$$\|\mathbf{W}\|^2 \leq \exp(2\mu_{\max} t),$$

where $\mu_{\max}$ denotes the maximum eigenvalue of the symmetric matrix $\mathcal{D}_n$. Consequently we require conditions to make sure that $\mu_{\max} < 0$. If this is so then for any $\eta > 0$ it is clear that for sufficiently large t,

$$\|\mathbf{W}(t)\| \leq \eta. \quad (17)$$

For any other choice of $\mathbf{Z}$ we may apply Young's inequality to (16) and derive

$$\frac{1}{2}\frac{\mathrm{d}}{\mathrm{dt}} \|\mathbf{W}\|^2 \leq (\mu_{\max} + \frac{\Gamma}{2}) \|\mathbf{W}\|^2 + \frac{1}{2\Gamma} \|\mathbf{k}\|^2,$$

where $\Gamma > 0$. Provided $\mu_{\max} < 0$ we can set $\Gamma = -\mu_{\max}$ and use Gronwall's lemma on the resulting inequation to get

$$\|\mathbf{W(t)}\| \cdot \leq \epsilon - \frac{\|\mathbf{k}\|}{\mu_{\max}} \tag{18}$$

for any $\epsilon > 0$ with t chosen sufficiently large. In passing we should observe that as the matrix B_n is dependent on $\mathbf{Z}$ then so too must $\mu_{\max}$.

We wish to establish that there exists some $\mathbf{Z}$ and a positive integer N such that for $n \geq N$ we have $\mu_{\max} < 0$. As it is only practical to find the eigenvalues of $\mathcal{D}_n$ analytically when $\mathbf{Z}$ is linear, we will deal with the problem numerically to allow different selections of $\mathbf{Z}$.

We continue our investigation into the stability of our discrete system by selecting the n-dimensional vector $\mathbf{Z}$ to be either A, B or C as given in Table 1. We will consider the cases when we have the boundary conditions $U_0 = 10$, $U_{n+1} = 0$ and also $U_0 = 0$, $U_{n+1} = 10$. Table 2 and Table 3 give the minimum value of n for $\mu_{\max}$ to be negative for each of the above choices of $\mathbf{Z}$ and ever-decreasing values of λ. For values of λ larger than those given in the tables we need only have $n \geq 2$.

The values for the discrete steady-state solutions referred to as Type A in Table 1, were found either by Newton iteration or a time integration into the steady-state. In each case the initial condition was taken to be the solution we wish to replicate, the discretised continuous steady-state solution (Type B). It should be noted that each of the values of n given in Tables 2 and 3 generates steady-state solutions to (8) that are poor discrete approximations to the continuous steady-state solution. The reason for this occurring is that there are not enough nodal values to resolve the boundary layer near $x = 1$ (recall Figure 3). The thickness of this layer for the boundary conditions that we use is approximately the value of λ, so the value of $n\lambda$ has to be considerably greater than unity for boundary layer resolution.

We shall now discuss the ramifications of the results shown in Table 2. Using (17) and (18) we can say that the values of n listed in the table are sufficient for convergence to within some ball about $\mathbf{Z}$. As mentioned in the previous paragraph the number of nodes given in the table is still much smaller than the value that would be used in practice and so this is not such a severe restriction. It should however be noted that these conditions on n are not necessary to achieve convergence. For example in the case of $\lambda = 0.02$, where our analysis suggests that we should take $n \geq 41$, it was found that steady-state solutions existed for values of n very much smaller than this. To examine their stability we numerically evaluated the eigenvalues resulting from substituting these solutions into the Jacobian of the system of equations (8). It was found that for $n \geq 8$, the maximum real part of these eigenvalues remained negative, verifying that they were stable solutions. For $n = 7$ we were unable to find a steady-state solution. For $n < 7$ there appeared to be an effect that depended on the parity of n. Further investigation into what is happening for these small values of n was not felt to be appropriate in this project: it is not relevant in the context

Table 1. Choice of **Z** for the calculations given in Table 2 and Table 3.

A	–	large-time solution of equation (8)
B_2	–	discrete sampling of u_0 in (3) at the nodes $\{x_j\}_{j=0}^{n+1}$
B_3	–	discrete sampling of u_1 in (6) at the nodes $\{x_j\}_{j=0}^{n+1}$
C	–	discrete sampling of u_s in (4) at the nodes $\{x_j\}_{j=0}^{n+1}$

Table 2. Minimum required value of n (> 1) for negative definiteness of $\mathcal{D}_n$ for specified **Z** and λ with $U_0 = 10$ and $U_{n+1} = 0$.

λ	0.4	0.3	0.2	0.1	0.09	0.08	0.07
A	2	2	2	6	8	10	10
B	2	4	7	15	17	19	22
C	2	4	7	15	17	19	22
λ	0.06	0.05	0.04	0.03	0.02	0.01	0.009
A	13	16	20	27	41	84	93
B	26	32	40	54	82	165	184
C	26	32	41	55	83	167	186
λ	0.008	0.007	0.006	0.005	0.004	0.003	0.002
A	105	120	141	169	212	283	425
B	207	237	276	332	415	554	832
C	209	239	279	335	419	560	841

Table 3. Minimum required value of n (> 1) for negative definiteness of $\mathcal{D}_n$ for specified **Z** and λ with $U_0 = 0$ and $U_{n+1} = 10$.

λ	0.06	0.05	0.04	0.03	0.02	0.01	0.009
A	2	2	7	10	16	33	36
B	2	2	2	2	2	2	2
λ	0.008	0.007	0.006	0.005	0.004	0.003	0.002
A	41	47	55	67	84	112	168
B	2	2	2	2	2	2	2

of solution approximation, although it may be of interest theoretically. In the cases $\lambda = 0.05$ and $\lambda = 0.005$ we were able to determine the existence of stable steady-state solutions down to $n = 3$ and $n = 17$ respectively, both very much below the estimates given in Table 2. As a consequence of this we can say that although our analysis leading to (17) and (18) gives lower bounds on n, for achievement of convergence it is not actually essential to select n on or above the appropriate bound. It is interesting to note that a relationship appears to exist between the choice of λ and the values of n derived from our analysis: the integer part of s/λ is a good approximation to the values of n that exceed 2 in Table 2 with $s = 0.84$ for Type A and $s = 1.66$ for Type B.

If we now turn our attention to Table 3, it is apparent that from any initial condition a solution will always enter a ball about the $\mathbf{Z}$ described by Type B, irrespective of the values of n or λ. The main reason for this is that the gradient of the continuous solution remains positive throughout the domain of x and the matrix B_n in (14) seems to contribute towards the negative definiteness of $\mathcal{D}_n$ in our subsequent analysis. The obvious question that this leaves us with is why we require a much larger choice of n to obtain a sufficient condition for convergence to the $\mathbf{Z}$ of Type A. As in the case of Table 2 this appears to be due to our analysis for the stability conditions imposing too great a restriction on n. Numerical experiments on selected values of λ with the boundary conditions as in Table 3 had no problems in finding steady-state solutions for all $n \geq 2$. Examinations of the eigenvalues of the associated Jacobians indicated that all these solutions were stable.

The existence of an IM is established by showing that our system satisfies the spectral gap condition [2]. It is shown in [8] that this condition held for the central difference discretisation of the KS equation. Since the eigenvalues of the dissipative operator in the KS equation are the squares of those that we would have to consider for Burgers' equation, it is relatively straightforward task to adapt this earlier proof to apply here. The nonlinear analysis of this section, together with the spectral gap condition, ensures the existence of an IM for the semi-discrete system (8).

4. Algorithm Illustration

Here we describe the two-grid method for the solution of an equation such as (1). We shall assume that time integration is carried out by means of a standard, implicit package for systems of ordinary differential equations. If $\mathbf{W}$ represents the discrete solution on the coarse grid then our objective is to solve a nonlinear dynamical system of the form $\dot{\mathbf{W}} = \mathbf{G}(\mathbf{W})$, with the AIM idea used to include the interplay between solutions on coarse and fine grids. The description that follows shows how $\mathbf{G}(\mathbf{W})$ is found using our two-grid method.

If h and H denote, respectively, the mesh spacings of the coarse and fine grids, we shall assume that the spacings satisfy the relation $H = h/2$. This choice of H

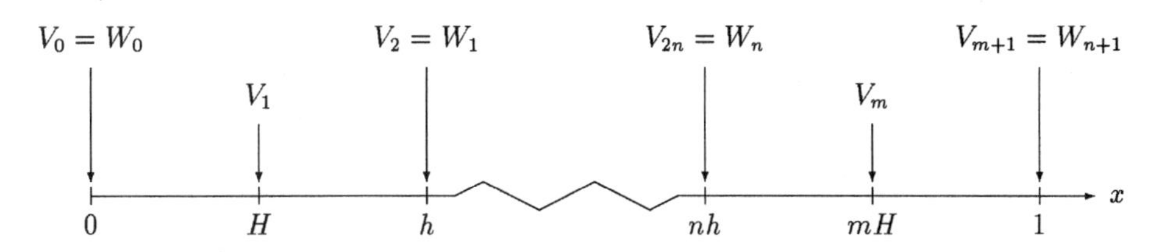

Fig. 4. **W** and **V** represent the coarse and fine solutions respectively.

as a function of h enables us to overlap the grids so that each coarse grid node is also on the fine grid. Consequently if we have n internal coarse grid points we must have $m = 2n + 1$ internal fine grid points. The idea behind our method is to improve the coarse grid approximation to $u(x, t)$ by applying the AIM properties that relate the coarse and fine grids. To be able to relate the solutions on the coarse grid and fine grid to one another we assume that they are equivalent at the overlapping nodes. This is illustrated in Figure 4, where the solution on the fine grid is represented by the m-dimensional vector $\mathbf{V}$. The improvement in accuracy produced by the two-grid method results from our ability to estimate — using the dynamics of the PDE — the solution at the non-common grid points using approximations that we possess on the coarse grid.

In the first stage of the solution process we use a second-order FD discretisation on the coarse grid to derive the set of n equations

$$\dot{V}_j = \frac{\lambda}{h^2}(V_{j+2} - 2V_j + V_{j-2}) - \frac{1-\theta}{2h}(V_{j+2} - V_{j-2})V_j - \frac{\theta}{4h}(V_{j+2}^2 - V_{j-2}^2), \qquad (19)$$

where $j = 2, 4, \ldots, 2n$ and, of course, $V_0 = L$ and $V_{m+1} = 0$.

Defining $F_j = \dot{V}_j$, for $j = 0, 2, \ldots, m + 1$, we are able to use (19) to obtain $F_2, F_4, \ldots, F_{2n}$, and since the boundary conditions associated with (1) are time independent we must have $F_0 = F_{m+1} = 0$. As in the spatially periodic case described in [7] we employ interpolation to estimate $F_1, F_3, \ldots, F_m$ at the intermediate fine grid nodes. If we choose to use a cubic interpolant to approximate the right hand side of (19) at the odd fine grid nodes, we find that

$$16F_j = \begin{cases} 15F_2 - 5F_4 + F_6 & j = 1, \\ -F_{j-3} + 9(F_{j-1} + F_{j+1}) - F_{j+3} & j = 3, 5, \ldots, m-2, \\ F_{m-5} - 5F_{m-3} + 15F_{m-1} & j = m. \end{cases}$$

The components of a vector $\mathbf{F}$ are now available on the fine grid. Consider for the moment the $m + 1$ equations resulting from a second-order FD discretisation of the fine grid, namely,

$$\dot{V}_j = \frac{\lambda}{H^2}(V_{j+1} - 2V_j + V_{j-1}) - \frac{1-\theta}{2H}(V_{j+1} - V_{j-1})V_j - \frac{\theta}{4H}(V_{j+1}^2 - V_{j-1}^2), \qquad (20)$$

with $j = 1, 2, \ldots, m$. Naturally the ultimate goal of our method is to be faster (in terms of computer processing time) and as accurate as the result that would be produced by integrating (20) in time. We attempt this by developing a relationship between the coarse and fine grids, reducing the dimension of our system to that of the coarse grid, but still retaining an element of the higher accuracy from the fine grid. This relation, or interaction, between the coarse and fine grids is carried out by using the AIM concept in a similar manner to that of the NLG methods [6].

The algorithms described in Wallace and Sloan [7] give the key ideas behind the approach. If $W(\cdot, t)$ and $V(\cdot, t)$ provide approximations to $u(\cdot, t)$ based on coarse and fine grid computations, respectively, then the difference $e(\cdot, t) = W(\cdot, t) - V(\cdot, t)$ will be dominated by high frequency components, since the low frequencies in W will match those in V. The difference, e, will behave like the high mode coefficients q of the NLG methods that we referred to in Section 1.The condition $\dot{q} = 0$ that is used to construct the algebraic relation $q = \phi(p)$ in the NLG method [6] is replaced here by $\dot{e} = 0$, and this is achieved by ensuring that the components of $\mathbf{F}$ based on coarse grid and fine grid computations are equal. This is implemented by assuming that the right hand side of (20) at $j = 1, 3, 5, \ldots, m$ is given by the value produced by interpolation from the coarse grid. This identification yields a system of $n + 1$ algebraic equations in $V_0, V_1, \ldots, V_{m+1}$ in which we treat the coarse grid values $V_0, V_2, \ldots, V_{m+1}$ as known, so that there are effectively $n + 1$ unknowns. This approach is really only supported by theory for the periodic spectral case: however, as in [5] we 'suppose that the physical basis for this assumption remains valid in the case of finite differences'. The numerical results presented later will lend support to the assumption.

We return now to the equations obtained by the prolongation process described above. With the coarse grid values treated as constants we obtain a set of $n+1$ linear equations in $V_1, V_3, \ldots, V_m$. The linearity of these equations becomes apparent when we introduce the constants K^+ and K^- to write them in the form

$$F_j = \frac{\lambda}{H^2}(K^+ - 2V_j) - \frac{1-\theta}{2H}V_j K^- - \frac{\theta}{4H}K^+K^-, \quad j = 1, 3, \ldots, m. \tag{21}$$

In the above $K^\pm = V_{j+1} \pm V_{j-1}$ which we know to be constant since both V_{j+1} and V_{j-1} are values on the *frozen* coarse grid. Note that if we had chosen a higher order FD method, these equations would be nonlinear and so a greater computational effort would be required for the interaction process.

Let us step aside to illustrate how discretisations of a higher order would result in a system of nonlinear equations in $V_1, V_3, \ldots, V_m$. As an illustrative example we consider a fourth-order FD method which would result in the following version of (21), namely,

$$\begin{aligned} F_j \;=\; & \frac{\lambda}{3H^2}(16K^+ - V_{j-2} - 30V_j - V_{j+2}) - \frac{\theta}{24H}(8K^+K^- + V_{j-2}^2 - V_{j+2}^2) \\ & -\frac{1-\theta}{12H}(8K^- + V_{j-2} - V_{j+2})V_j, \qquad j = 1, 3, \ldots, m. \end{aligned}$$

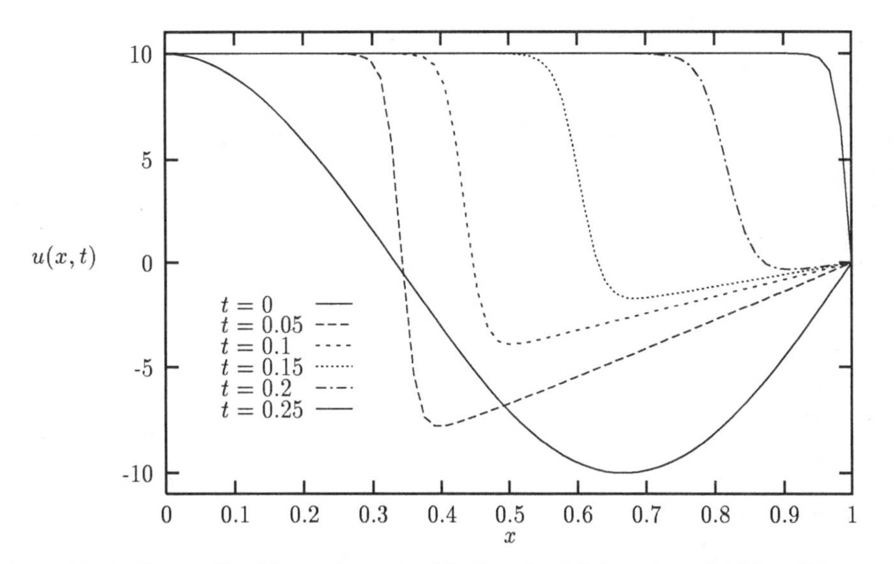

Fig. 5. Time profile of Burgers' equation (1) where $\lambda = 0.1$, from shown initial condition.

We have kept the same constants as equation (21); however, now we have a system of nonlinear equations to solve. The linear nature of (21) is a special case resulting from the discretisation and at some juncture we might be able to take advantage of this for use with higher order methods. In particular, it would be worthwhile to consider how retaining (21) as a representation of the interaction process would affect the accuracy when a higher order method is used for (20). Basically, would the computational saving made by solving the linear system (21) be worth the loss in accuracy by not using the same high discretisation order throughout? This question remains unanswered, but it should perhaps be considered in some later work. We now leave the issue of higher order methods and return to the explanation of our second-order discretisation.

Thanks to the interaction process given by (21) we now possess a full set of values for $\mathbf{V}$ on the fine grid which we can use to recalculate $\mathbf{F}$ on the coarse grid using differences based on a grid spacing H rather than h. Consequently, by applying (20) at $j = 2, 4, \ldots, 2n$ we can improve the estimate for $\mathbf{F}$ as it now incorporates information from the fine grid.

It does seem rather wasteful however to carry out the above procedure in regions where the coarse mesh alone can cope adequately with the profile of the PDE solution. As an example, consider the evolution of the dynamical system whose solution is given in Figure 5: clearly there are no features away from the developing boundary layer likely to cause difficulty. In the light of this we investigate to see if applying our algorithm only where "the action" occurs results in an increase in computational efficiency with no corresponding decrease in accuracy. To get information on the range over which we should apply our two-grid algorithm, we follow a profile from some initial starting point using a flat method (no fine grid interaction) with a sufficient number of internal nodes to achieve an accurate solution. By observing how this solution evolves we can determine the spatial range over which application of the two-grid method is likely be most beneficial. We can then repeat this integration using a flat method with grid interaction included only in regions where the previous integration has suggested that the two-grid algorithm might be helpful. This development may be thought of as an extremely crude adaptive method. It is simplistic in the sense that we are using a priori knowledge of the solution. The usefulness of both the *two-grid* and *adaptive two-grid* methods will again depend on how they measure up against the *flat method.* We now illustrate the effects of using the flat, two-grid and adaptive two-grid methods on the time integration of our discrete approximation to Burgers' equation.

5. Numerical Results

Since our stability analysis in the previous section took $\theta = 2/3$, all the numerical calculations to follow also use this particular discretisation of the nonlinear term. In Figure 5 we see the evolution in time from the initial condition $u(x,0) = L\cos(3\pi x/2)$, of (1), when we have $L = 10$ and $\lambda = 0.1$. By comparing Figure 5 with Figure 3 we observe that after $t = 0.25$ our solution appears to be very close to the steady-state solution of the continuous problem. Due to the simple structure of the PDE (1) and our knowledge of its continuous steady-state solution it was felt that by beginning an integration from the initial condition given above, we may be sufficiently close to the attractor to derive a benefit from the two-grid method. To test this we carried out integrations using the two-grid and flat methods to $t = 0.1$. It was found that in terms of accuracy the two-grid method out-performed the flat by approximately a factor of two, so we are clearly getting an improvement, but as yet we do not know the associated cost.

It should be stated that as our baseline a fourth-order FD discretisation possessing a high degree of spatial accuracy was used to provide us with an exact solution at $t = 0.1$. We chose an external routine to perform the time integration from the NAG (Numerical Algorithms Group) library. The routine D02EAF that was chosen is particularly well suited for the efficient integration of *stiff* systems. It uses the BDF to perform the discretisation in time with variable time steps, and is thus more accurate than the Crank-Nicolson method used in Wallace and Sloan [7]. The choice of time integrator is akin to that used by García-Archilla and de Frutos [3]. The disadvantage of using the NAG routine is that we have little control of its internal workings. With regard to the accuracy we are able to preside over a parameter, TOL say, that exerts control over the time-stepping restrictions imposed on the integrator

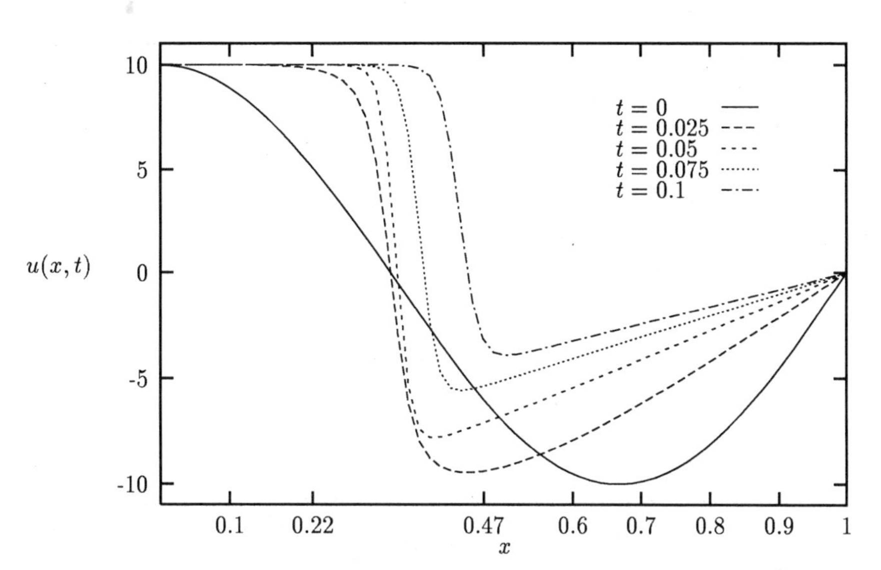

Fig. 6. Path of the sharp front in the integration to $t = 0.1$ for the system described in 5.

used in the routine. Unfortunately the link between the accuracy and this parameter is rather complex: a decrease in TOL will not necessarily lead to an increase in accuracy and computing time. For any one of the spatial discretisation methods, the computed solution can be no more accurate than the spatial discretisation error that is determined by the choice of n. To achieve a specified computational error at minimum cost we select n so that the spatial error falls marginally below the specified acceptable error, with the parameter TOL as large as possible whilst maintaining a total error dominated by spatial discretisation. Numerical experiments suggest that this is not an unreasonable strategy. If n is increased beyond this value — using an increased value of TOL that results in time-related errors being present in the total error — the specified error is reached only at an increased computational cost. When we compare the various methods (in which the same choice of n results in different spatial errors) we ensure that our selection of n for each of them results in a similar spatial error.

As stated at the end of Section 4we will further consider an adaptive two-grid method for this particular integration. If we look at Figure 6 we see that the steepest part of the solution profile remains between $x = 0.22$ and 0.47 when we travel from $t = 0$ to $t = 0.1$. Thus it was decided to let the coarse and fine grids interact at the node points lying within the smallest domain that contains this spatial range. This method gave results that were as accurate as those produced by the two-grid method, and we would naturally expect it to be computationally cheaper since the interaction takes place on a quarter of the grid.

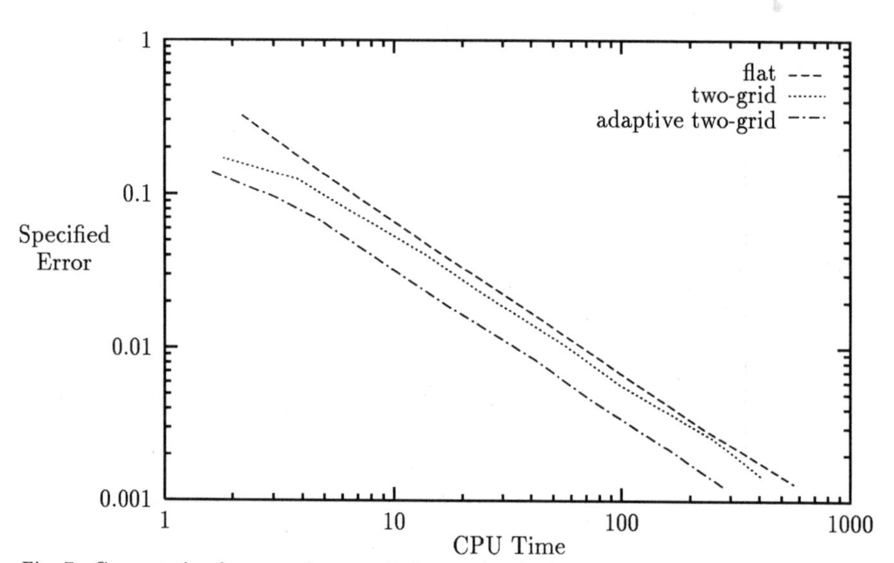

Fig. 7. Computational cost against specified error for the integration from $t = 0$ to $t = 0.1$ with $\lambda = 0.1$.

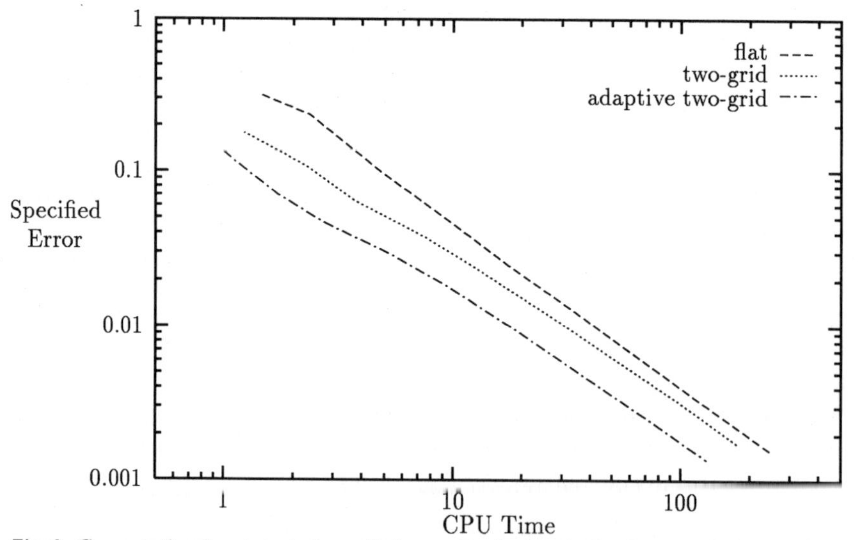

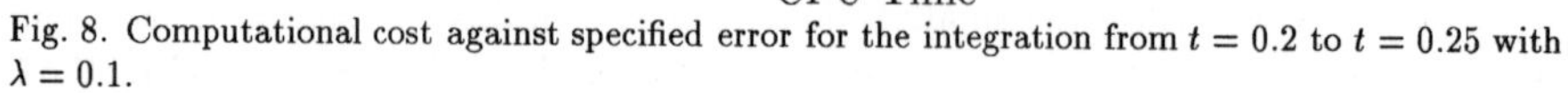

Fig. 8. Computational cost against specified error for the integration from $t = 0.2$ to $t = 0.25$ with $\lambda = 0.1$.

It was found that the two-grid methods gave approximately half the error of the flat method when applied on the same coarse grid. Figure 7 illustrates what this additional accuracy costs us: we compare the computing time required to meet a specified error for the flat, two-grid and adaptive two-grid methods. This is in contrast to the results obtained in Wallace and Sloan [7] when considering pseudospectral approximations to the KS equation, in which the two-grid method fared relatively poorly. However by far the most distinctive feature of this graph is the excellent performance of the adaptive two-grid method. It would appear that very little, if anything at all, has been done in incorporating the AIM idea with adaptivity. As we stated earlier, the approach that we have adopted here has been very simplistic in nature and could certainly be advanced upon. This brings about the hope that further increases in the sophistication of our adaptive method will bring about even greater improvements.

Since Devulder *et al.* [1] and García-Archilla and de Frutos [3] state that normally one would not necessarily expect much of an improvement when integrating through a transient phase, we decided to carry out further experiments. These involved integrating from the initial condition used above with a flat method on a coarse grid until $t = 0.2$ (selecting TOL so that the time discretisation error is effectively eliminated). From this point we continue to $t = 0.25$ using one of the three methods on the same coarse grid. The difference in accuracy between the methods was found to be similar to that obtained in the previous integration. The graph for the cost versus accuracy for the integrations between $t = 0.2$ to 0.25 (they, of course, are using the same method up to $t = 0.2$) is given in Figure 8. The results themselves are not greatly different from Figure 7; however, we do see that the two-grid method is performing better relative to the flat. This could be due to the fact that when the interaction is included the solution is much closer to the steady-state solution and hence to the attractor.

6. References

1. C. Devulder, M. Marion and E. S. Titi, *On the rate of convergence of the nonlinear Galerkin methods*, Math. Comput., **60** (1993), 495–514.
2. C. Foias and E. S. Titi, *Determining nodes, finite difference schemes and inertial manifolds*, Nonlinearity, **4** (1991), 135–153.
3. B. García-Archilla and J. de Frutos, *Time integration of the nonlinear Galerkin method*, IMA J. Numer. Anal., **15** (1995), 221–244.
4. J. D. Lambert, *Numerical methods for ordinary differential equations: the initial value problem*, John Wiley & Sons, Chichester, second ed., 1991.
5. L. G. Margolin and D. A. Jones, *An approximate inertial manifold for computing Burgers' equation*, Physica D, **60** (1992), 175–184.
6. M. Marion and R. Témam, *Nonlinear Galerkin methods*, SIAM J. Numer.

Anal., **26** (1989), 1139–1157.

7. R. Wallace and D. M. Sloan, *Numerical solution of a nonlinear dissipative system using a pseudospectral method and inertial manifolds*, SIAM J. Sci. Comput., **16** (1995), 1049–1070.
8. R. Wallace, *Numerical solutions of nonlinear dissipative systems using inertial manifolds*, PhD thesis, University of Strathclyde, Glasgow, Scotland, 1995.

THE EFFECTIVENESS OF DROP-TOLERANCE BASED INCOMPLETE CHOLESKY PRECONDITIONERS FOR THE CONJUGATE GRADIENT METHOD

I. C. SMITH*, R. WAIT and C. ADDISON
Institute of Advanced Scientific Computation
University of Liverpool, Liverpool, L69 3BX, U.K.
E-mail: ismith@supr.scm.liv.ac.uk, wait@liv.ac.uk, cliff@liv.ac.uk

ABSTRACT

Incomplete Cholesky factorisation based preconditioners have been widely used in association with the Conjugate Gradient Method. The standard incomplete Cholesky factorisation, or $IC(0)$ factorisation, forces the Cholesky factor $\mathbf{L}$ (where $\mathbf{A} \approx \mathbf{LL}^T$) to have the same sparsity as $\mathbf{A}$. Two main strategies exist for controlling the amount of fill; namely control based upon position (or fill level) and control based upon magnitude (or drop tolerance). The latter has received little popular attention and is the focus of this report.

1. Introduction

Incomplete Cholesky factorisation based preconditioners have been widely used in association with the Conjugate Gradient Method (CG) as a means of improving the convergence rate of the solver. The standard incomplete Cholesky factorisation, or $IC(0)$ factorisation, forces the Cholesky factor $\mathbf{L}$ (where $\mathbf{A} \approx \mathbf{LL}^T$) to have the same sparsity as $\mathbf{A}$. Whilst this is effective in reducing the number of CG iterations required for convergence, further improvement is possible by allowing a certain amount of fill in $\mathbf{L}$. In the extreme case, were the fill is unconstrained, the factorisation is exact ($\mathbf{A} = \mathbf{LL}^T$) and all that is left is essentially a direct method where CG merely provides iterative improvement.

Two main strategies exist for controlling the amount of fill; namely control based upon position (or fill level) and control based upon magnitude (or drop tolerance). The former has been widely cited [1,2] and was examined earlier in this project however, the latter has received less popular attention and is the focus of this report.

The drop tolerance based incomplete Cholesky factorisation may be written as:

$$a_{k,k} \leftarrow \sqrt{a_{k,k}} \tag{1}$$

$$\text{for each } k, \quad i > j > k \quad a_{i,j} \leftarrow \begin{cases} a_{i,j} - a_{i,k}a_{k,k}^{-1}a_{k,j} & \text{if } |a_{i,j}| > tol \\ a_{i,j} & \text{otherwise} \end{cases} \tag{2}$$

where *tol* is the specified drop tolerance. In other words, any fill at position (i, j) is discarded or dropped if it is less than or equal to the drop tolerance in magnitude. Since the test is an absolute one, it is to be expected that the amount of fill will vary between matrices of similar sparsity. To overcome this difficulty, it is instead possible

*This work is funded by EPSRC Grant GR/K/13028

to test the magnitude of the fill element against the corresponding diagonal entries thus:

$$\text{retain } a_{i,j} \text{ if } |a_{i,j}| > tol \times a_{i,i}a_{j,j} \tag{3}$$

In a strict physical sense, this is dimensionally incorrect and a further refinement may be to use:

$$\text{retain } a_{i,j} \text{ if } |a_{i,j}| > tol \times \sqrt{a_{i,i}a_{j,j}} \tag{4}$$

However since the square root operation is generally achieved through a library function call, this test may be less computationally appealing than (3) and a more efficient implementaion is:

$$\text{retain } a_{i,j} \text{ if } (a_{i,j})^2 > (tol)^2 \times a_{i,i}a_{j,j} \tag{5}$$

2. Effect of drop tolerance on fill and CG convergence

As a starting point, the Sherman1 matrix was selected as a suitable test candidate from the Harwell-Boeing Sparse Matrix Collection [4]. This system is based on a finite difference approximation to a p.d.e., discretized over a regular $10 \times 10 \times 10$ grid, with 1 equation per grid point. The sparsity pattern of the resulting 1000×1000 system is shown in fig. 1. Initially the absolute drop tolerance rule, given in (2), was employed and the resulting Cholesky factor used as a preconditioner for a standard CG iterative solver. In all cases in this report, the CG error tolerance was 1×10^{-3}.

Fig. 2(a) illustrates the effect of drop tolerance on fill. Although the relationship is not as linear as that seen when using position based fill [3], the transition from incomplete to complete factorisation takes place over a useful range of drop tolerance (approximately five orders of magnitude). The effect on CG performance is dramatic and can be seen in fig. 2(b). It is apparent that the number of CG iterations diminishes rapidly for only a small increase in fill. Clearly, increasing the amount of fill effectively trades off solver cost against preconditioner cost so that the ultimate goal should be to minimise the aggregate cost, a least in a sequential environment. Fig. 2(c) reveals this broader context and suggests that a drop tolerance in the range $10^{-2} - 10^{-4}$ is useful whilst a value of 10^{-3} is optimal for this system. For completeness, the above experiments were repeated using the relative drop tolerance rule given in (4). The results, shown in fig. 3, are not appreciably different from those above and this rule was used in all subsequent experiments.

Having examined a system which had previously performed well using position based fill, the next step was to test a matrix which has previously performed poorly. BUS494 ($n = 494$) is again drawn from the Harwell-Boeing collection and is far more irregularly structured than Sherman1, being akin to a nested dissection ordering (see

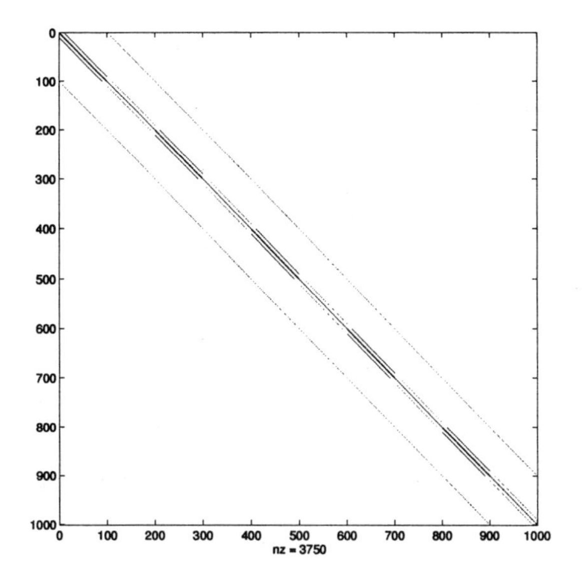

Figure 1: Sherman1 sparsity pattern

fig. 4). The results obtained for this system are shown in fig. 5. Here again the degree of fill in can be controlled over a useful range of drop tolerance values with only a small amount of fill giving near instant CG convergence. The aggregate preconditioner/solver cost is minimised by a drop tolerance of 10^{-4} and rises by only 20% at 10^{-3}. It is interesting to note that even for large amounts of fill, the aggregate cost does not rise appreciably. This is in agreement with a previous belief that the system is small enough to favour direct solution.

3. Comparison between drop tolerance based Cholesky and standard Cholesky factorisations

In order to judge how drop tolerance based Cholesky preconditioners perform compared with the $IC(0)$ and full Cholesky types, a further twelve matrices were drawn from the Harwell-Boeing collection for testing. The matrices vary widely in size, order and sparsity pattern and should therefore be representative of a broad class of problems (details of the individual systems are given in Appendix A). Of the complete set of thirteen, matrices 1 and 2 did not converge within n iterations using the $IC(0)$ preconditioner. Although convergence was achieved using the drop tolerance based preconditioners, the number of iterations required ($> 0.9n$) was thought to be too great for the systems to be amenable to iterative solution. For this reason,

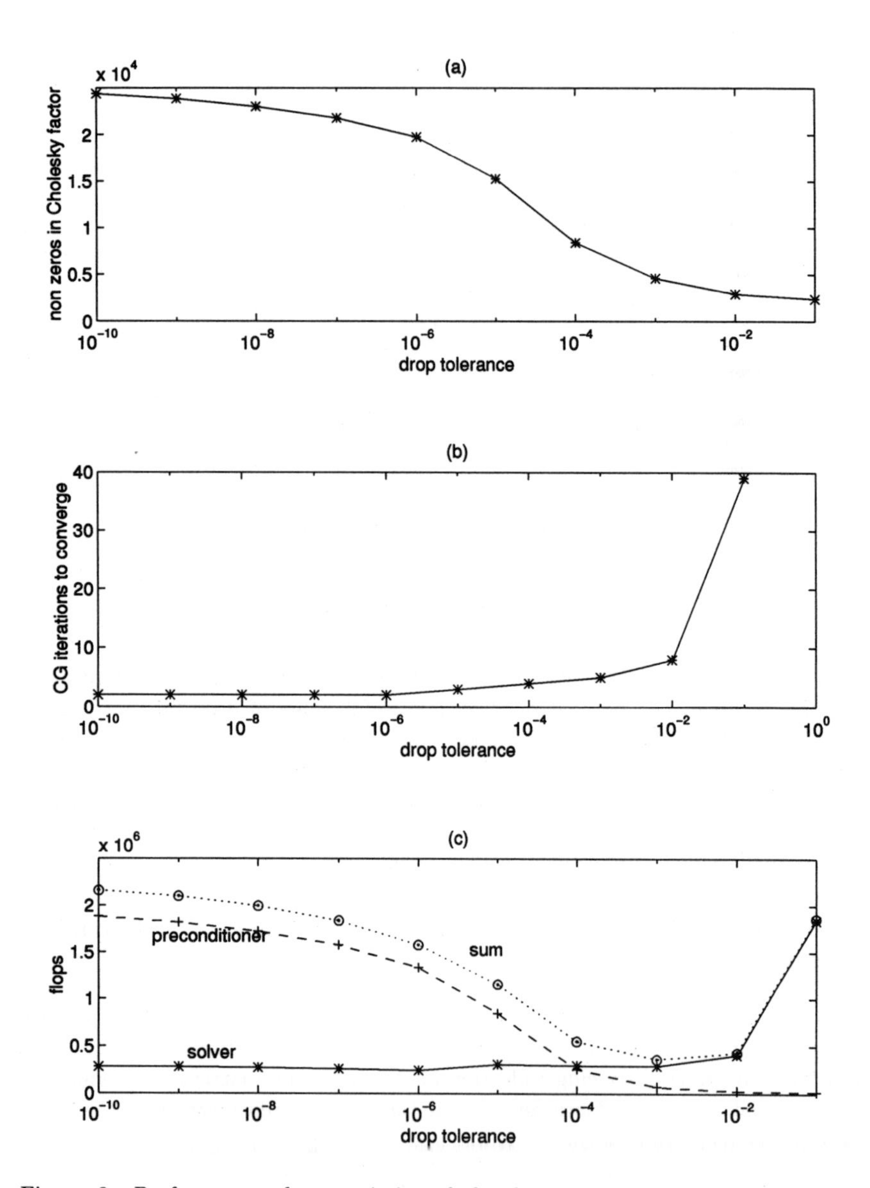

Figure 2: Performance characteristics of the drop tolerance based fill incomplete Cholesky preconditioner when applied to Sherman1 (absolute drop tolerance rule)

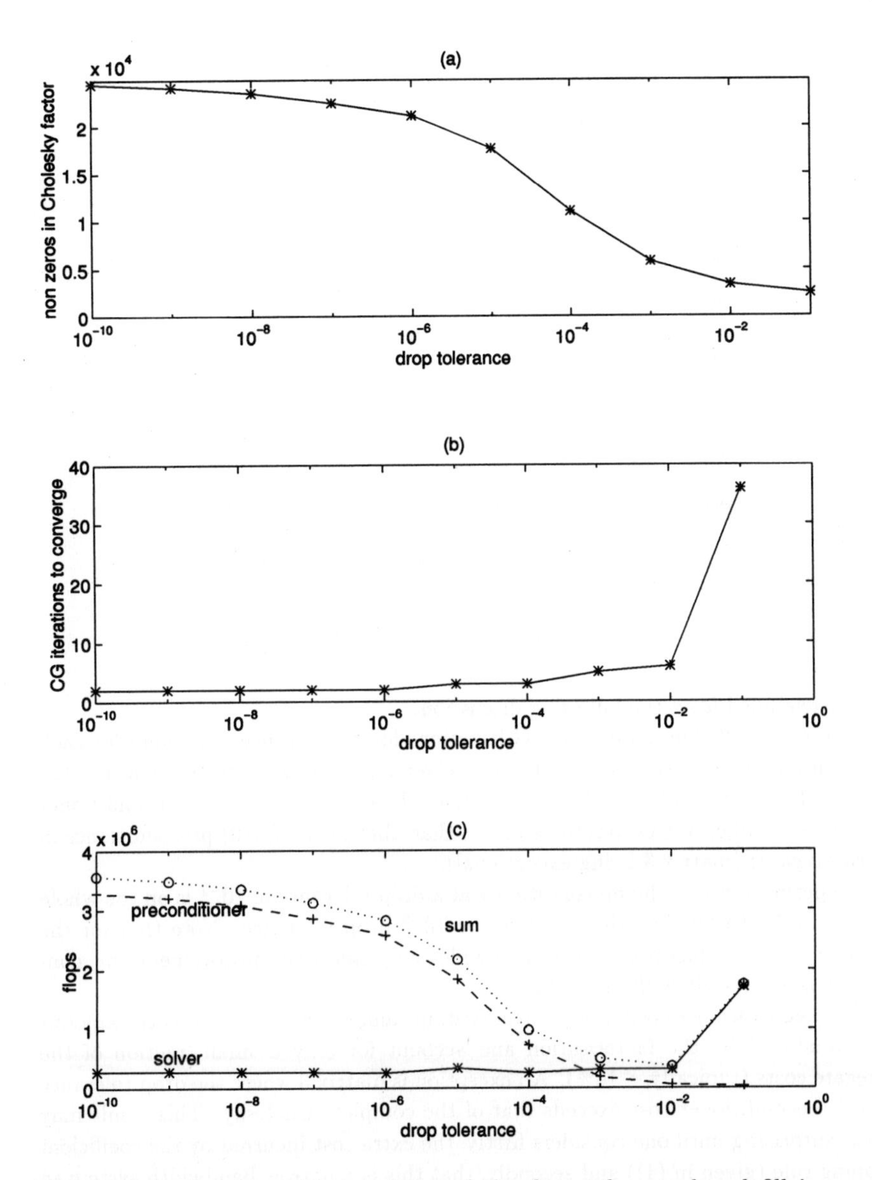

Figure 3: Performance characteristics of the drop tolerance based fill incomplete Cholesky preconditioner when applied to Sherman1 (relative drop tolerance rule).

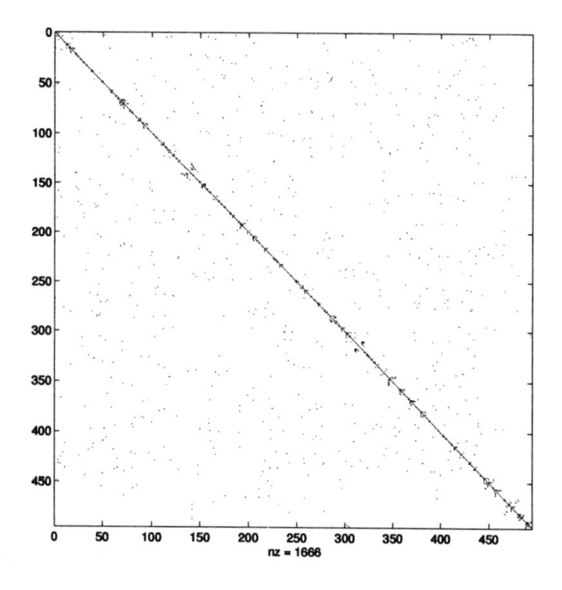

Figure 4: BUS494 sparsity pattern

they are excluded from the following discussion.

As in section 2, the number of CG iterations for convergence was found for each matrix in addition to the preconditioner, solver and aggregate costs. The results, shown in fig. 6(a), indicate that for a drop tolerance of 10^{-3}, the preconditioner reduces the number of CG iterations to less than that of the $IC(0)$ preconditioner in almost all cases (matrix 3 being exceptional).

The performance of the preconditioner at a drop tolerance of 10^{-2} is on the whole superior to $IC(0)$ but less than that of the 10^{-3} drop tolerance. Note that for the complete Cholesky factorisation, CG provides only iterative improvement and convergence is always virtually instantaneous.

The preconditioner costs are, as expected, in rank order according to the amount of fill produced by the factorisation and account for only a small fraction of the aggregate costs (typically $< 10\%$). An exception is matrix 3 where the drop tolerance $= 10^{-3}$ preconditioner cost exceeds that of the complete Cholesky. This result may appear surprising until one considers firstly, the extra cost incurred by the coefficient dropping rule (given in (4)) and secondly, that this is a narrow bandwidth system so that the amount of fill in a full Cholesky factorisation is relatively small.

The solver costs are somewhat varied; in some instances the drop tolerance preconditioner has improved the iteration matrix condition (and hence reduced the number

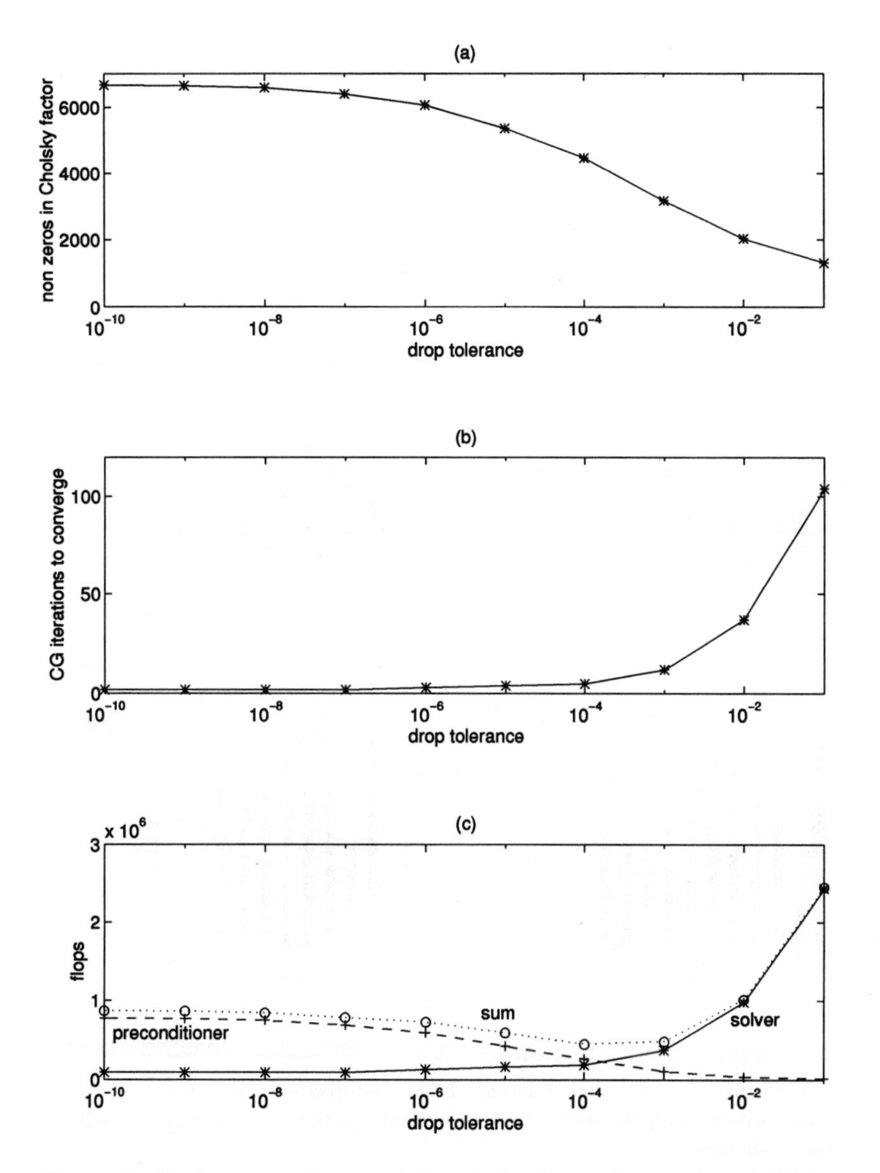

Figure 5: Performance characteristics of the drop tolerance based fill incomplete Cholesky preconditioner when applied to BUS494 (relative drop tolerance rule).

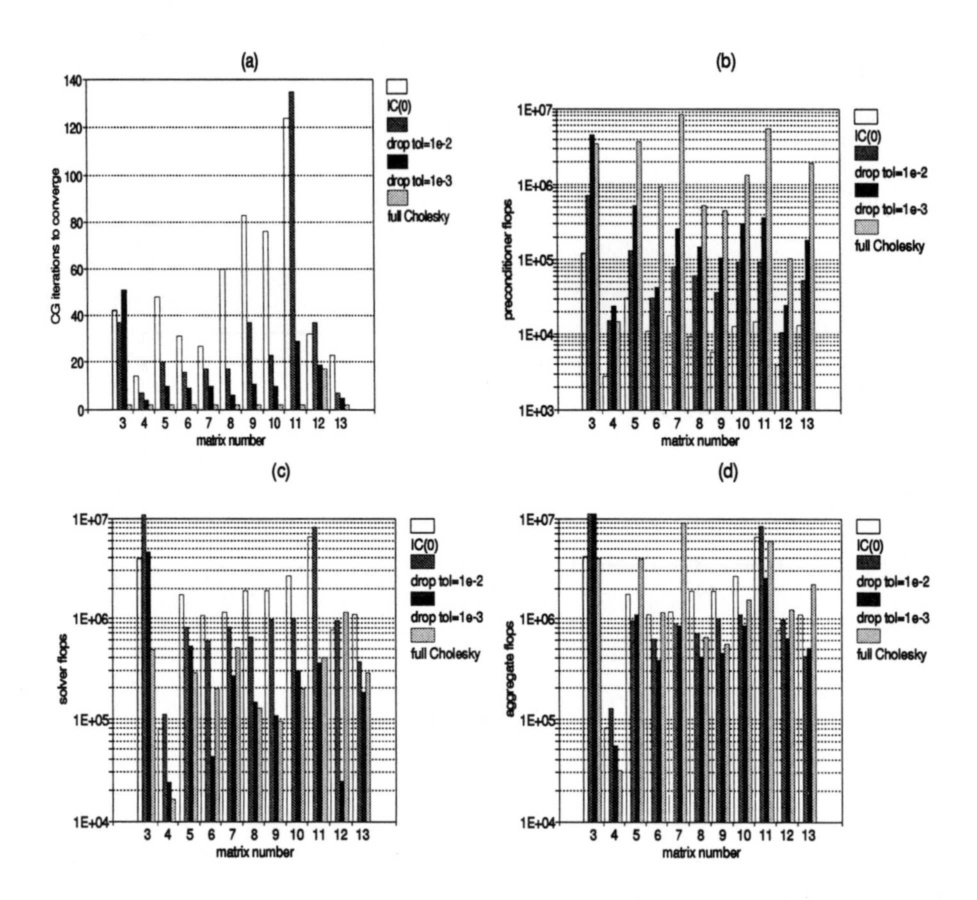

Figure 6: Performance of the drop tolerance based factorisation compared with the standard incomplete ($IC(0)$) and full Cholesky factorisations for various Harwell-Boeing sparse matrices.

of CG iterations) enough to reduce the cost below that of either an $IC(0)$ or full Cholesky preconditioner (e.g. matrices 6, 7, 11 and 12) whilst in other cases, the increase in work per iteration has increased the overall cost (e.g. matrices 3, 5 and 10). The broader pattern only becomes clear when the aggregate costs are examined. Here the drop tolerance $= 10^{-3}$ preconditioner produces the lowest total cost in all but two instances (matrices 3 and 4). However, when a drop tolerance of 10^{-2} is employed the total cost may exceed that of an $IC(0)$ or full Cholesky factorisation.

To summarise, it appears that correct choice of drop tolerance can give a preconditioner which reduces the overall cost, at least for a sequential implementation, to below that of either an $IC(0)$ or full Cholesky preconditioner in the majority of cases. The choice of 10^{-3} as a relative drop tolerance seems to be a good heuristic even though it may not produce optimal results. A value greater than this reduces the number of CG iterations although the extra work per iteration may outweigh the reduction in total cost. Another useful heuristic seems to be to err on the side of caution and, if in doubt, choose a drop tolerance of $< 10^{-3}$ (say $10^{-3} - 10^{-4}$) so that the work load is shifted toward the preconditioner construction phase. The motivation being that for small amounts of fill, the preconditioner cost is generally much less than that of the solver.

4. Combining position and magnitude based fill

The results presented so far have compared drop tolerance, $IC(0)$ and full Cholesky factorisation based preconditioners. Since the latter represent the two extremes of position based fill, neither of which may come close to the optimum in terms of aggregate cost, it is instructive to examine the effect of combining fill based on both fill level and drop tolerance. The algorithm used here first allows unconstrained fill up to some level k and then rejects any further fill on the basis of magnitude. It is immediately evident that reducing the drop tolerance is a far more effective means of reducing the required number of CG iterations than increasing fill level.

The results for Sherman1 are presented in fig 7. Here the contour, fill level = 0, corresponds to a factorisation based entirely on drop tolerance whilst the contour, drop tolerance = 0.1, approximates a factorisation based entirely on fill level.

Fig. 7(a) seems to suggest that any drop tolerance based fill leads to a reduction in CG iterations which outweighs the increase in flops per iteration, hence the almost monotonic decrease in solver cost for constant fill level. This is not the case when the drop tolerance is held constant however. Here the number of solver flops initially increases with fill, as the work per iteration increases, then falls again as the reduction in iteration count takes precedence. The small ridge along the drop tolerance $= 10^{-4}$ contour is accountable by a single CG iteration increase, attributable almost certainly to round-off error.

The variation in preconditioner cost (fig. 7(b)) is at first sight suprising. Clearly

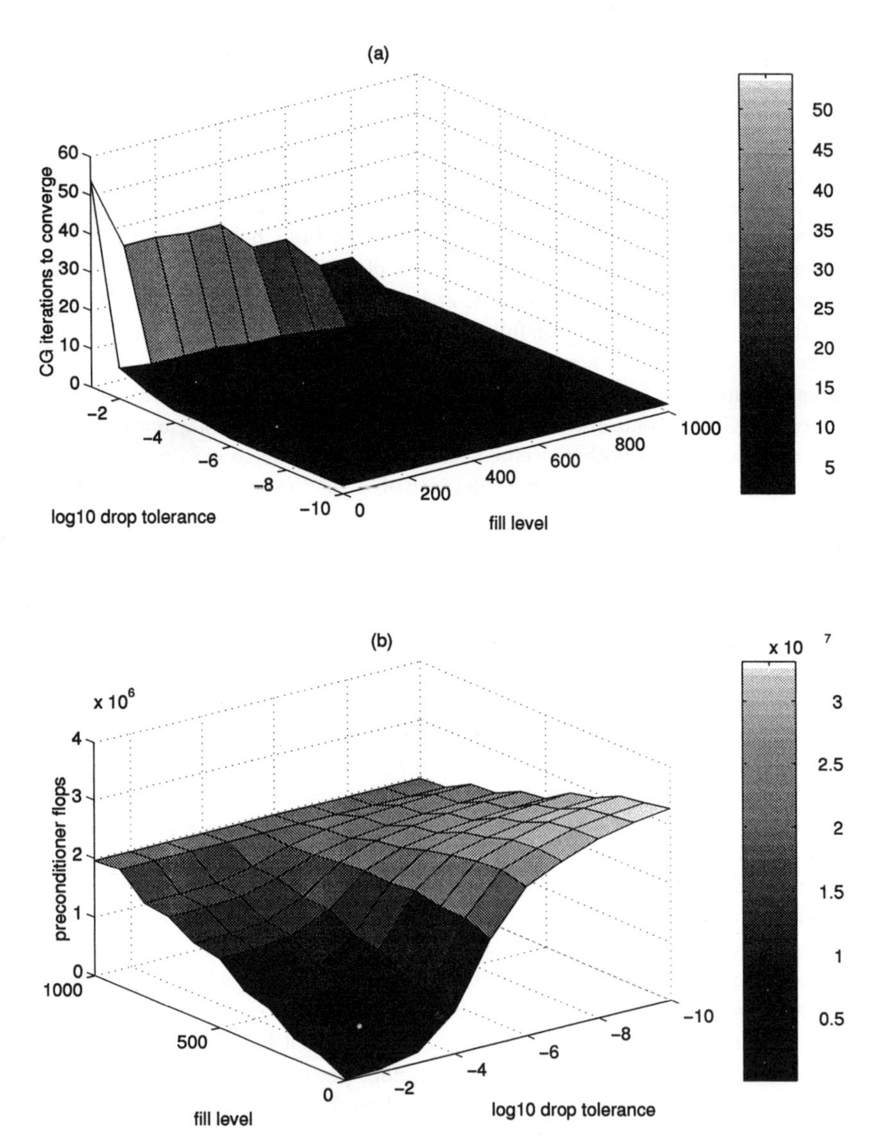

Figure 7: Effect of combining magnitude and position based fill in the incomplete Cholesky preconditioner applied to Sherman1.

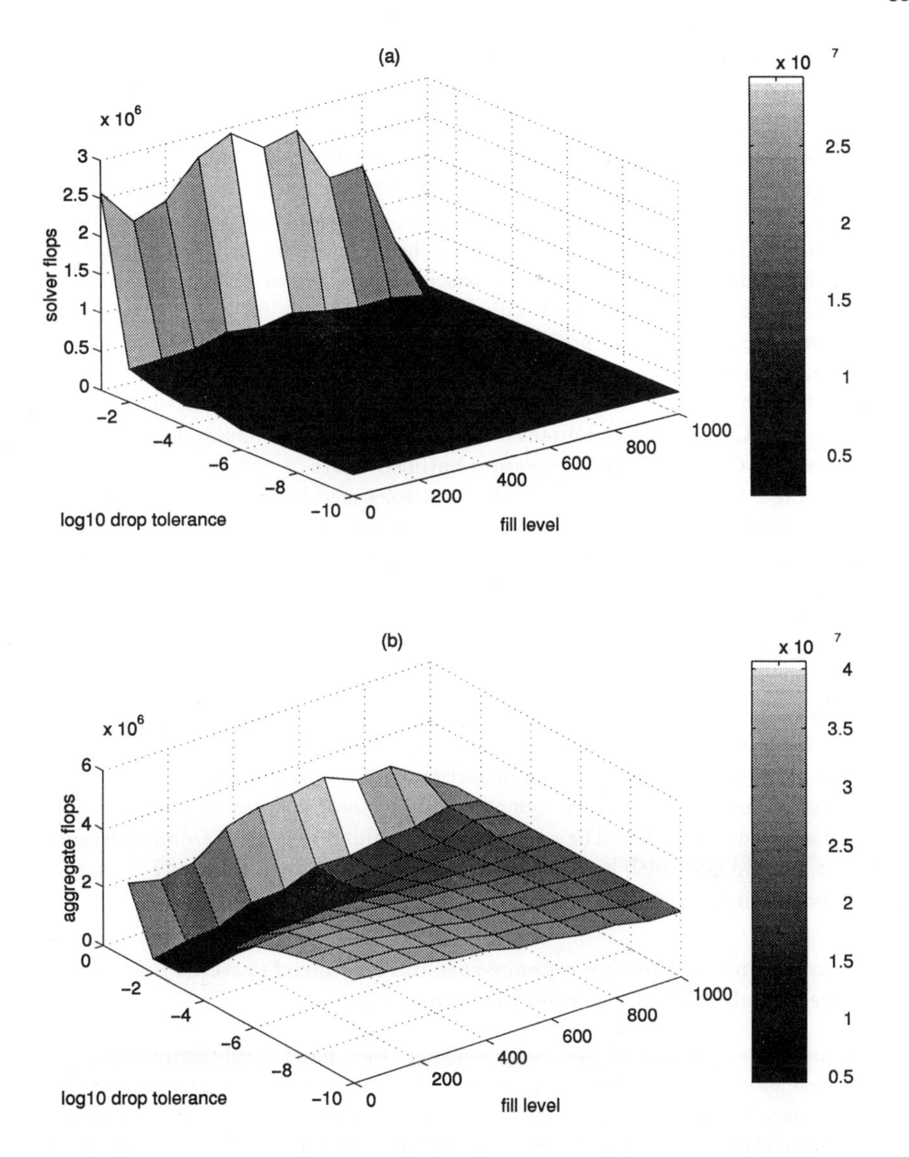

Figure 7: (continued) Effect of combining magnitude and position based fill in the incomplete Cholesky preconditioner applied to Sherman1.

for a fixed fill level the same general trend is seen as in fig. 2(c); however there seems to be a net *decrease* in cost at high fill levels for a fixed drop tolerance. The reason behind this lies in the fact that, for each fill element which is created, a magnitude based fill requires additional floating point operations to position based fill; namely those required by the additional multiply and, more significantly, the additional square root operations. For a low drop tolerance (say 10^{-10}), an almost complete factorisation already exists at zero fill level so that increasing the fill level effectively trades position and magnitude based fill. Since position based fill is acquired more 'cheaply' than magnitude based fill the overall cost will therefore fall as the fill level is raised and the position based fill starts to predominate.

If the aggregate costs are examined, it is evident that the overall cost is minimised by a drop tolerance of 10^{-3} at zero fill level. At high values of drop tolerance, adding position based fill appears to be disadvantageous whilst at low values, increasing the fill level reduces the overall cost for reasons outlined above.

In section 2 the effect of using solely drop tolerance based fill in the BUS494 system factorisation was examined. Whilst the results looked promising, the question naturally arises as to whether any further performance increase may be accrued by combining position based fill. Fig. 8(a) shows that there is a monotonic decrease in CG iterations for both increasing position and increasing magnitude based fill; however the effect is most dramatic in the latter case. As with Sherman1 there is a faint ridge, in this case along the drop tolerance $= 10^{-5}$ contour, which amounts to a 1 iteration increase and is accountable for by round-off error. This ridge also manifests itself in the solver cost characteristics.

Whilst the preconditioner cost characteristics follow the same general trends as for Sherman1, the solver costs fall monotonically for both increasing position based fill and magnitude based fill. The aggregate cost is minimised with a drop tolerance of 10^{-4} at zero fill level with high values of fill level again only leading to a reduction in overall costs at low values of drop tolerance.

5. Improvements in the convergence rate of re-ordered systems using drop tolerance based Cholesky preconditioners

Symmetric re-ordering of systems, resulting from p.d.e. approximations, has widely been used in an attempt to increase the amount of work which can be performed in parallel during iterative solution. Unfortunately, it can sometimes be the case that the number of CG iterations required to solve the re-ordered system is far in excess of that required to solve the original system (see [5] for examples). and consequently much, if not all, of the parallel speed up is lost. To investigate whether the use of drop tolerance based preconditioners could help ameliorate this situation, experiments were performed on red-black (a.k.a. 2 colour) and symmetric minimum degree re-orderings of some of the Harwell-Boeing matrices. The drop tolerance was

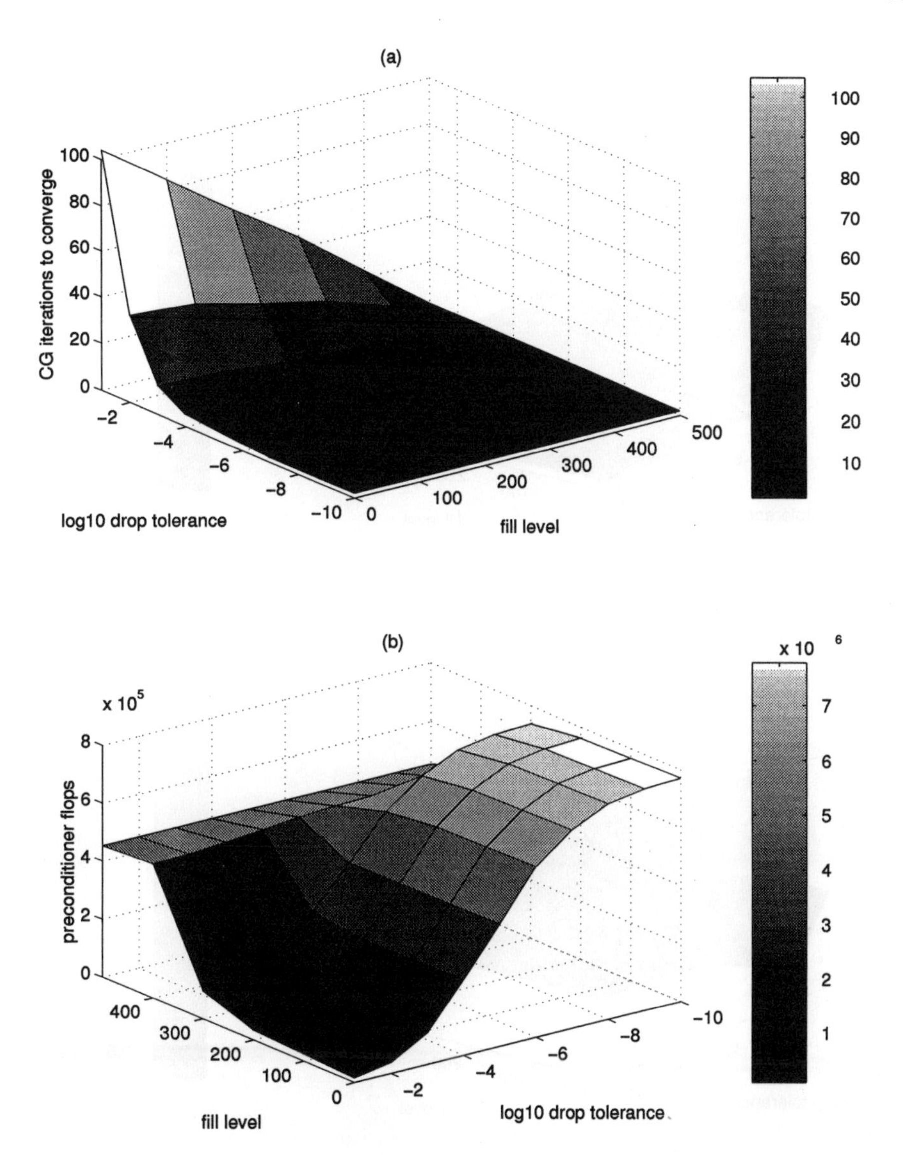

Figure 8: Effect of combining magnitude and position based fill in the incomplete Cholesky preconditioner applied to BUS494.

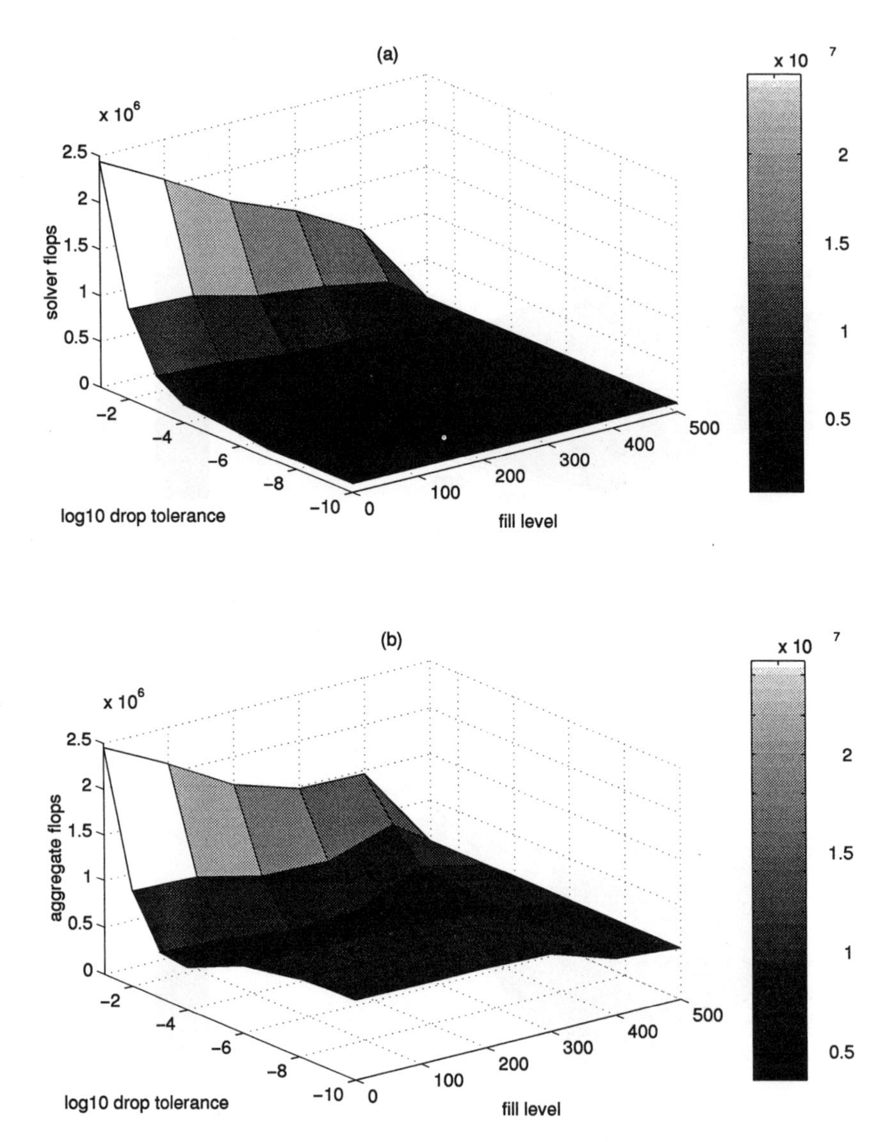

Figure 8: (continued) Effect of combining magnitude and position based fill in the incomplete Cholesky preconditioner applied to BUS494.

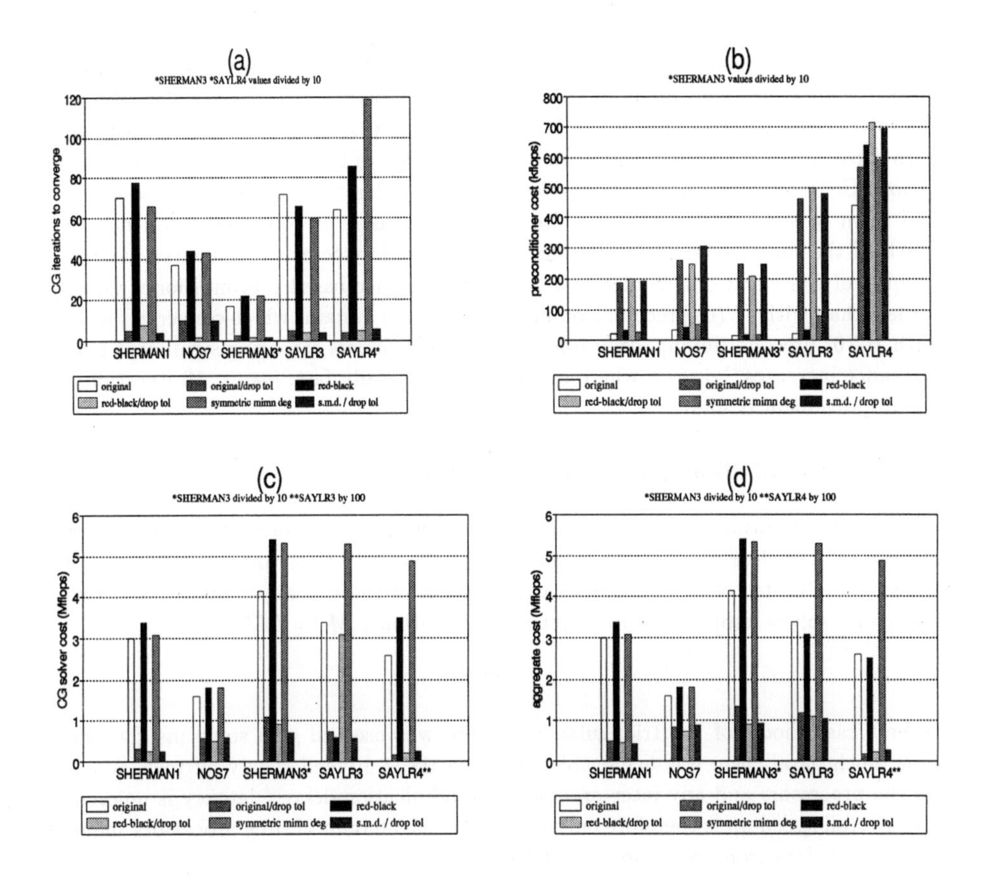

Figure 9: Performance of the drop tolerance based Cholesky factorisation preconditioner for various re-ordered Harwell-Boeing sparse matrices.

set at 10^{-3} and the $IC(0)$ preconditioner was used as a bench mark.

With only a small set of systems to look at, it is perhaps inevitable that the more extreme degradations in CG convergence would not be present. In fact, by examining fig. 9(a), it is clear that the re-ordered systems perform less well than the original only in the case of SAYLR4.

However what is also clear is the large reduction in iteration count gained through drop tolerance preconditioning in all cases. A corresponding reduction is also seen in the solver costs, shown in fig. 9(b). The preconditioner costs are small in comparison with those of the solver and reflect merely the amount of fill produced by the factorisation. Consequently, the aggregate costs differ only slightly from those of the solver and seem to suggest that use of a drop tolerance based preconditioner can negate much of the increase in CG iterations for re-ordered systems. It appears that the number of CG iterations required for the drop tolerance based preconditioner varies far less than with $IC(0)$ preconditioning. Therefore, in systems which converge slowly with $IC(0)$ preconditioning, drop tolerance preconditioners can bring significant performance increases, however where convergence is already fast, (say < 10 iterations) the improvement is slight.

6. The effect of fill on parallel implementations

Since the motivation behind re-ordering is to enhance the potential parallelism of the solver, it is clearly important to look at the effect fill has on any possible parallel speed up. The factor which limits parallelisation in the CG algorithm are the two triangular solves implicit in applying the preconditioner M $(= LL^T)$ thus:

$$s \Leftarrow \mathbf{M}^{-1} r$$

A best case model of the triangular solve step was used to gain some insight into how much parallelism was achievable. The matrix is divided evenly by rows across all the processors and any inter-processor communication delay / latency is ignored. The overall speed up is therefore only limited by the amount of time processors spend waiting for data from other processors (see [6] for details).

In the case of the Sherman1 system (fig. 10), the amount of inherent parallelism is small and it is only when the number of processors divides evenly into the number of rows, leading to an even load balance, that any real parallel speed up is seen. The 2-colour re-ordering (fig. 10), however provides near perfect scaling although recall that this is a best case model. In the case of the symmetric minimum degree re-ordering (fig. 10), the scaling is not as good and the efficiency asymptotically approaches 0.4 for greater than 40 processors.

The degradation in parallel performance caused by fill in is severe, as can be judged by fig. 11. Even in the case of the 2-colour re-ordering, the maximum achievable speed up is approximately 2.5. This figure is exceeded only slightly, albeit at much lower efficiency, in the case of the 2-colour re-ordering, whilst the effect of fill on the original system is to inhibit almost any potential parallelism.

Although the parallel performance results are disappointing for Sherman1, recent work tentatively suggests that for larger systems of similar origin, greater speed ups may be possible. In the case of a system derived from a 7-point p.d.e. approximation discretized over a $40 \times 40 \times 40$ grid ($n = 64000$) a speed up of 15 was predicted on 40 processors using the 2-colour re-ordering compared with 3.5 for Sherman1.

References

1. J. A. Meijerink and Van der Vorst, *Guidelines for the Usage of Incomplete Decompositions in Solving Sets of Linear Equations as They Occur in Practical Problems*, J. Comp. Phys., **44**, 134-155 (1981).
2. O. Axelsson, *A Survey of Preconditioned Iterative Methods for Linear Systems of Algebraic Equations*, BIT **25**, 166-187 (1985).
3. I. C. Smith, *Survey of Preconditioners for the Conjugate Gradient Method*, Internal report, (1995).
4. I. S. Duff, R. G. Grimes and J. G. Lewis, *Users' Guide for the Harwell-Boeing Matrix Collection (Release I)*, available via anonymous ftp from **orion.cerfacs.fr** in directory **/pub/harwell_boeing**, (1992).
5. I. Duff and G. Meurant, *The Effect of Ordering on Preconditioned Conjugate Gradients*, BIT **29**, 635-657 (1989).
6. Y. Saad, *Krylov Subspace Methods on Supercomputers*, SIAM J. Sci. Stat. Comp. **10**, 1200-1232 (1989).

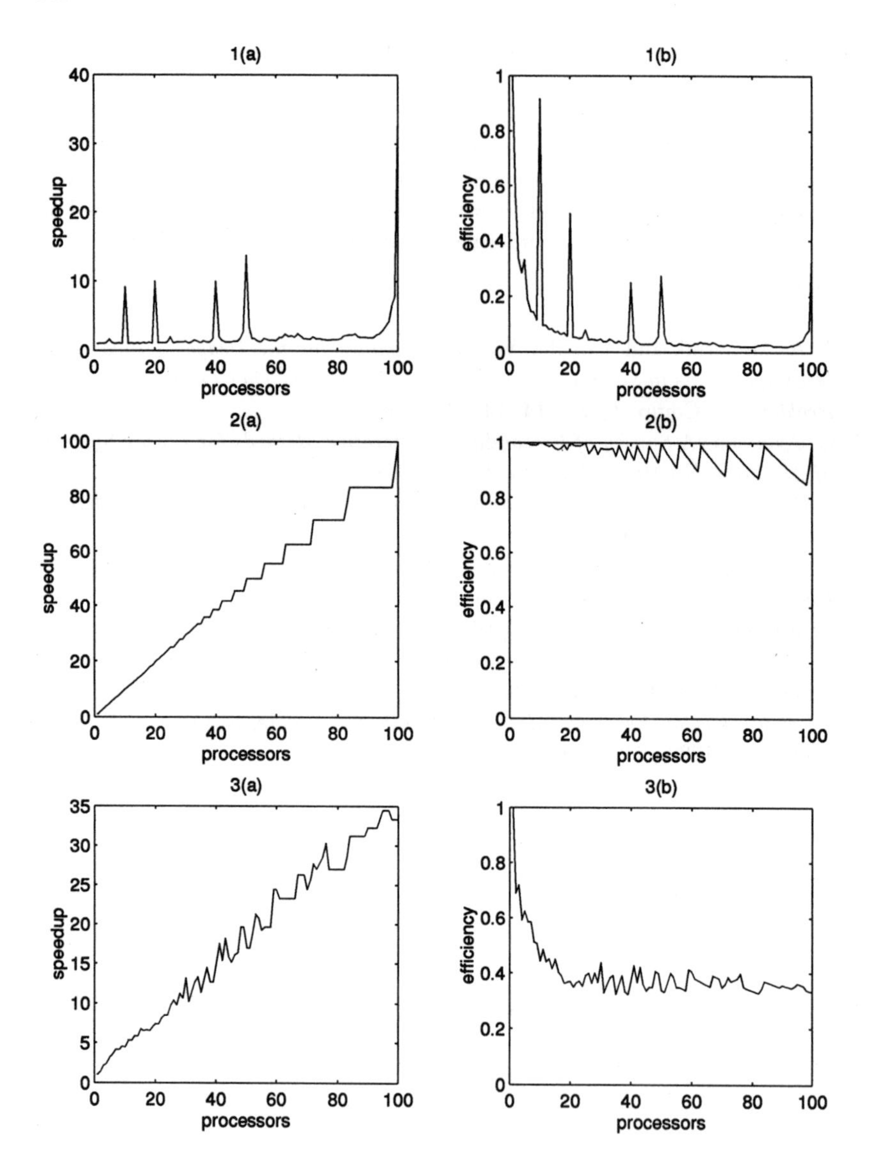

Figure 10: Theoretical maximum parallel performance using Sherman1: **(1)** original matrix, **(2)** 2 colour re-reordering, **(3)** symmetric minimum degree re-ordering.

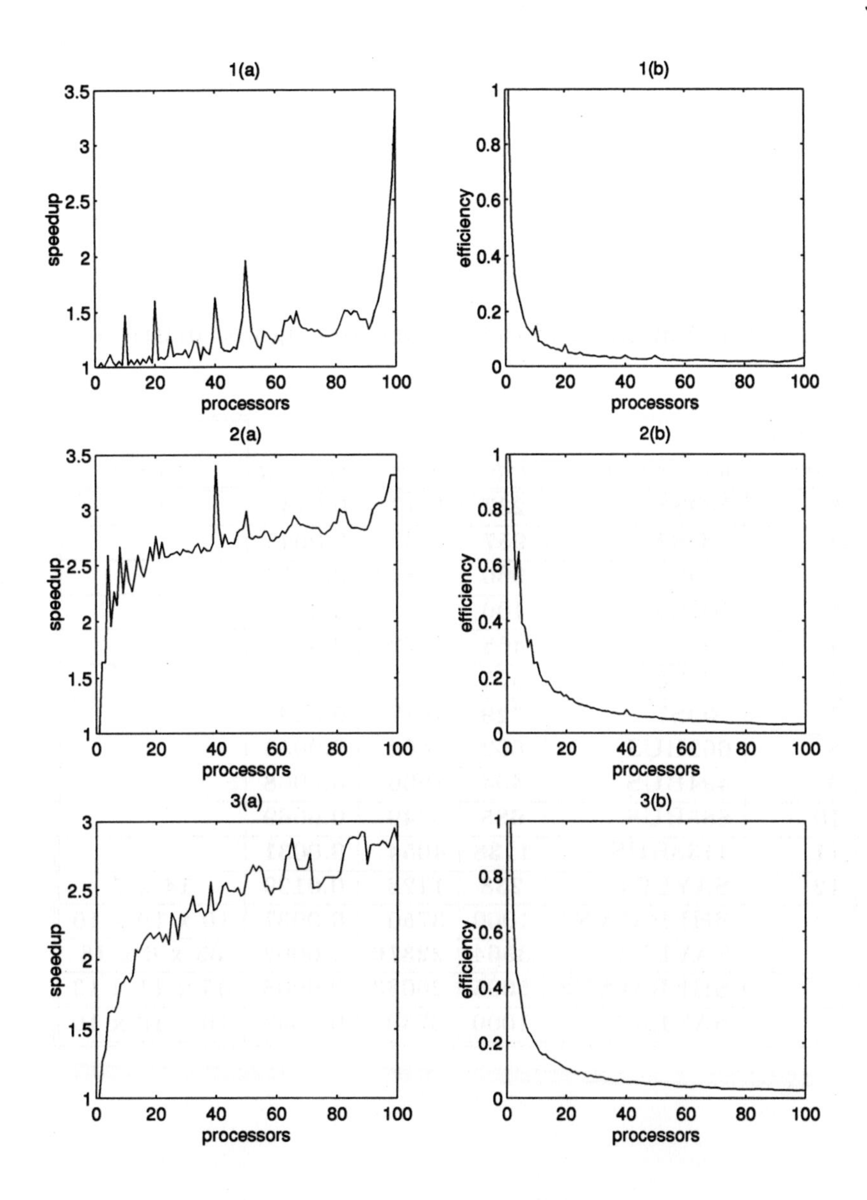

Figure 11: Theoretical maximum parallel performance using Sherman1: incomplete Cholesky factorisation with 10^{-3} drop tolerance **(1)** original matrix, **(2)** 2 colour re-reordering, **(3)** symmetric minimum degree re-ordering.

Appendix A 7. Summary of the Harwell-Boeing test matrices

matrix number	*matrix name*	*order*	*non zeros*	*sparsity*	*grid size*
1	**NOS1**	**237**	**1017**	**0.0142**	
2	**NOS2**	**957**	**4137**	**0.0045**	
3	**NOS3**	**960**	**15844**	**0.0172**	
4	**NOS4**	**100**	**594**	**0.0594**	
5	**NOS5**	**468**	**5172**	**0.0236**	
6	**NOS6**	**675**	**3255**	**0.0071**	
7	**NOS7**	**729**	**4617**	**0.0087**	
8	**662BUS**	**662**	**2464**	**0.0056**	
9	**494BUS**	**494**	**1666**	**0.0068**	
10	**685BUS**	**685**	**3249**	**0.0069**	
11	**1138BUS**	**1138**	**4054**	**0.0031**	
12	**SAYLR1**	**238**	**1128**	**0.0199**	**14 x 7**
13	**SHERMAN1**	**1000**	**3750**	**0.0037**	**10 x 10 x 10**
-	**SAYLR4**	**3564**	**22316**	**0.0002**	**33 x 6 x 18**
-	**SHERMAN3**	**5005**	**20083**	**0.0008**	**35 x 11 x 13**
-	**SAYLR3**	**1000**	**3750**	**0.0037**	**10 x 10 x 10**

CREATING AND COMPARING WAVELETS

GILBERT STRANG
Department of Mathematics, Massachusetts Institute of Technology
Cambridge, Massachusetts 02139, USA
E-mail: gs@math.mit.edu

ABSTRACT

This paper emphasizes two points about the design and application of filters and filter banks and wavelets:

- The algebra behind wavelet design is now quite simple.
- The comparison of two competing wavelets is still experimental and empirical.

As example we consider two particular 9/7 constructions (nine coefficients in analysis and seven in synthesis). Both are symmetric, so neither is orthogonal. Both have important advantages. We are unable to say which construction is better. The paper begins with the conditions on the filter coefficients for perfect reconstruction. That part is a brief summary of the exposition in Strang and Nguyen [1].

1. Introduction

This paper discusses two-channel filter banks, leading to wavelets. The structure involves four filters (which are just convolutions, see below). The analysis bank has a lowpass filter H_0 and a highpass filter H_1. The outputs y_0 and y_1 from those filters are "downsampled" by keeping only the even-numbered components:

$$\begin{aligned} y_0 &= H_0 x \quad \text{and} \quad v_0(n) = (\downarrow 2) y_0(n) = y_0(2n) \\ y_1 &= H_1 x \quad \text{and} \quad v_1(n) = (\downarrow 2) y_1(n) = y_1(2n). \end{aligned}$$

The full-length input vector x yields two half-length vectors v_0 and v_1. *That analysis step is inverted by the synthesis step.* The v's are "upsampled" to put zeros into their odd-numbered components. The results are filtered by F_0 and F_1, and their sum is the output $\hat{x}$:

$$\hat{x} = F_0(\uparrow 2)v_0 + F_1(\uparrow 2)v_1.$$

The intention is that $\hat{x} = x$.

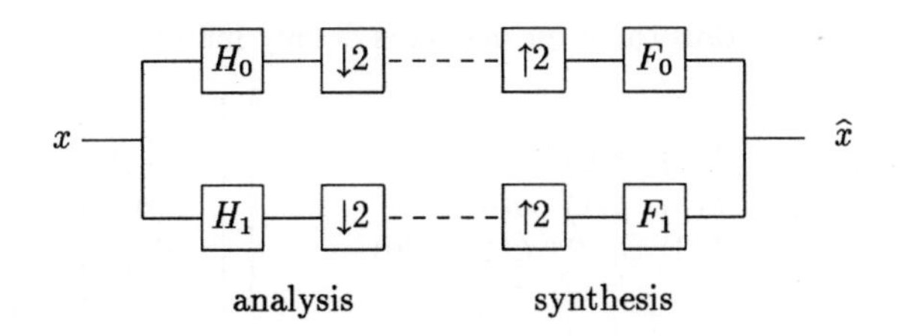

In that case the filter bank gives *perfect reconstruction.* The analysis wavelets and synthesis wavelets will be *biorthogonal.* So are the scaling functions (biorthogonality means $\int \tilde{\phi}(t-k)\phi(t-\ell)dt = \delta_{k\ell}$). All these functions come from solving two-scale equations, which involve t and $2t$. This second scale $2t$ corresponds to the $2n$ in downsampling.

We emphasize particularly how the properties of these functions in continuous time follow from the properties of the filters in discrete time. The heart of the construction is the choice of H_0 and F_0. These are given by the impulse responses h_0 and f_0, which are the vectors of filter coefficients: $h_0 = (h_0(0), \ldots, h_0(8))$ and $f_0 = (f_0(0), \ldots, f_0(6))$. Those choices determine H_1 and F_1 by a pattern of "alternating signs", $h_1(n) = (-1)^n f_0(n)$ and $f_1(n) = (-1)^{n+1} h_0(n)$.

We first determine the conditions on H_0 and F_0 (thus on h_0 and f_0) for perfect reconstruction. Then we explain the special importance of "zeros at $z=-1$" in the transfer functions, which are simply polynomials built from the filter coefficients:

$$H_0(z) = \sum_0^8 h_0(n)z^{-n} \qquad \text{and} \qquad F_0(z) = \sum_0^6 f_0(n)z^{-n}.$$

The algebra is all straightforward. We have a change of basis, produced by a *wavelet transform.* The components of v_0 and v_1 express the input vector x in the new basis. This transform can be applied again to the lowpass output v_0 that is normally most important. Scaling functions and wavelets appear in the limit of an infinite iteration. Four or five levels give a typical tree, in practice.

If successful, many components of the v's will be small. The signal is *compressed* by setting small coefficients to zero (not invertible!). The reconstructed output from the synthesis bank will no longer agree exactly with x. But if x and $\hat{x}$ are close, the input signal is now represented by a small number of components. In the new basis, the signal can be efficiently transmitted and stored. One important and unresolved difficulty is the meaning of the word "close".

2. The Conditions for Perfect Reconstruction

A filter is a convolution: $y(n) = \sum h(k)x(n-k)$. This linear transformation is represented by a Toeplitz matrix (meaning constant diagonals). The coefficient $h(k)$ appears along the kth subdiagonal. The input vector x is very long in practice and infinitely long in theory— thus the filter matrix is doubly infinite:

$$Hx = \begin{bmatrix} \cdot & \cdot & \cdot & \cdot & & & & \\ & h(3) & h(2) & h(1) & h(0) & & & \\ & & h(3) & h(2) & h(1) & h(0) & & \\ & & & h(3) & h(2) & h(1) & h(0) & \\ & & & & \cdot & \cdot & \cdot & \cdot \end{bmatrix} \begin{bmatrix} \cdot \\ x(-1) \\ x(0) \\ x(1) \\ \cdot \end{bmatrix} = \begin{bmatrix} \cdot \\ y(-1) \\ y(0) \\ y(1) \\ \cdot \end{bmatrix}.$$

Downsampling removes $y(-1)$ and $y(1)$. In the product $(\downarrow 2)H$, all the odd-numbered rows of H are removed:

$$(\downarrow 2)H = \begin{bmatrix} \cdot & \cdot & \cdot & \cdot & & & & \\ & h(3) & h(2) & h(1) & h(0) & & & \\ & & & h(3) & h(2) & h(1) & h(0) & \\ & & & & & \cdot & \cdot & \cdot & \cdot \end{bmatrix}.$$

Notice the *double shift* between rows. $(\downarrow 2)H$ is the fundamental operator in wavelet analysis (1×2 block Toeplitz matrix). When the two analysis filters $(\downarrow 2)H_0$ and $(\downarrow 2)H_1$ are combined, by interleaving rows of the two matrices, we get the block Toeplitz matrix (with 2×2 blocks) that represents the analysis bank:

$$H_b = \begin{bmatrix} h_0(3) & h_0(2) & h_0(1) & h_0(0) & & \\ h_1(3) & h_1(2) & h_1(1) & h_1(0) & & \\ & & h_0(3) & h_0(2) & h_0(1) & h_0(0) \\ & & h_1(3) & h_1(2) & h_1(1) & h_1(0) \\ & & & & \cdot & \cdot \end{bmatrix}$$

The inverse of H_b is the synthesis matrix F_b. *The key feature of these matrices is that both are banded.* In the language of signal processing, all filters are FIR (finite impulse response). Banded Toeplitz matrices with banded inverses are possible only because these are *block* matrices. The inverse of a polynomial $1 + z^{-1}$ is not a polynomial. But a matrix polynomial can have a polynomial inverse: for example

$$H_p(z) = \frac{1}{2}\begin{bmatrix} 1+z^{-1} & 1-z^{-1} \\ 1-z^{-1} & 1+z^{-1} \end{bmatrix}$$

has determinant z^{-1} and inverse

$$\frac{1}{2}\begin{bmatrix} z+1 & -z+1 \\ -z+1 & z+1 \end{bmatrix}.$$

This example illustrates two further points. First, the determinant is a monomial. Since we divide by the determinant, this monomial is the key to a polynomial inverse. Second, the inverse is anticausal (powers of z) when the original filters are causal (powers of z^{-1}). To keep all filters causal and all matrices lower triangular, the convention is to separate out the monomial determinant:

$$\begin{aligned} F_p(z) &= z^{-1}H_p^{-1}(z) \qquad &&\text{is causal,} \\ F_p(z)H_p(z) &= z^{-1}I &&\text{is a one–step delay.} \end{aligned}$$

This means that the output $\hat{x}(n)$ from the filter bank agrees with $x(n-1)$, not with $x(n)$. We still call this perfect reconstruction! In general the filters produce ℓ delays and the output is $\hat{x}(n) = x(n-\ell)$. The product of the lower triangular block

Toeplitz matrices F_b and H_b is the shift matrix that has identity blocks on the ℓth subdiagonal. Our real task is to find the conditions on the coefficients $f_0(n)$ and $h_0(n)$ for this to happen.

The algebra of convolution is summed up in the convolution rule. This transforms the matrix equation $y = Hx$ into a simpler equation for the associated polynomials. These polynomials are just multiplied: $Y(z) = H(z)X(z)$ is

$$\left(\sum y(n)z^{-n}\right) = \left(\sum h(k)z^{-k}\right)\left(\sum x(\ell)z^{-\ell}\right).$$

The terms z^{-k} and $z^{-(n-k)}$ in the factors give z^{-n} in the product. Thus the coefficient in $y(n)z^{-n}$ is the convolution $\sum h(k)x(n-k)$ that comes from matrix multiplication.

The algebra of $(\downarrow 2)$ and $(\uparrow 2)$ is almost as neat. For vectors,

$$(\uparrow 2)(\downarrow 2)\begin{bmatrix} y(-2) \\ y(-1) \\ y(0) \\ y(1) \\ y(2) \end{bmatrix} = (\uparrow 2)\begin{bmatrix} y(-2) \\ y(0) \\ y(2) \end{bmatrix} = \begin{bmatrix} y(-2) \\ 0 \\ y(0) \\ 0 \\ y(2) \end{bmatrix}.$$

For the z-transform $Y(z) = \sum y(n)z^{-n}$, only even powers remain. The result of downsampling and upsampling is the even part

$$\frac{1}{2}\Big(Y(z) + Y(-z)\Big) = \frac{1}{2}\Big(H(z)X(z) + H(-z)X(-z)\Big).$$

That term with $-z$ reflects aliasing. Two inputs can give the same output. The constant vector $y(n) \equiv 1$ and the alternating vector $y(n) = (-1)^n$ have the same even components, and therefore they look the same after downsampling.

The conditions for perfect reconstruction $\hat{x}(n) = x(n-\ell)$ come by following the signal through the filter bank. We do it in the z-domain, starting with $X(z)$. The lowpass channel yields $Y_0(z) = H_0(z)X(z)$. Then it takes the even part. Then it multiplies by $F_0(z)$. The highpass channel has 1 in place of 0, and we add:

$$\hat{X}(z) = \begin{matrix} \frac{1}{2}F_0(z)\Big(H_0(z)X(z) + H_0(-z)X(-z)\Big) \\ + \\ \frac{1}{2}F_1(z)\Big(H_1(z)X(z) + H_1(-z)X(-z)\Big) \end{matrix} = z^{-\ell}X(z).$$

The coefficient of $X(z)$ is $z^{-\ell}$ (no distortion, only a delay). The coefficient of $X(-z)$ must be zero (no aliasing in the final output). These are the PR conditions:

$$F_0(z)H_0(z) + F_1(z)H_1(z) = 2z^{-\ell} \quad (1)$$

$$F_0(z)H_0(-z) + F_1(z)H_1(-z) = 0. \quad (2)$$

It is the anti-aliasing equation (2) that leads to the "alternating sign" constructions, $h_1(n) = (-1)^n f_0(n)$ and $f_1(n) = (-1)^{n+1} h_0(n)$. In terms of polynomials, this is $H_1(z) = F_0(-z)$ and $F_1(z) = -H_0(-z)$. Then (2) is automatically satisfied, and (1) reduces to an equation for the *product filter* $P_0(z) = F_0(z)H_0(z)$:

$$P_0(z) - P_0(-z) = 2z^{-l}. \tag{3}$$

This is the key equation for perfect reconstruction.

Note that the left side of (3) is an odd function, so ℓ must be odd. The equation says that *the only odd term in* $P_0(z)$ *is* $z^{-\ell}$ *with coefficient 1.* We can separate the design of a PR filter bank into three steps:

1. Choose a polynomial $P_0(z)$ that satisfies (3).
2. Factor $P_0(z) = F_0(z)H_0(z)$.
3. Choose $H_1(z) = F_0(-z)$ and $F_1(z) = -H_0(-z)$.

This simplicity is what was meant in our first point in the abstract. It is deceptive, because it does not indicate what makes one design better than another. Part of the answer (only part!) is in the number of zeros at $z = -1$.

3. Zeros at -1: Approximation and Vanishing Moments

For a lowpass filter, the polynomial

$$H(z) = \sum h(n)z^{-n} = \sum h(n)e^{-in\omega}$$

is near zero at the highest frequency $\omega = \pi$. In the z-plane, with $z = e^{i\omega}$, this is the point $z = e^{i\pi} = -1$. Then the multiplication $Y(z) = H(z)X(z)$ ensures that high frequencies in x (at or near $\omega = \pi$) are nearly annihilated in y. The lowest frequency $\omega = 0$ (or $z = +1$) passes through the filter provided $H(1) = 1$. Hence the name "lowpass".

This condition $H(-1) = 0$ is fundamental in wavelet theory. It must hold exactly, not just approximately, to have any chance of continuous scaling functions and wavelets. The scaling function $\phi(t)$ solves the *dilation equation* or *refinement equation:*

$$\phi(t) = 2\sum h(k)\phi(2t-k). \tag{4}$$

The Fourier transform of this equation is

$$\hat{\phi}(\omega) = \sum h(n)e^{-i\omega n/2}\hat{\phi}(\omega/2) = H(e^{i\omega/2})\hat{\phi}(\omega/2). \tag{5}$$

Periodicity gives $H(e^{ik\pi}) = 0$ for all odd integers k. Then equation (5) yields $\hat{\phi}(2\pi n) = 0$ for every $n \neq 0$. This is the first of the so-called "Strang–Fix conditions", implying that the translates of $\phi(t)$ add to a constant, which we may normalize to 1:

$$\sum_{-\infty}^{\infty} \phi(t-n) \equiv 1. \tag{6}$$

This conclusion could also be reached directly from the dilation equations

$$\phi(t-n) = 2\sum h(k)\phi(2t-2n-k)$$

by summing on n. The sum $S(t)$ is 1-periodic, and we restrict to $0 \le t < 1$. Separating even from odd k leads to

$$S(t) = \left(2 \sum_{\text{even } k} h(k)\right) S(2t) + \left(2 \sum_{\text{odd } k} h(k)\right) S(2t-1). \tag{7}$$

The coefficients are $H(1) \pm H(-1)$. Thus both coefficients are 1. The sum satisfies the "Haar equation" $S(t) = S(2t) + S(2t-1)$, whose solution on the period interval $[0,1)$ is $S(t) \equiv 1$.

We often use the simplified notation $H(\omega)$ to replace $H(e^{i\omega})$. Note that the transform of $\phi(2t)$ is $\frac{1}{2}\hat{\phi}(\omega/2)$. Then (4) applies recursively to $\omega/2, \omega/4, \ldots$ and leads to an infinite product formula for $\hat{\phi}(\omega)$:

$$\hat{\phi}(\omega) = H(\frac{\omega}{2})H(\frac{\omega}{4})\hat{\phi}(\frac{\omega}{4}) = \cdots = \prod_{1}^{\infty} H(\frac{\omega}{2^j}). \tag{8}$$

This product converges for each ω. But is it the transform of a smooth $\phi(t)$?

Summary

The special zero $H(-1) = 0$ leads to $S(t) \equiv 1$. Constant polynomials can be produced from translates of $\phi(t)$. By standard results in approximation theory, these translates give at least first-order approximation to any smooth function $f(t)$:

$$\|f(t) - \sum a_k \phi(t-k)\| \le C\|f'(t)\|,$$

for suitable a_k. The particular choice $a_k = f(k)$ is the "quasi–interpolate". When the mesh size changes from 1 to h, by rescaling t, the familiar factor h appears on the right side and the approximation error is $O(h)$.

All this followed from a simple zero $H(-1) = 0$. *Suppose that the zero at $z = -1$ has higher multiplicity p.* Then the corresponding steps lead to the Strang–Fix condition of order p: the derivatives are $\hat{\phi}^{(j)}(2\pi n) = 0$ for $n \ne 0$ and $j < p$. This is equivalent to pth-order approximation by the translates of $\phi(t)$.

Theorem 1 *A pth-order zero of $H(z)$ at $z = -1$ implies these properties, provided $\phi(t)$ exists in L^2:*

1. *The translates $\phi(t-n)$ can reproduce all polynomials of degree less than p.*

2. *The translates give pth-order approximation of a smooth $f(t)$:*

$$\|f(t)-\sum a_k\phi(t-k)\| \le C_p\|f^{(p)}(t)\| \tag{9}$$

for suitable a_k. Again the rescaling of t produces the factor h^p in (9).

3. *The wavelets that are orthogonal to the scaling functions have p vanishing moments.*

Orthogonality to the scaling functions $\phi(t-k)$ means orthogonality to their combinations $1, t, \ldots, t^{p-1}$. This gives the vanishing moments:

$$\int_{-\infty}^{\infty} t^m \tilde{w}_{jk}(t)dt = \int_{-\infty}^{\infty} t^m 2^{j/2}\tilde{w}(2^j t-k)dt = 0 \qquad \text{for } m < p.$$

Note! The convention is that $\phi(t)$ without the tilde is the *synthesis* function. Thus the coefficients in (4) and (5) should be written $f_0(k)$ instead of $h(k)$. The *analysis* functions $\tilde{\phi}(t)$ and $\tilde{w}(t)$ are constructed from

$$\begin{aligned}\tilde{\phi}(t) &= 2\sum h_0(k)\tilde{\phi}(2t-k)\\ \tilde{w}(t) &= 2\sum h_1(k)\tilde{\phi}(2t-k).\end{aligned}$$

Biorthogonality between tilde and non-tilde follows from the perfect reconstruction conditions, at every step of the iteration that solves the dilation equation. This iteration is the "cascade algorithm":

$$\tilde{\phi}^{(i+1)}(t) = 2\sum h_0(k)\tilde{\phi}^{(i)}(t). \tag{10}$$

The initial $\tilde{\phi}^{(0)}(t)$ is the box function on [0,1]. *But the iteration may not converge.* There is a condition on the eigenvalues of the matrix $T = (\downarrow 2)2HH^T$:
Condition E: All eigenvalues satisfy

$$|\lambda(T)| < 1$$

except for a simple eigenvalue $\lambda = 1$.

This gives L^2 convergence of the cascade algorithm [1]. The scaling function basis and the wavelet basis are stable [2]. Furthermore the smoothness of $\tilde{\phi}(t)$ and $\tilde{w}(t)$ are determined [3] by the spectral radius $\rho = |\lambda_{\max}(T)|$, when we exclude the special eigenvalues $\lambda = 1, \frac{1}{2}, \frac{1}{4}, \ldots, (\frac{1}{2})^{2p-1}$ that are automatic from the pth order zero of $H(z)$ at $z=-1$. The number of derivatives of $\tilde{\phi}(t)$ and $\tilde{w}(t)$ in L^2 is given by

$$s_{\max} = -\log\rho/\log 4. \tag{11}$$

Each additional factor $(1+z^{-1})/2$ in $H(z)$ increases p by 1 and $s_{\max}$ by 1. The new $\tilde{\phi}(t)$ is just the convolution of the old $\tilde{\phi}(t)$ with the box function. The perfect examples are *B-splines*, which come from the "pure" filter

$$H(z) = \left(\frac{1+z^{-1}}{2}\right)^p.$$

The B–spline of degree $p-1$ has $s_{\max} = p - 1/2$. Of course the B–splines are not biorthogonal to themselves (except for the box function when $p = 1$). The product $F_0(z)H_0(z)$ is allowed only one odd power $z^{-\ell}$. We now create filters with this perfect reconstruction property, and compare the functions that come out of the cascade algorithm.

4. Three Choices of 9/7 Symmetric Biorthogonal Filters

The most popular filters are constructed by factoring a Daubechies polynomial of degree $4p-2$:

$$D_{4p-2}(z) = (1+z^{-1})^{2p} Q_{2p-2}(z).$$

$Q(z)$ is needed so that $D(z)$ will have only one odd power (with coefficient 1). The unique polynomial Q_{2p-2} of lowest degree $2p-2$ is the Daubechies choice. Examples are

$$\begin{aligned} D_6(z) &= (1+z^{-1})^4\,(-1+4z^{-1}-z^{-2})/32 \\ D_{14}(z) &= (1+z^{-1})^8\,(-5+40z^{-1}-131z^{-2}+208z^{-3}-131z^{-4}+40z^{-5}-5z^{-6})/2^{12}. \end{aligned}$$

The polynomial Q_{2p-2} comes directly from the binomial expansion of $(1-y)^{-p}$, truncated after p terms [1]. This polynomial of degree $(p-1)$ in $y = (2-z-z^{-1})/4$ becomes a polynomial of degree $2p-2$ in z^{-1} (after shifting by z^{1-p}).

Here are three filter banks with interesting properties. Many others are interesting too!

1. The FBI 9/7 filters were constructed by Daubechies and chosen by the FBI in digitizing fingerprints. $H_0(z)$ and $F_0(z)$ are factors of $D_{14}(z)$, each with $p = 4$ zeros at $z = -1$. (Thus $(1+z^{-1})^8$ is split down the middle.) The other factors of degree 4 in H_0 and 2 in F_0 are chosen to preserve the symmetry of $Q_6(z)$. Real reciprocal roots z and $1/z$ go into F_0, and complex reciprocal roots $z, \bar{z}, 1/z, 1/\bar{z}$ go into H_0. *The coefficients are not rational*, and we give (inadequately) two decimals of the filter coefficients:

$$\begin{aligned} h_0 &= [\;\; 0.03 \;\; -0.02 \;\; -0.08 \;\; 0.27 \;\; 0.60 \;\; 0.27 \;\; -0.08 \;\; -0.02 \;\; 0.03] \\ f_0 &= [\; -0.05 \;\; -0.03 \;\; 0.30 \;\; 0.56 \;\; 0.30 \;\; -0.03 \;\; -0.05]. \end{aligned}$$

These filters are frequently used in image compression. Our normalization is

$$\sum h_0(k) = \sum f_0(k) = 1.$$

Then an extra $\sqrt{2}$ is needed in all four filters $H_0,\ H_1,\ F_0,\ F_1$, by Eq.(1).

2. The spline 9/7 filters come from a different factorization of $D_{14}(z)$, in which $F_0(z) = (1+z^{-1})^6/64$. Then $\phi(t)$ is the extremely smooth B–spline of degree 5 (with 4 continuous derivatives). But this leaves only two zeros at -1 for $H_0(z)$, which must swallow $Q_6(z)$ whole:

$$\begin{aligned} h_0 &= [\ -5 \quad 30 \quad -56 \quad -14 \quad 154 \quad -14 \quad -56 \quad 30 \quad -5]/64 \\ f_0 &= [\ 1 \quad 6 \quad 15 \quad 20 \quad 15 \quad 6 \quad 1]/64. \end{aligned}$$

This is a foolish choice. We will see that there is no L_2 solution to the dilation equation involving h_0. Condition E fails.

3. The "binary" 9/7 filter bank selects $F_0(z) = D_6(z)$. This choice (by the physicist Tomas Arias of M.I.T.) surprised the author. In itself it has one odd power z^{-3}—which leads to the useful interpolating property $\phi(n) = \delta(n-3)$. But it is the product F_0H_0 that must have one odd power, and what is H_0? The biorthonormal filter is needed in compression, if not in physics.

The lowest degree is 8 for a symmetric H_0 with a zero at $z=-1$. A direct calculation gives

$$\begin{aligned} h_0 &= [\ 1 \quad 0 \quad -8 \quad 16 \quad 46 \quad 16 \quad -8 \quad 0 \quad 1]/64 \\ f_0 &= [\ -1 \quad 0 \quad 9 \quad 16 \quad 9 \quad 0 \quad -1]/32. \end{aligned}$$

Notice that all coefficients are integers divided by powers of 2. This means perfect arithmetic, and fast execution on a chip.

We communicated this construction to Wim Sweldens who responded that he had already found the same binary filters. His method of "lifting" is extremely useful[5]. Starting from one admissible pair, in this case $H_0 \equiv 1$ and $F_0 = D_6$, the choice $H_{\text{new}}(z) = H_0(z) + F_0(-z)S(z^2)$ also gives perfect reconstruction. The unrestricted S allows our filter (Wim's filter) to have two zeros at $z=-1$. With four zeros the lengths become 13/7, and our compression experiments were less satisfactory—we don't know why.

5. Smoothness and Behavior of the Filters

We come now to a *comparison* of the three examples: FBI, spline and binary. Various measures are easy to compute. They give partial information:

Zeros at $z=-1$: 4/4, 2/6, 2/4 (thus binary loses)
Smoothness $s_{\max}$ of $\tilde{\phi}(t)$ and $\phi(t)$: 1.4/2.1, −2.2/5.5, 0.59/2.44.

The largest non–special eigenvalues of the matrix T for the binary h_0 and f_0 were 0.4394 and 0.0339. The smoothness $s_{\max} = 0.59$ and 2.44 came directly from Eq. (11).

The largest non-special eigenvalue for the foolish choice h_0, biorthonormal to the quintic B-spline filter, was 21.314. The function $\tilde{\phi}(t)$ is a wild distribution (−2.2 derivatives in L_2). This example is *not* to show that splines are a poor choice—they are often very good. But we must keep enough zeros at $z = -1$ in both $H_0(z)$ and $F_0(z)$. When f_0 with binomial coefficients gives a spline, stability and smoothness may require a longer h_0 than would be needed for perfect reconstruction.

A third important quantity is the *coding gain*, to give the expected compression for inputs that have Markov correlation 0.95 between neighboring pixels. None of those measures is totally consistent with human visual perception. Therefore filters are generally chosen for good looks on well-known images.

We were able to compare the FBI 9/7 with the binary 9/7 on a "boats" image [1].

25:1 compression			50:1 compression	
FBI	binary		FBI	binary
40.55	40.14	MEAN SQUARE ERROR	84.55	85.92
32.05	32.10	PEAK SIGNAL/NOISE RATIO	28.86	28.79
43.71	45.07	MAXIMUM ERROR	68.81	81.99
OK	better	PERCEPTUAL QUALITY	OK	better

Note the slight advantage of the FBI, objectively. Note the equally slight advantage of the binary filters, subjectively. We believe that the binary property and the interpolation property may be significantly useful in applications.

Acknowledgement I would like to add a less conventional tribute to Ron Mitchell by thanking him for teaching my sons to play "football".

6. References

1. Gilbert Strang and Truong Nguyen, *Wavelets and Filter Banks*, Wellesley–Cambridge Press (1996).
2. Albert Cohen and Ingrid Daubechies, A stability criterion for biorthogonal wavelet bases and their related subband coding schemes, Duke Math. J., 68:313–335 (1992).
3. Lars Villemoes, Wavelet analysis of refinement equations, SIAM J. Math. Anal., 25:1433–1466 (1994).
4. Gilbert Strang, Eigenvalues of $(\downarrow 2)H$ and convergence of the cascade algorithm, IEEE Trans. on Signal Processing (1996).
5. Wim Sweldens, The lifting scheme: a custom–design construction of biorthonormal wavelets, Appl. Comput. Harm. Anal., (To appear).

THE NUMERICAL ANALYSIS OF SPONTANEOUS SINGULARITIES IN NONLINEAR EVOLUTION EQUATIONS

YVES TOURIGNY
School of Mathematics, University of Bristol
Bristol BS8 1TW, United Kingdom
E-mail: y.tourigny@bristol.ac.uk

ABSTRACT

The paper describes an extension to differential equations of Lyness' algorithm for the approximation of the Taylor coefficients of an analytic function. The computational basis of this new procedure is provided by standard time-stepping schemes employed in the complex domain. It is shown that the resulting "discrete" coefficients are exponentially accurate approximations of the true coefficients. By combining this procedure with the ratio test and its variants, one obtains a flexible and reliable tool for the investigation of singularities arising from evolution equations. An application to a semilinear parabolic problem is presented.

1. Introduction

Consider the Cauchy problem

$$\frac{dx}{dt} = f(x,t), \quad x(0) = x_0, \tag{1.1}$$

where f is analytic at $(x_0, 0)$. One of the remarkable features of such nonlinear evolution equations is the fact that solutions often develop singularities whose location cannot be determined a priori by a mere inspection of the data (cf. Bender and Orszag[2] for an elementary introduction to the topic). Numerical investigations of such singularities has become increasingly important, both in the theory of differential equations (characterization of integrable systems[11], blow-up[10]) and in applications such as Fluid Dynamics (vortex sheets, free-surface flows[1]).

For solutions that are analytic in time, one way of "deciphering" the singularities is to process the Taylor coefficients of the solution. Thus, if the solution of eq. (1.1) is given locally by the expansion

$$x(t) = \sum_{n=0}^{\infty} x_n t^n, \quad |t| < R,$$

one can for example determine the radius of convergence R (or, equivalently, the distance from the origin to the nearest singularity in the complex plane) by using Hadamard's formula

$$\frac{1}{R} = \overline{\lim_{n\to\infty}} \sqrt[n]{|x_n|}.$$

This formula is admittedly somewhat unwieldy! For solutions with non-vanishing real coefficients and a singularity $t_\star$ along the real axis, the ratio test

$$t_\star = \lim_{n\to\infty} \frac{x_{n-1}}{x_n},$$

is commonly used. Many variants of such formulas may be found in Tourigny and Grinfeld[13] and in the references therein.

This approach was used very successfully by Chang and Corliss in their implementation of the Taylor series method for ordinary differential equations.[3,4] ATOMFT, the latest version of this software, comprises a translator that accepts a statement of the differential problem and produces a Fortran object code. This code consists of recurrence relations that are used to generate the Taylor coefficients at each integration step. Importantly, the method includes a subroutine that uses ratio-test-like estimates in order to determine the type and position of the singularities on the circle of convergence. This information may be used to estimate the local error at each step, or even to vault over the singularities if those lie on the path of integration.[5]

While the Taylor series method is competitive for small systems of equations, many problems are treated more effectively by other discretization techniques. In such circumstances, although Taylor coefficients would be most useful in providing information on the singularities of the solution, the idea of using a translator to produce the recurrence relations for the coefficients, as in the ATOMFT software, is impractical.

This paper is devoted to the description, analysis and application of a new procedure for the calculation of approximate Taylor coefficients. This procedure may be viewed as a means of obtaining a "discrete version" of the recurrence relation satisfied by the true coefficients of the solution x of the evolution equation (1.1). While the technique retains the asymptotic exponential accuracy associated with the classical Taylor series method, its implementation requires nothing beyond the use of standard time-stepping schemes and the evaluation of f (and, possibly, of its Jacobian matrix) at points in the complex plane.

In order to motivate the method, let us assume that an approximation of x is required in the range $|t| \le T < R$. The Taylor coefficients of x may be conveniently expressed by making use of Cauchy's formula for a circular contour of radius T. This yields

$$x_n = \frac{x^{(n)}}{n!}(0) = \frac{1}{T^n}\left(\frac{1}{2\pi}\int_0^{2\pi} x(Te^{i\theta})\, e^{-in\theta} d\theta\right). \tag{1.2}$$

The expression within brackets is the nth Fourier coefficient of the periodic function $x(Te^{i\theta})$. This observation forms the basis of an algorithm due to Lyness[9] which yields, via the Fast Fourier Transform, approximations of the Taylor coefficients of an analytic function from a sample of its values at equally-spaced points along the complex circle $|t| = T$.

In our context however, the values assumed by the function x on that circle are not known beforehand. Standard time-stepping schemes, employed along the complex circle, could possibly be used to provide the missing values. Unfortunately, the analysis of Tourigny and Grinfeld[13] reveals that a naive implementation of this idea yields an approximation of the (normalized) Taylor coefficients that achieves a poor *relative* accuracy. More precisely, for a time-stepping scheme of order p with a time step of size Δt, only $O(p \log \frac{1}{\Delta t})$ of the normalized coefficients may be approximated in this way to within a specified accuracy tolerance as $\Delta t \to 0$. This is a serious drawback because many more coefficients may typically be required to extract reliable information on the singularities.

We shall instead aim at setting up a discrete equation in order to define the approximate coefficients implicitly. Let us therefore introduce the notation

$$u(\theta) = x(Te^{i\theta}) = \sum_{n=0}^{\infty} u_n e^{in\theta}$$

where, according to (1.2),

$$u_n = T^n \frac{x^{(n)}}{n!}(0).$$

Using Cauchy's formula for the function $f(x(t), t)$, we readily infer from (1.1) that

$$u_0 = x_0 \tag{1.3}$$

and, for $n = 1, 2, \ldots$,

$$n\, u_n = T \frac{1}{2\pi} \int_0^{2\pi} f(u(\theta), Te^{i\theta})\, e^{-i(n-1)\theta} d\theta. \tag{1.4}$$

In the remainder of this paper, we shall analyse a discretization of (1.3), (1.4) that provides an approximation of the first N Fourier coefficients of u, and hence of the normalized Taylor coefficients of x. It will be shown that the resulting procedure is stable, convergent and, in a sense that will be made clear, exponentially accurate as $N \to \infty$. The analysis relies on a stability theory expounded by López–Marcos and Sanz–Serna[8]. The method and results of this paper can cater for systems of more than one differential equation. However, for notational convenience, the analysis will consider only the scalar case. In §2, we define the discrete scheme and prepare the ground for the derivation of the main results in §3. §4 contains a brief discussion of some important practical aspects of the method. In particular, we explain how the discrete solution may be found via Newton's method at a computational cost of $O(M^2 N^2)$ operations per iteration, where M is the number of differential equations that make up the system. Finally, in §5, the procedure is applied to a semilinear parabolic problem featuring solutions that blow up in finite time.

2. The discretization

We introduce the discrete space

$$S_h = \text{Span}\{1, e^{i\theta}, \ldots, e^{i(N-1)\theta}\},$$

where $h = \frac{1}{N}$ and N is a positive integer. For an arbitrary element $v_h \in S_h$, we shall use the representation

$$v_h = \sum_{n=0}^{N-1} v_{hn} e^{in\theta}.$$

The mapping $\Phi_h : S_h \to S_h$ is defined by

$$(\Phi_h(v_h))_0 = v_{h0} - x_0 \tag{2.1}$$

and

$$(\Phi_h(v_h))_n = n\, v_{hn} - T\frac{1}{N}\sum_{j=0}^{N-1} f(v_h(\theta_j), Te^{i\theta_j})\, e^{-i(n-1)\theta_j} \tag{2.2}$$

for $n = 1, \ldots, N-1$. We propose to construct an approximation $U_h \in S_h$ of u by solving the discrete equation

$$\Phi_h(U_h) = 0. \tag{2.3}$$

This construction yields approximate "normalized" Taylor coefficients which may be used for a variety of purposes. In order to discuss the convergence properties of the discretization, it is convenient to introduce the discrete Fourier transform $u_h \in S_h$ of u. Its coefficients, usually referred to as the *discrete Fourier coefficients* of u, are given explicitly by

$$u_{hn} = \frac{1}{N}\sum_{j=0}^{N-1} u(\theta_j)\, e^{-in\theta_j}.$$

It should be observed that u_h interpolates u at the N equally-spaced points $\theta_j = j2\pi/N$, $j = 0, \ldots, N-1$. The relationship between exact and discrete Fourier coefficients may be expressed by the relation

$$u_{hn} - u_n = \sum_{k=1}^{\infty} u_{n+kN},$$

where the series on the right-hand side is the so-called aliasing error.[12] The following result, whose proof follows immediately from Cauchy's formula, will often be used in the course of the analysis.

Lemma 1 *Let z be a function analytic in the disk $|t| \leq r$, where $r > T$. Let w_n and w_{hn} denote respectively the exact and discrete Fourier coefficients of the periodic function $w(\theta) = z(Te^{i\theta})$. Then, for $n = 0, 1, \ldots, N-1$, we have*

$$|w_n - w_{hn}| \leq \frac{1}{1-\lambda} \max_{|\tau|=r} |z(\tau)| \, \lambda^{n+N},$$

where $\lambda = T/r$.

This lemma suggests the choice of discrete norm

$$\|v_h\| = \max_{0 \leq n \leq N-1} \lambda^{-\alpha n} |v_{hn}| \tag{2.4}$$

for elements of S_h. In this expression, $\alpha \in (0,1)$, $\lambda = T/r$ and r is an arbitrary number in (T, R). It follows from Lemma 2.1 that, in this discrete norm, the difference between u_h and the Nth partial Fourier sum of u is $O(\lambda^N)$. Hence, the convergence of the discretization (2.3) may be investigated by considering the difference $U_h - u_h$. The analysis will rely on the general theory developed by López–Marcos and Sanz–Serna, which may thus be summarised:

Theorem 1 *For $R_h \in (0, \infty]$, let $B(u_h, R_h) \in S_h$ denote the ball of radius R_h centered at u_h. Assume that the following conditions hold:*

(1) There exists positive constants h_0 and L such that, for $h \leq h_0$, the inverse of the Fréchet derivative $\Phi'_h(u_h)$ exists and

$$\|\Phi'_h(u_h)^{-1}\| \leq L.$$

(2) There exists a constant Q, with $0 \leq Q < 1$, such that, for h sufficiently small, and for each $v_h \in B(u_h, R_h)$, we have

$$\|\Phi'_h(v_h) - \Phi'_h(u_h)\| \leq Q/L.$$

(3)

$$\lim_{h \to 0} \|\Phi_h(u_h)\| = o(R_h).$$

Then, for h sufficiently small, the discrete equation (2.3) is uniquely solvable in $B(u_h, R_h)$ and

$$\|u_h - U_h\| \leq L/(1-Q) \|\Phi_h(u_h)\|.$$

The essence of this result is that a consistent discretization with a continuous derivative and a stable linearization is convergent. It is thus closely related to the work of Keller.[7] The distinguishing feature, which we shall draw upon, is that, subject to condition (3), the present theory allows the size of the ball $B(u_h, R_h)$ to shrink to zero as $h \to 0$.

3. Error bounds

In this section, we simply work through each of the three conditions which together form the hypothesis of Theorem 2.1. For ease of exposition, we shall restrict our attention to the scalar case $M = 1$ although it should be clear that the extension of our results to the general case is completely straightforward. In the remainder, the symbol C will be used to denote a generic positive constant, not necessarily the same at each occurrence, independent of h but which may at times depend on the solution x of Eq. (1.1) and on the parameters $\alpha \in (0,1)$ and $\lambda \in (0,1)$ in (2.4).

Proposition 1 (Stability of the Linearization) *There exists a constant $L = L(r,\alpha)$ and an integer $N_0 = N_0(r,\alpha)$ such that, for $N > N_0$, we have*

$$\|v_h\| \le L\|\Phi_h'(u_h)v_h\| \quad \forall\, v_h \in S_h.$$

Proof. Let $v_h \in S_h$. By definition of the Fréchet derivative, we have

$$(\Phi_h'(u_h)v_h)_0 = v_{h0}$$

and, for $n = 1, \dots, N-1$,

$$(\Phi_h'(u_h)v_h)_n = n\, v_{hn} - T\sum_{m=0}^{N-1} v_{hm}\left(\frac{1}{N}\sum_{j=0}^{N-1} f_x(u(\theta_j), Te^{i\theta_j})\, e^{-i(n-m-1)\theta_j}\right),$$

where we have used the fact that u_h is the interpolant of u. Let $b_h \in S_h$ denote the discrete Fourier transform of the function $f_x(u(\theta), Te^{i\theta})$. We may rewrite the above equality as

$$n\, v_{hn} = (\Phi_h'(u_h)v_h)_n + T\left(\sum_{m=0}^{n-1} v_{hm} b_{h(n-1-m)} + \sum_{m=n}^{N-1} v_{hm} b_{h(N+n-1-m)}\right).$$

Multiplying this equality by $\lambda^{-\alpha n}$ and using the estimate

$$|b_{hn}| \le \frac{1}{1-\lambda}\max_{|\tau|=r}|f_x(x(\tau),\tau)|\,\lambda^n, \quad n = 0,1,\dots,N-1,$$

we infer that

$$n\,\lambda^{-\alpha n}|v_{hn}| \le \lambda^{-\alpha n}|(\Phi_h'(u_h)v_h)_n| +$$
$$+ C\left(\max_{0\le m\le n-1}\lambda^{-\alpha m}|v_{hm}|\sum_{m=0}^{n-1}\lambda^{(1-\alpha)(n-m)} + \lambda^N\|v_h\|\sum_{m=n}^{N-1}\lambda^{(1-\alpha)(n-m)}\right).$$

Note that, since $\alpha \in (0,1)$, we have

$$\sum_{m=n}^{N-1} \lambda^{(1-\alpha)(n-m)} \leq \int_0^N \lambda^{(\alpha-1)s} ds \leq \frac{\lambda^{(\alpha-1)N}}{(\alpha-1)\log\lambda}.$$

Hence, we obtain

$$n\,\lambda^{-\alpha n}|v_{hn}| \leq \lambda^{-\alpha n}|(\Phi_h'(u_h)v_h)_n| + C_\star \lambda^{\alpha N}\|v_h\| + C_\star \max_{0\leq m\leq n-1} \lambda^{-\alpha m}|v_{hm}|. \tag{3.1}$$

We shall henceforth reserve the particular symbol $C_\star$ for the constant appearing in (3.1). We introduce the sequence $\{L_n\}$ which is defined recursively via $L_0 = 1$ and

$$L_n = \max\left\{L_{n-1}, \frac{1}{n}(1 + C_\star L_{n-1})\right\}. \tag{3.2}$$

Let us prove by induction that, for $n = 0, 1, \ldots, N-1$, we have

$$\max_{0\leq m\leq n} \lambda^{-\alpha m}|v_{hm}| \leq L_n \left(\max_{0\leq m\leq n} \lambda^{-\alpha m}|(\Phi_h'(u_h)v_h)_m| + C_\star \lambda^{\alpha N}\|v_h\|\right). \tag{3.3}$$

The result is obviously true for $n = 0$. Let us assume that it holds for $n-1$. For n, we have, by virtue of (3.1) and the induction hypothesis,

$$\begin{aligned} n\,\lambda^{-\alpha n}|v_{hn}| &\leq \left(\lambda^{-\alpha n}|(\Phi_h'(u_h)v_h)_n| + C_\star \lambda^{\alpha N}\|v_h\|\right) + \\ &+ C_\star L_{n-1}\left(\max_{0\leq m\leq n-1} \lambda^{-\alpha m}|(\Phi_h'(u_h)v_h)_m| + C_\star \lambda^{\alpha N}\|v_h\|\right) \leq \\ &\leq (1 + C_\star L_{n-1})\left(\max_{0\leq m\leq n} \lambda^{-\alpha m}|(\Phi_h'(u_h)v_h)_m| + C_\star \lambda^{\alpha N}\|v_h\|\right). \end{aligned}$$

(3.3) therefore holds and the induction proof is complete. For $n = N-1$, (3.3) becomes

$$\|v_h\| \leq L_{N-1}\left(\|\Phi_h'(u_h)v_h\| + C_\star \lambda^{\alpha N}\|v_h\|\right).$$

In order to complete the proof of the proposition, it suffices to show that the sequence $\{L_n\}$ is bounded. Clearly, it is non-decreasing. Thus, there exists a positive integer $n_0 = n_0(r, \alpha)$ such that

$$(n - C_\star)L_{n-1} \geq 1 \quad \forall n \geq n_0.$$

It follows that $L_n = L_{n_0}$ for all $n \geq n_0$, as required □

Remark 1 The proof simplifies considerably if one makes the assumption that T is sufficiently small. The main difficulty has been to avoid this undesirable restriction.

Proposition 2 (Continuity of the derivative) *Assume that*

$$R_h = o(h^2).$$

Then,

$$\sup_{v_h \in B(u_h, R_h)} \|\Phi_h'(u_h) - \Phi_h'(v_h)\| = o(1).$$

Proof. Let $v_h \in B(u_h, R_h)$. For any $z_h \in S_h$, we have

$$((\Phi'_h(u_h) - \Phi'(v_h))z_h)_0 = 0$$

and, for $n = 1, \dots, N-1$,

$$((\Phi'_h(u_h) - \Phi'_h(v_h))z_h)_n = T \sum_{m=0}^{N-1} z_{hm} c_{h(n-1-m)},$$

where $c_h \in S_h$ is the discrete Fourier transform of the function $(f_x(v_h(\theta), Te^{i\theta}) - f_x(u_h(\theta), Te^{i\theta}))$. Multiplying this equality by $\lambda^{-\alpha n}$, we readily infer that

$$\|(\Phi'_h(u_h) - \Phi'_h(v_h))z_h\| \le CN\|c_h\|\|z_h\|. \tag{3.4}$$

We proceed to estimate the norm of c_h. Expanding $f_x(\cdot, Te^{i\theta})$ in a Taylor series about $u(\theta_j)$ and posing $e_h = u_h - v_h$, we may write

$$c_{hn} = \sum_{k=1}^{\infty} \frac{1}{k!} \left(\frac{1}{N} \sum_{j=0}^{N-1} \frac{\partial^{k+1} f}{\partial x^{k+1}} (u(\theta_j), Te^{i\theta_j})(e_h(\theta_j))^k e^{-in\theta_j} \right) =$$

$$= \sum_{k=1}^{\infty} \frac{1}{k!} \sum_{l=0}^{N-1} f_{hl}^{(k+1)} \frac{1}{N} \sum_{j=0}^{N-1} (e_h(\theta_j))^k e^{-i(n-l)\theta_j},$$

where $f_h^{(k)} \in S_h$ denotes the discrete Fourier transform of the function $\frac{\partial^k f}{\partial x^k}(u(\theta), Te^{i\theta})$. Multiplying this equality by $\lambda^{-\alpha n}$, we are led to

$$\lambda^{-\alpha n}|c_{hn}| \le \sum_{k=1}^{\infty} \frac{1}{k!} \sum_{l=0}^{N-1} \lambda^{-\alpha l} |f_{hl}^{(k+1)}| \, \lambda^{-\alpha(n-l)} |(e_h^k)_{h(n-l)}|, \tag{3.5}$$

where $(e_h^k)_h \in S_h$ is the discrete Fourier transform of e_h^k (the kth power of e_h). It is easily verified by induction that

$$\|(e_h^k)_h\| \le N^{k-1} \|e_h\|^k.$$

Reporting this in (3.5), we find

$$\lambda^{-\alpha n}|c_{hn}| \le \sum_{k=1}^{\infty} \frac{1}{k!} (N\|e_h\|)^k \|f_h^{(k+1)}\|. \tag{3.6}$$

Now,

$$\|f_h^{(k+1)}\| \le \frac{1}{1-\lambda} \max_{|\tau|=r} |\frac{\partial^{k+1} f}{\partial x^{k+1}}(x(\tau), \tau)| \le$$

$$\le \frac{1}{1-\lambda} \max_{|\tau|=r+\epsilon} |f_x(x(\tau), \tau)| \, k! \, \epsilon^{-k}$$

by Cauchy's formula, with $0 < \epsilon < R - r$. We therefore conclude from (3.6) that

$$\|c_h\| \le C \sum_{k=1}^{\infty} \left(\frac{N\|e_h\|}{\epsilon} \right)^k \le CN\|e_h\| \tag{3.7}$$

for N sufficiently large, since $\|e_h\| = o(N^{-2})$. Thus, returning to (3.4), we find, for N large enough,

$$\|\Phi'_h(u_h) - \Phi'_h(v_h)\| \le CN^2\|e_h\| \le CN^2 R_h,$$

and the proof is complete □

The introduction by López–Marcos and Sanz–Serna of h-dependent "stability thresholds" R_h was designed to cater for partial differential problems.[8] Their occurrence in the present context is due to the well-known fact that the nonlinear recurrence relation (1.4) is unstable in the strict sense. Indeed, consider the example given by Corliss and Chang[5], namely

$$\frac{dx}{dt} = x^2, \quad x(0) = x_0.$$

This leads to the recurrence relation

$$n\, u_n = T \frac{1}{2\pi} \int_0^{2\pi} u^2(\theta)\, e^{-i(n-1)\theta} d\theta = T \sum_{j=0}^{n-1} u_j u_{n-1-j} \tag{3.8}$$

for $n = 1, \ldots, N-1$, with $u_0 = x_0$. The solution is $u_n = x_0(Tx_0)^n$. If one denotes by $\tilde{u}_n$ the solution corresponding to the perturbed initial datum $(1+\epsilon)x_0$, then the relative difference is

$$\max_{0 \le n \le N-1} \left| \frac{u_n - \tilde{u}_n}{u_n} \right| = (1+\epsilon)^{N-1} - 1. \tag{3.9}$$

In particular, if ϵ is independent of N, then this relative difference grows without bound as $N \to \infty$. The concept of stability threshold expresses the fact that, when computing the first N coefficients, perturbations up to a certain maximum size may be tolerated. Thus, (3.9) shows that the recurrence is stable under perturbations of size $\epsilon = o(N^{-1})$ since

$$\max_{0 \le n \le N-1} \left| \frac{u_n - \tilde{u}_n}{u_n} \right| \le C\,\epsilon\, N = o(1) \quad \text{as } N \to \infty.$$

It is reasonable to conjecture that such stability features are inherited by the discrete scheme defined by Eq. (2.1). Note that the permissible size of the perturbation in this example is larger than the stability threshold R_h allowed by the above proposition. This indicates that the condition $R_h = o(h^2)$ may be unduly restrictive. However, the important point is that, as the above example illustrates, it is necessary to cater for the fact that R_h does shrink to zero as $N \to \infty$. As we shall see, the consistency error decays much faster than the stability threshold, thus making convergence possible.

Proposition 3 (Consistency) *There exists a constant $C = C(r, \alpha) > 0$ such that*

$$\|\Phi_h(u_h)\| \leq C\lambda^N.$$

Proof. By virtue of Eqs. (1.3) and (1.4), we have

$$(\Phi_h(u_h))_0 = u_{h0} - u_0$$

and, for $n = 1, \ldots, N-1$,

$$(\Phi_h(u_h))_n = n\,(u_{hn} - u_n) - T(f_{h(n-1)} - f_{n-1}),$$

where $f_h \in S_h$ and f_n denote respectively the discrete Fourier transform and the nth Fourier coefficient of the function $f(u(\theta), Te^{i\theta})$. Thus, applying Lemma 2.1,

$$\lambda^{-\alpha n}|(\Phi_h(u_h))_n| \leq C\,n\,\lambda^{(1-\alpha)n+N} \leq C\lambda^N$$

since $\alpha \in (0,1)$ □

It remains only to invoke Theorem 2.1 to obtain the

Theorem 2 *Let $r \in (T, R)$, $\lambda = T/r$ and $\alpha \in (0,1)$. Assume that*

$$R_h = o(N^{-2}).$$

Then, for N large enough, the discrete equation (2.3) admits a unique solution in the ball $B(u_h, R_h)$. Further, we have

$$\max_{0 \leq n \leq N-1} \lambda^{-\alpha n}|u_n - U_{hn}| \leq C\lambda^N.$$

Remark 2 Given the coefficients U_{hn}, one may approximate the true solution $x(t)$ of Eq. (1.1) by the "discrete" Taylor polynomial

$$X_h(t) = \sum_{n=0}^{N-1} U_{hn}(t/T)^n.$$

It is a simple matter to show that, for $T < R$, Theorem 3.1 implies the bound

$$\max_{0 \leq t \leq T} |x(t) - X_h(t)| = O(\lambda^N).$$

In this way, one obtains a new discretization method for ordinary differential equations which, while it requires only the evaluation of f at points in the complex plane, achieves the same infinite order of accuracy as the classical Taylor series method.

4. Solving the discrete equation

The computation of U_h requires the solution of a nonlinear system of equations. The choice of iterative procedure is very much dependent on what value of the parameter T is adequate for the application in hand. Note that

$$u_n = T^n \frac{x^{(n)}}{n!}(0) \sim (T/R)^n \quad \text{as } n \to \infty,$$

where R is the distance to the nearest singularity in the complex plane. On a computer with finite precision arithmetic, a large enough value of T should be chosen in order to ensure that a sufficient number of coefficients may be obtained to a good relative accuracy. If only a few coefficients are needed, it is often adequate to choose T small. In that case, Picard's method is convenient. Otherwise, Newton's method or its variants should be used.

4.1. Picard's method

Let $U_h^{(k-1)} \in S_h$ denote an approximation of U_h. In Picard's method, the next approximation $U_h^{(k)} \in S_h$ is given by

$$U_{h0}^{(k)} = x_0$$

and, for $n = 1, \dots, N-1$,

$$U_{hn}^{(k)} = U_{hn}^{(k-1)} + \frac{1}{n} T \frac{1}{N} \sum_{j=0}^{N-1} f(U_h^{(k-1)}(\theta_j), Te^{i\theta_j})\, e^{-i(n-1)\theta_j}.$$

A straightforward analysis reveals that the iteration sequence converges linearly provided that T is "small enough". The reduction factor at each iteration is, roughly speaking, $T\|b_h\|$, where b_h is the discrete Fourier transform of $f_x(u(\theta), Te^{i\theta})$. The cost of each iteration is $O(N \log N)$ function evaluations. Note, however, that $O(N)$ iterations are needed, at least in principle, in order to attain the exponential accuracy of the discrete solution.

4.2. A simplified form of Newton's method

In Newton's method, the next iterate is found by solving the linear system of equations

$$\Phi_h'(U_h^{(k-1)})(U_h^{(k)} - U_h^{(k-1)}) = -\Phi_h(U_h^{(k-1)}). \tag{4.1}$$

Let $\delta_h = U_h^{(k)} - U_h^{(k-1)}$. We may write (4.1) componentwise as

$$\delta_{h0} = 0$$

and, for $n = 1, \ldots, N-1$,

$$n\,\delta_{hn} - T\sum_{m=0}^{N-1} b_{h(n-1-m)}^{(k-1)}\delta_{hm} = -(\Phi_h(U_h^{(k-1)}))_n. \tag{4.2}$$

In this notation, $b_{hn}^{(k-1)}$ is an $M \times M$ matrix with entries

$$(b_{hn}^{(k-1)})_{\mu,\nu} = \frac{1}{N}\sum_{j=0}^{N-1}(f_x(U_h^{(k-1)}(\theta_j), Te^{i\theta_j}))_{\mu,\nu}\, e^{-in\theta_j},$$

where f_x denotes the Jacobian matrix of f. (4.2) constitutes a linear system of order MN which can be solved in $O(N^3M^3)$ operations by Gaussian elimination. The entries of the associated matrix and of the right-hand side vector may be computed in $O(M^2N\log N)$ operations via the Fast Fourier Transform. Therefore, the overall cost of each Newton iteration is $O(N^3M^3)$ operations.

We now propose a slightly modified form of Newton's method which entails only $O(N^2M^2)$ operations per iteration without, as we shall argue, significant loss of accuracy. Multiplying (4.2) by $\lambda^{-\alpha n}$, we find

$$n\,\lambda^{-\alpha n}\delta_{hn} - T\lambda^{-\alpha}\sum_{m=0}^{n-1}\lambda^{-\alpha(n-1-m)}b_{h(n-1-m)}^{(k-1)}\lambda^{-\alpha m}\delta_{hm} =$$

$$= -\lambda^{-\alpha n}(\Phi_h(U_h^{(k-1)}))_n + \epsilon_h,$$

where

$$\epsilon_h = \lambda^{\alpha(N-1)}T\sum_{m=n}^{N-1}\lambda^{-\alpha(n-1-m+N)}b_{h(n-1-m+N)}^{(k-1)}\lambda^{-\alpha m}\delta_{hm}.$$

Now

$$|\epsilon_h| \le TN\|b_h^{(k-1)}\|\|\delta_h\|\lambda^{\alpha(N-1)},$$

where $|\cdot|$ denotes some norm on M-dimensional vectors and $\|b_h^{(k-1)}\|$ has the obvious meaning in terms of the subordinated matrix norm. Consequently, it makes sense to disregard the contribution from the off-diagonal upper triangular part of the matrix associated with (4.1) and to replace (4.2) by

$$n\,\delta_{hn} - T\sum_{m=0}^{n-1} b_{h(n-1-m)}^{(k-1)}\delta_{hm} = -(\Phi_h(U_h^{(k-1)}))_n. \tag{4.3}$$

With this new system, $U_h^{(k)}$ may be obtained by forward substitution and the cost of each iteration is $O(N^2M^2)$ operations. This is the same computational complexity as in Corliss and Chang's implementation of the Taylor series' method. Of course, a further simplification of Newton's method consists of iterating with a fixed Jacobian matrix f_x evaluated at $U_h^{(0)}$.

4.3. The starting scheme

Theorem 3.2 only guarantees the unique solvability of the discrete equation in a ball of radius $R_h = o(h^2)$ centered at u_h. It is therefore important to show that the initial approximation $U_h^{(0)}$ may be constructed in such a way that it actually lies in that ball.

For definiteness, let $R_h = h^{2+\epsilon}$ with $\epsilon > 0$. For N sufficiently large, there exists an integer $N_\star$ depending only on x, α and λ such that

$$\lambda^{-\alpha n}|u_{hn}| \leq \frac{1}{1-\lambda}\max_{|\tau|=r}|x(\tau)|\,\lambda^{(1-\alpha)n} \leq R_h$$

for all $N_\star < n \leq N-1$. Hence, we may choose

$$U_{hn}^{(0)} = 0, \quad \text{for } N_\star < n \leq N-1. \tag{4.4}$$

On the other hand, the first $N_\star$ approximate coefficients may be obtained by combining Lyness' algorithm with a standard time-stepping method as in Tourigny and Grinfeld.[13] Suppose that a method of order $p > 2+\epsilon$ is used to numerically integrate the differential equation along the complex circle $Te^{i\theta}$ using N equally-spaced steps. This yields approximate values V_j of $u(\theta_j)$ which satisfy the bound

$$\max_{0\leq j\leq N-1}|V_j - u(\theta_j)| \leq Ch^p.$$

Applying the Fast Fourier Transform to the set $\{V_j\}_{j=0}^{N-1}$, we obtain approximations V_{hn} of the discrete Fourier coefficients u_{hn} and it is readily seen that

$$\max_{0\leq n\leq N-1}|V_{hn} - u_{hn}| \leq Ch^p.$$

Hence, for $0 \leq n \leq N_\star$, we have

$$\lambda^{-\alpha n}|V_{hn} - u_{hn}| \leq Ch^p\lambda^{-\alpha N_\star} \leq h^{2+\epsilon}$$

as $h \to 0$. We may thus choose $U_{hn}^{(0)} = V_{hn}$ for $0 \leq n \leq N_\star$. This, together with (4.4) ensures that

$$\|U_h^{(0)} - u_h\| \leq R_h$$

for N sufficiently large, as required.

The integer $N_\star$ in the above construction is, of course, not known a priori. In practice, we have found that the choice $U_{hn}^{(0)} = V_{hn}$ for $n = 0, \ldots, N-1$, is usually sufficient to ensure convergence.

5. Application to a semilinear parabolic problem

The partial differential problem

$$\frac{\partial w}{\partial t} = 4\,\frac{\partial^2 w}{\partial x^2} + w^2, \quad (x,t) \in (-1,1) \times (0,\infty), \tag{5.1}$$

with $w(\pm 1, t) = 0$ for $t \geq 0$ and the initial condition

$$w_0(x) = \alpha \cos(\pi x/2), \quad -1 < x < 1,$$

belongs to a class that has been studied by many authors since Fujita's early paper[6]. It can be shown that a global solution exists provided that α is "sufficiently small". On the other hand, there are values of α for which the solution blows up at a finite time.

In order to compute the blow-up time $t_\star$, we first discretize the equation with respect to the space variable x. The Chebyshev collocation method leads to the system of ordinary differential equations

$$\frac{dW_j}{dt} = 4\,(AW)_j + W_j^2, \tag{5.2}$$

for $j = 1, \ldots, J-1$, with $W_0(t) = W_J(t) = 0$ for $t \geq 0$ and $W_j(0) = \alpha \cos(\pi x_j/2)$. In this notation, $x_j = \cos(j\pi/J)$, A is the square of the $(J+1) \times (J+1)$ Chebyshev collocation derivative matrix[12] and $W_j(t)$ approximates $w(x_j, t)$. Note that, in compact time intervals excluding $t_\star$, the error $W_j - w(x_j, \cdot)$ decays exponentially as $J \to \infty$.

The case $J = 2$ will serve as a test for the theoretical results of §3. The system (5.2) then reduces to the scalar equation

$$\frac{dW_1}{dt} = -8\,W_1 + W_1^2$$

with $W_1(0) = \alpha$. It is readily found that W_1 has a simple pole at

$$t = t_\star^{(2)} = \frac{1}{8} \ln\left(\frac{\alpha}{\alpha - 8}\right).$$

It follows that W_1 exists for all time if $\alpha < 8$ (since $t_\star^{(2)}$ is in this case "off" the real axis) and only locally if $\alpha > 8$.

A fourth-order diagonally implicit Runge–Kutta method was used as a starting scheme for the modified Newton version of the algorithm described in §4. The calculations were carried out on a Sun computer using double precision complex arithmetic.

Table 1: Decay of the relative error for increasing values of N.

N	$\max_{0\leq n<\tilde{N}} \left\|\frac{V_{hn}-u_n}{u_n}\right\|$	rate	$\max_{0\leq n<\tilde{N}} \left\|\frac{U_{hn}-u_n}{u_n}\right\|$	rate
2	0.18271493D+00	–	0.13433256D+00	–
4	0.60284048D+00	-1.72	0.78154643D-01	0.78
8	0.69091723D+02	-6.84	0.48748401D-02	4.00
16	0.24144151D+00	8.16	0.20812793D-05	11.2
32	0.11614600D-01	4.38	0.14712231D-12	23.8
64	0.67912801D-03	4.10	0.49250555D-14	4.90

Table 2: The ratio test with the computed coefficients U_{hn} ($N = 32$).

n	1	2	4	8	12	16
$T\frac{U_{h(n-1)}}{U_{hn}}$	.27777778	.13157895	.14648030	.14625888	.14625891	.14625891

Table 1 shows the maximum relative error for the first $\tilde{N} = \max\{N, 8\}$ coefficients, as N increases. In that calculation, $\alpha = 11.6$ and $T = 0.05$. The second column displays the corresponding error for the starting approximation V_{hn} (cf. §4.3). For both approximations, the local convergence rate may be defined as $\ln(e_N/e_{2N})/\ln 2$, where e_N is the error. The results provide clear evidence of the exponential accuracy of the discrete coefficients U_{hn} as N increases, thus confirming the validity of Theorem 3.2.

Note that, for this choice of α, we have

$$t_\star^{(2)} = 0.14625890658\ldots.$$

Using the ratio test with the discrete coefficients corresponding to $N = 32$, we obtain the results shown in Table 2.

In order to learn about the singularities associated with the parabolic problem (5.1), it is of course necessary to choose larger values of J. For the case $\alpha = 11.6$, Table 3 lists a sample of parameter values that were used to compute estimates of the location $t_\star^{(J)}$ of the positive singularity of $W_{\frac{J}{2}}$ nearest to the origin.

In that table, the fifth column shows the least index n for which all the computed normalized coefficients with index greater than n were less that 10^{-7} in modulus. This

Table 3: Typical parameter values for estimating $t_\star^{(J)}$ as J increases.

J	expansion point	T	N	n	ratio-test estimate
4	0.20D+00	0.500D-01	64	41	0.27867806D+00
8	0.44D+00	0.100D-01	128	12	0.48735295D+00
16	0.45D+00	0.155D-02	256	6	0.48829244D+00
32	0.45D+00	0.100D-03	256	3	0.48837677D+00

particular threshold value was chosen because, in double precision arithmetic, smaller coefficients cannot be computed to a sufficiently good relative accuracy. The ratio-test estimate shown in the last column is obtained from the computed coefficients with indices n and $n-1$.

When $J \geq 8$, it was found that, for larger values of T, the starting scheme would not yield a sufficiently good initial guess to ensure the convergence of the Newton iteration sequence. The situation was not significantly improved by taking more than N steps along the complex circle. In fact, those preliminary numerical experiments suggest that the effect of stiffness is to create a line of complex singularities very near to the real axis. Yet, if this were the case, one would expect the normalized coefficients to grow as T approaches its limiting value. Further research is needed in order to elucidate this phenomenon.

An obvious, though unattractive, way to deal with the presence of stiffness is to make the ratio T/R (and hence the normalized coefficients) sufficiently large by computing very near to the singularity of interest. An alternative, and more satisfying approach, is based on the observation that, in theory (cf. Theorem 3.2), the accuracy of the method actually improves as $T \to 0$. Hence, instead of seeking to increase T, one should cater for the rapid decay of the normalized coefficients by working in extended precision arithmetic.

6. References

1. G. Baker, M. Siegel and S. Tanveer, *J. Comput. Phys.* **120** (1995), 348.
2. C. M. Bender and S. A. Orszag, *Advanced Mathematical Methods for Scientists and Engineers* (McGraw–Hill, New York, 1978).
3. G. Corliss and Y. F. Chang, *ACM Trans. Math. Software* **8** (1982), 114.
4. Y. F. Chang, *ATOMFT: User Manual* (Technical Report 385, Marquette University, Wisconsin, 1993).
5. G. Corliss, *Math. Comput.* **35** (1980), 1181.
6. H. Fujita, *J. Faculty Sc. Univ. of Tokyo* **13** (1966) 109.
7. H. B. Keller, *Math. Comput.* **29** (1975), 464.
8. J. C. López–Marcos and J. M. Sanz–Serna, *IMA J. Numer. Anal.* **8** (1988), 71.
9. J. N. Lyness, *Math. Comput.* **22** (1968), 352.
10. C. Sulem, P. L. Sulem and H. Frisch, *J. Comput. Phys.* **50** (1983) 138.
11. M. Tabor, *Chaos and Integrability in Nonlinear Dynamics: an Introduction* (Wiley, New York, 1989).
12. E. Tadmor, *SIAM J. Numer. Anal.* **23** (1986), 1.
13. Y. Tourigny and M. Grinfeld, *Math. Comput.* **62** (1994), 155.